SEMICONDUCTOR MANUFACTURING TECHNOLOGY

Advanced Series in Electrical and Computer Engineering – Vol. 13

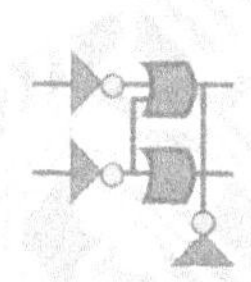

Chue San YOO

Former Associate Professor at Tamkang University and
National Tsing Hua University, Taiwan

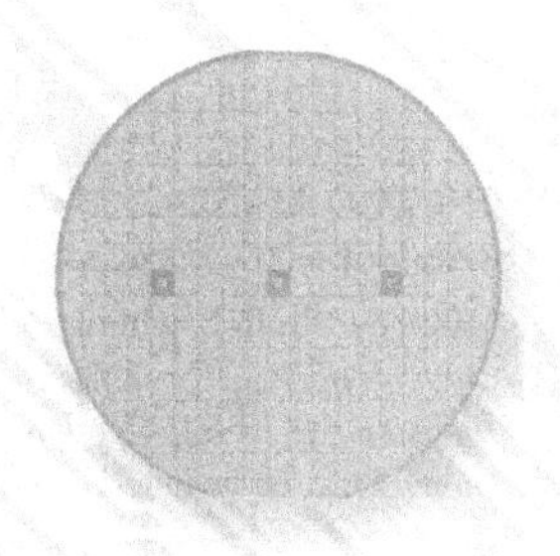

SEMICONDUCTOR MANUFACTURING TECHNOLOGY

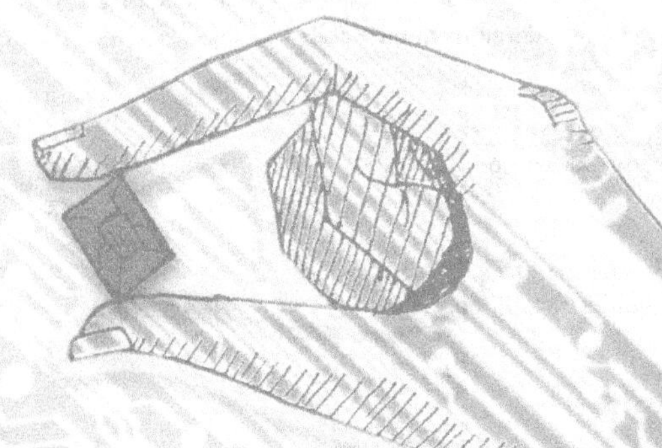

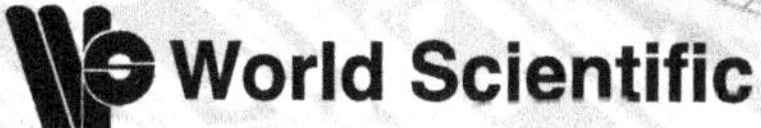

World Scientific

NEW JERSEY • LONDON • SINGAPORE • BEIJING • SHANGHAI • HONG KONG • TAIPEI • CHENNAI

Published by

World Scientific Publishing Co. Pte. Ltd.
5 Toh Tuck Link, Singapore 596224
USA office: 27 Warren Street, Suite 401-402, Hackensack, NJ 07601
UK office: 57 Shelton Street, Covent Garden, London WC2H 9HE

British Library Cataloguing-in-Publication Data
A catalogue record for this book is available from the British Library.

SEMICONDUCTOR MANUFACTURING TECHNOLOGY
Advanced Series in Electrical and Computer Engineering — Vol. 13

ISBN-13 978-981-256-823-6
ISBN-10 981-256-823-9

Typeset by Stallion Press
Email: enquiries@stallionpress.com

Printed in Singapore.

Preface

Over the last 20 years, I have been intensively involved in semiconductor process technology research, development, and manufacturing. Alongside my professional career, I have also enjoyed teaching semiconductor processing at Tamkang and Chinghwa universities as an adjunct professor. The contents of this book have largely evolved from the handouts that I prepared over the years for these courses. I also revised the course material for training junior engineers at various semiconductor companies. My experience indicates that it is relatively difficult for people who first embrace the semiconductor world to comprehend the essence of semiconductor processing technology. This is largely due to the fact that most people have a single engineering discipline in schools, while semiconductor processing is interdisciplinary in nature. It even takes a person with an engineering degree a few years before he or she can gain a good grasp of the overall semiconductor processing technology.

This book aims to provide readers with an easy-to-understand view of semiconductor industry, technology, and manufacturing. The first two chapters provide an overview of the industry. This includes an introduction to semiconductor processing and what the industry is focusing on. In addition, the integrated manufacturing flow, from raw material to device passivation, is explained, and each involved process module is defined. It also explains what ICs are and what they consist of. The following chapters further elaborate each individual process module technology. By and large, the sequence of the process module introduction follows typical manufacturing flows. The manufacturing processes are divided into two parts: the front-end and back-end processes. The former includes oxidation, gas kinetics, plasma physics, CVD, plasma CVD and etching, photolithography, mask making, and doping technology. The pattern generation or mask making is included since it is relatively unfamiliar to those who work in wafer processing. Furthermore, it is widely believed that the photomask technology has to be closely integrated into wafer processing to ensure wafer production success as technologies migrate beyond 0.13 μm. The back-end process introduction, beginning with contact formation, primarily focuses on metallization and planarization and their technology evolution. Various silicide formation processes are also included in the metallization due to its inevitable role in advanced semiconductor manufacturing. Following the conventional back-end technology discussion, copper and low dielectric constant materials are also introduced as they are widely used in prevailing nanometer device manufacturing.

While most text is self-explanatory, some mathematics modeling is included occasionally to better explain the theories. Those who are in an undergraduate program or who do not have enough time to check into the mathematics can skip the modeling sessions and still gain insight into semiconductor manufacturing principles.

Acknowledgments

I am grateful to my late father, Ching mu Yoo, who gave me an unforgettable childhood. I would also like to thank my wife for her support over the past several years, when I was writing this book. To my 10-year-old son, Gordon. I like to tell you I am so proud of your drawing on the cover page. Thank you so much. I am indebted to the engineers I have taught in various semiconductor companies and all the students I have taught in the past 15 years at Tamkang University and Chinghwa University. They have helped greatly in shaping the contents of this book.

Contents

Chapter 1

OVERVIEW

This chapter provides an introduction to the semiconductor industry. It begins with the classification of solids in terms of resistivity and introduces the semiconductor industry's evolution from its early stage to the prevailing planar technology. It then illustrates each of the six major chip manufacturing stages: design, mask making, wafer substrate making, wafer processing, back-end service, and qualification. The substrate and wafer processes are elaborated with schematic process flows. The last two sections are dedicated to explaining what the semiconductor industry is focusing on and where it is devoting its efforts.

1.1. Classification of Materials

A semiconductor device is made up of a wide variety of materials, ranging from insulators, such as silicon oxides or silicon nitrides, to excellent conductors, such as aluminum and copper. Most important, it is based on the semiconductor. What is a semiconductor? This can be explained from a resistivity viewpoint. The resistance of a material is a function of its geometry:

$$R = \rho \frac{L}{A},$$

where R is the resistance of the material, L is the length of the material, A is the cross-sectional area perpendicular to the current flow, and ρ is the resistivity of the material, which is a physical property of the material.

Table 1.1. Resistivities of materials used in semiconductor manufacturing.

Materials	Resistivity (Ω-cm)
SiO_2, Si_3N_4	$> 10^5$
Si	5×10^4
Ge	50
Heavily doped Si	10^{-4}
Cu	10^{-6}

Conductors have a resistivity of less than 0.01 Ohm-cm, while insulators have a resistivity of greater than 100 000 Ohm-cm. Materials with resistivities ranging between those of the conductors and insulators are called semiconductors. Table 1.1 shows the resistivities of different materials that are often used in semiconductor manufacturing. Silicon oxide and silicon nitride are good insulators that are often used for isolation between conductors and passivation layers. Silicon and germanium are typical wafer substrates on which devices are made. Copper, on the other hand, is used for interconnects.

Resistivities of solid materials are related to the way the atoms or molecules are bonded together. The intermolecular or interatomic forces characterize the solids, which are classified into four groups (molecular, ionic, covalent, and metallic). Table 1.2 shows the materials that are used in semiconductor manufacturing as well as their classifications.

The elements in columns II to VI in the periodic table are considered semiconductor materials. Table 1.3 shows the electronic configuration and some of the properties of each element in column IV. Silicon and germanium are elemental semiconductors, and each atom has four valence electrons. On the other hand, elements in columns II and VI, e.g., Zn and S, or columns III and V, e.g., Ga and As, can form compound semiconductors such as ZnS and GaAs. The common feature of the semiconductor materials is that each atom has effectively four valence electrons. When bonded to other atoms, they form covalent bonds. Germanium was the material of choice for major device fabrication in the early years of semiconductor production. It has lower resistivity than silicon. However, owing to its

Table 1.2. Bonding natures of solids and examples in semiconductor manufacturing.

	Molecular	Ionic	Covalent	Metallic
Units occupying lattice sites	Molecules	Ions	Atoms	Positive ions
Interunit forces	van der Waals	Electrostatic	Shared electrons	Ions–electrons
General properties	Soft	Hard	Very hard	Varies
Melting point	Low	High	Very high	Varies
Conductivity	Insulators	Insulators	Semiconductors	Good conductors
Examples in semiconductor industry	N_2, H_2, CO_2*	NaOH, HCl†	C, Si, Ge	Al, Cu, W

*N_2, H_2, CO_2: when condensed to solid.
†NaOH, HCl: in water solution for wafers or masks cleaning process.

Table 1.3. Elements in column IV of the periodic table.

Elements	Z	Electronic configuration	M.P.,°C	B.P.,°C	Ionization potential, eV
C	6	(2) $2S^22P^2$	3500	4000	11.26
Si	14	(10) $3S^23P^2$	1400	2400	8.15
Ge	32	(28) $4S^24P^2$	937	2800	8.13
Sn	50	(48) $5S^25P^2$	232	2260	7.33
Pb	82	(78) $6S^26P^2$	327	1700	7.42

limited sources, it is much more expensive than silicon. Germanium is brittle and its oxide is soluble in water, which limits its applications in realizing planar processing technology. As semiconductor device production volume increased and its applications multiplied, germanium was soon replaced by silicon.

Silicon is the second richest element on earth, next to oxygen. Therefore it is relatively cheap and easy to access compared to other elements. The fact that silicon dioxides are not soluble in water gave rise to planar processing technology in 1959. By far, silicon is the most commonly used elemental material in the semiconductor industry. In addition, GaAs is the most commonly used compound semiconductor material. The biggest advantage of using GaAs is that the electron mobility of GaAs is ten times faster than that of silicon.

One can imagine that a calculation task can be accomplished by GaAs devices a few times faster than by silicon devices. However, the manufacturing technology for GaAs has been lagging behind that of silicon. Until recent years, device manufacturing using GaAs has remained a small-scale production. One of the major limitations is the thermal budget. As processing temperature increases, arsenic (As) may outgas from the substrate material. Arsenic is extremely toxic in nature. Furthermore, as more arsenic atoms outgas from the substrate, the substrate becomes richer in Ga, and therefore it loses its intrinsic semiconducting characteristics.

1.2. Evolution of Integrated Circuit (IC) Industry

The first half of the twentieth century can be considered the vacuum tube era. It was inspired by J. A. Fleming, who invented the vacuum tube diode in 1904 by using the Edison effect. The vacuum tube functioned as a valve, conducting current in only one direction. In 1906, Lee De Forest inserted a third electrode into the diode to obtain a triode, which amplified the input signals. After that, all electronic devices ranging from radios and TV sets to computers were made of vacuum tubes. However, the vacuum tubes were big, fragile, and power consuming. Electronic appliances that were made up of vacuum tubes needed to be interrupted after being used for a period of time to avoid overheating. The world's first successful computer — The University of Pennsylvania's ENIAC — comprised some 18 000 vacuum tubes. The footprint of the ENIAC was big enough to fill up a huge room. Whenever it shuts down due to overheating, the maintenance engineers often needed to search for the burnt-out tube out of tens of thousands of tubes. Complicated systems with vacuum tubes were very difficult to maintain.

With the demonstration of transistor functions accomplished by Walter H. Brattain, John Bardeen, and William Shockley in 1947 and the invention of the semiconductor integrated circuit (IC) by Jack Kilby in 1958, the use of vacuum tubes began to be replaced rapidly by devices made of semiconductor materials. Today, the applications of semiconductor devices, ranging from everyday appliances

Table 1.4. Milestones in electronic industry evolution.

Milestones	Inventors
Vacuum tube diode	J. A. Fleming, 1904
Vacuum tube triode	Lee De Forest, 1906
Vacuum tube computer, ENIAC	University of Pennsylvania, 1946
Semiconductor transistor	William Shockly *et al.*, 1947
Integrated circuit (IC)	Jack Kilby, 1958
Silicon planar technology	J. A. Hoerni, 1959
Microprocessor, $> 1\,\text{K}$ transistors	Intel 4004, 1971
1 Kb DRAM	Intel's 1103, 1970
Microprocessor, $> 9.5\,\text{M}$ transistors	Intel's Pentium III, 1999
Microprocessor, $> 42\,\text{M}$ transistors	Intel's Pentium 4, 2000

to communication devices and computers, are much broader than those made of vacuum tubes. Table 1.4 shows the major milestones in the human electronic evolution. It is evident that the technology advanced with an astonishing speed. With the invention of integrated circuits and planar processing technology, people began to put a large number of devices on the same chip (or die). The number of devices per chip increases with advances in processing technology. Over the last 30 years, the packing density has increased almost 10 000 times. The level of integration has evolved from large-scale integration (LSI $> 10^3$ devices/chip) to very large scale integration (VLSI $> 10^5$ devices/chip) to today, with ultra large scale integration (ULSI $> 10^7$ devices/chip). Nonetheless, the technology still has a lot of room for improvement in terms of size, speed, and cost.

The high packing density is made possible by the silicon planar processing technology, which has evolved over the past generations. Using a diode manufacturing process as a vehicle, one can gain insight into the silicon processing technology evolution, as shown in Fig. 1.1. The grown junction method basically dips a *p*-type substrate into an *n*-type silicon melt. As the *p*-type substrate is pulled up, the *n*-type melt crystallizes next to the *p*-type substrate. Therefore a *p*–*n* junction is formed. The alloy junction, on the other hand, is formed by placing an *n*-type pellet onto a *p*-type substrate. As the temperature is raised above the silicon melting point and then cooled

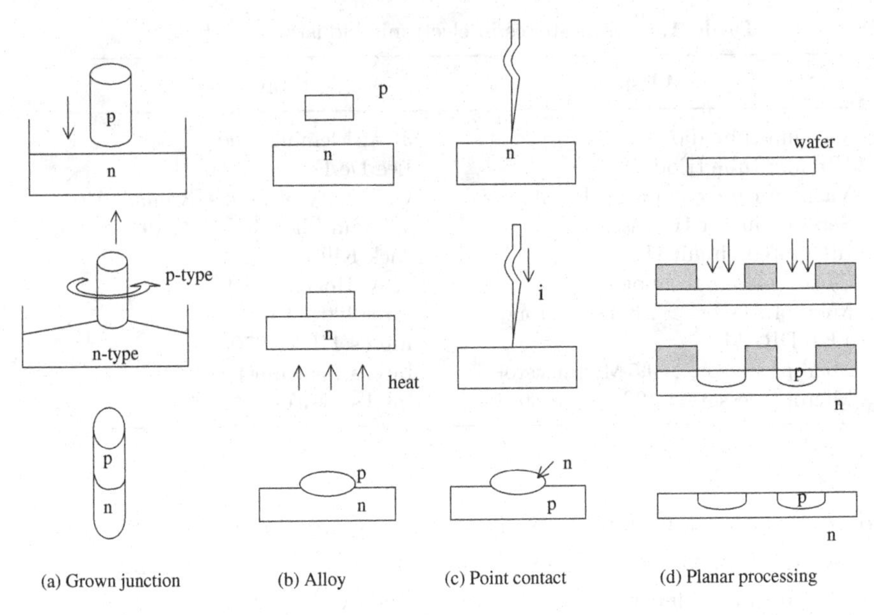

Fig. 1.1. Junction formation approaches, illustrating the evolution of semiconductor processing technology.

down, an alloy junction can be formed. These two approaches result in junctions with large capacitances, owing to large junction areas.

The point contact junction approach brings an n-type whisker in contact with a p-type substrate surface, and when a high current is passed through the whisker, the local heating brings about a small area p–n junction. Therefore a small capacitance junction diode can be obtained. The technologies mentioned earlier are either difficult to control or too difficult to mass produce. Silicon planar processing technology (described by J. A. Hoerni in 1960), which employs a silicon oxide mask to define the size of a diode, enables a huge number of devices to be made simultaneously on the same silicon wafer surface. Planar processing technology results in capacitors with different capacitances and excellent controllability. This is made possible by two unique features of silicon: (1) silicon can be oxidized to obtain silicon dioxide, which is a stable insulator, and (2) silicon and silicon dioxide have a good chemical durability during wafer cleaning.

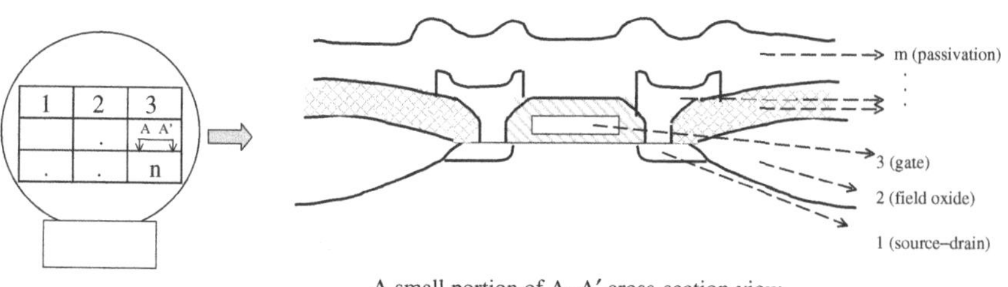

Fig. 1.2. A single silicon wafer contains n discrete dice. Each die is made of m layers of different materials defined by m layers of masks.

Prevailing silicon planar processing technology is far more complicated than it used to be. Figure 1.2 shows the cross-section of a simplified, single-layer metal circuit structure. It defines a device structure into m layers. Each layer corresponds to a photomask. The layers are made up of different materials and are manufactured layer by layer. The pattern within each layer is divided into n regions. Each region designates an individual chip or die. The planar technology enables device manufacturing to transform from a discrete to an integrated level. The combination of planar technology and integrated circuit design greatly accelerated the evolution of the semiconductor industry. As the technology advances, the mask number, m, increases to increase the packing density. One example of this is the use of multilayer metal to increase the level of integration. Meanwhile, the number of dice, n, can be increased to reduce the cost per die. This is accomplished by using a smaller geometry technology, assuming the chip function stays unchanged.

The silicon planar processing technology has been evolving since 1960. Today, hundreds of millions of transistors can be packed onto a tiny die. Figure 1.3 shows the volume production of 1 M DRAM during 1991, while the current technology is capable of producing 256 Mb DRAM. The 1-Gb DRAM is now in volume production. The level of integration is expected to increase a thousand fold by the year 2016. The memory size of 1 Gb DRAM corresponds to 64 000 sheets of newspaper. For the microprocessor, the level of integration is expected to increase ten-fold by the year 2010. With the prevailing semiconductor processing technology, the ENIAC computer could be

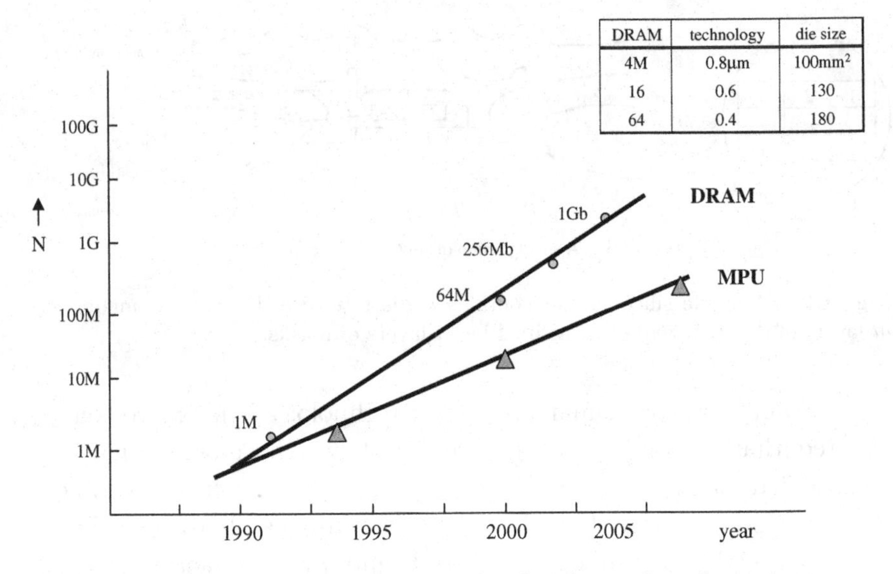

DRAM	technology	die size
4M	0.8μm	100mm²
16	0.6	130
64	0.4	180

Fig. 1.3. As semiconductor technology shrinks, the number of transistors per die (either on DRAM or MPU) increases dramatically, but the die size increases disproportionately (N: the number of transistors per die).

shrunk onto a tiny chip smaller than 1 mm^2, and yet it will be even more powerful and will perform faster computing functions.

1.3. From Design to Chips

The process of manufacturing silicon chips begins with high-quality sand as a raw material and ends with the resulting silicon chips, which are used in end product such as PCs, TVs, or other electronic appliances. This entire process can be divided into six different stages, as illustrated in Fig. 1.4. These subprocesses are as follows: (1) circuit design; (2) mask making; (3) silicon substrate manufacturing; (4) wafer processing; (5) back-end services, including testing; and packaging; and (6) qualification. The circuit design defines a circuit that meets the specified functions, performance, and cost. A circuit design primarily includes two phases. One phase is the logical design, while the other phase is the physical layout design. The logical design determines the circuit structures needed to implement the prescribed

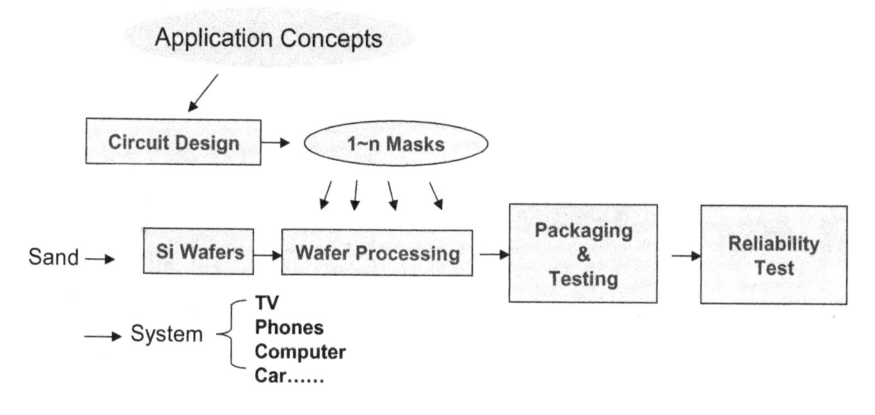

Fig. 1.4. From design to packaged chips.

function. The physical layout design concerns the implementation of the circuit structure on the silicon substrate. In this phase, the circuit is separated into different levels, and each level is represented by a large number of polygons, as illustrated in Fig. 1.5. The output of the product circuit design is the physical layout pattern of a number of layers. A mask shop takes each layer's pattern information and generates the layout patterns on a mask, chrome patterns on quartz, as demonstrated in Fig. 1.6. A circuit or product generally consists of a number of masks, depending on the complexity of the product and wafer process technology. The masks must be used sequentially

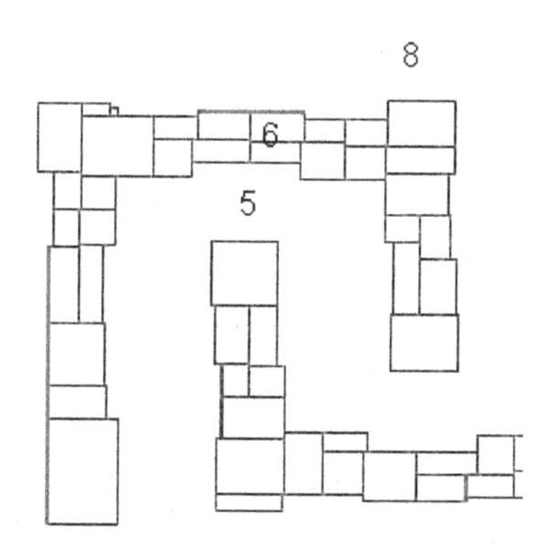

Fig. 1.5. Examples of circuit patterns composed of polygons.

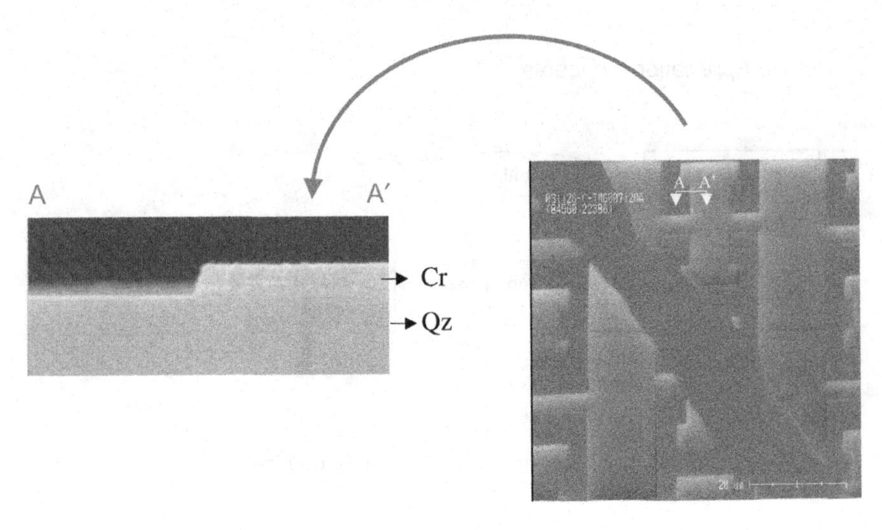

Fig. 1.6. Mask patterns with chrome on quartz and the cross-sectional view.

in the designated order to produce a functional product. Each mask is used in the wafer's photolithographic process step, followed by an etching for selective or local material removal for pattern delineation on wafers.

The circuit patterns are transferred from the masks to the wafer surface layer by layer, using the photolithographic process. The wafer substrate making and wafer processing will be discussed in detail in the following sections. The wafer processing is used to make a large number of transistors on a planar silicon surface, connecting them together to render a functional circuit. The wafers have to undergo electrical testing to determine whether the electrical functions comply with the original design requirements. If they do not meet the requirements, the circuit must be revised and retested until the function is optimized, as illustrated in Fig. 1.7. After the testing is completed, a product qualification procedure is required. The purpose of the product qualification is to ensure statistically that the manufactured chips are functioning properly under extreme environmental conditions for a certain lifetime. To accomplish this product qualification, each wafer is sawed into individual dice. It is then packaged and tested in various severe environmental and operational condition such

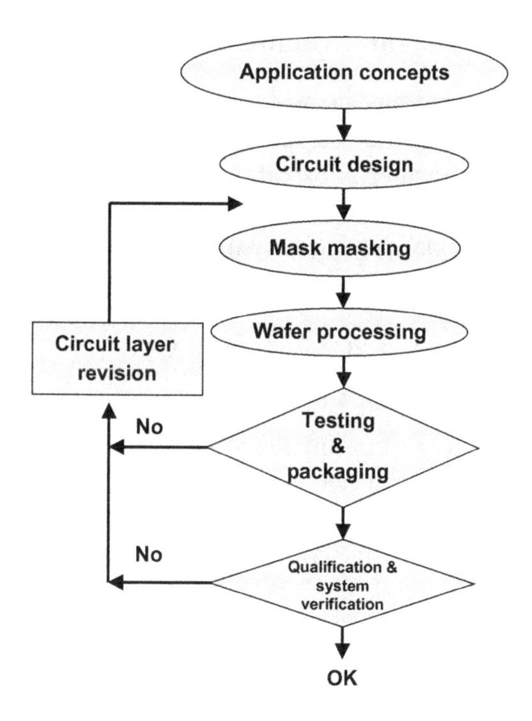

Fig. 1.7. Iterative procedure for circuit design.

Table 1.5. Product qualification items and stress conditions.

Items	Stress condition
(1) High-temperature operating life test	125°C, 1000 h
(2) Temperature and humidity bias test	85°C, 85% RH, 1000 h
(3) Pressure cook test	121°C, 2 atm, 100% RH, 200 h
(4) Temperature cycling (air to air)	– 65°C 10 min to 50°C 10 min, 1000 cycles
(5) Thermal shocks test (liq. to liq.)	– 65°C 5 min to 150°C 5 min, 200 cycles

as extreme humidity, temperatures, and operating voltages. Table 1.5 shows some typical product qualification items and their corresponding stress conditions. A functional chip design is not achieved unless the final qualification and system verification have been completed successfully.

1.4. The Wafer Substrate Manufacturing Flow

To achieve a high yield in the wafer manufacturing process, it is essential for the process to begin on a high-quality single crystalline silicon substrate. The wafer substrate manufacturing process plays a critical role in this regard. To achieve this goal, high-quality sand (primarily silicon dioxide) is heated to above 1600°C to molten silicon and then reduced with carbon to form polycrystalline silicon:

$$\mathrm{SiO_2} + 2\mathrm{C} \longrightarrow \mathrm{Si} + 2\mathrm{CO_2}\,.$$

The polycrystalline silicon at this stage is a metallurgical-grade silicon that contains excessive impurities such as heavy metals, carbon, and oxygen in the range of several hundreds of parts per million. It is not qualified for making devices on its surface. Therefore further refining processes, as illustrated in Fig. 1.8, are required. The metallurgical-grade silicon is fed into a fluidized bed, where it reacts with HCl to produce various chlorosilanes such as monochlorosilane, dichlorosilane, and tricholorosilane (TCS):

$$\mathrm{Si} + 3\mathrm{HCl} \longrightarrow \mathrm{SiHCl}_x + \mathrm{H_2}\,,$$

where x can be 1, 2, or 3.

These products are fed into a distillation column. The distillation is a mass separation process that takes advantage of the different boiling points of solvents in a solution. The lower boiling constituent TCS is obtained from the top of the column. It is then further purified using an absorption tower. The absorption is a mass separation process that relies on the fact that solutes can have different solubility levels in a solvent. The absorbing solvent and the solutes flow in opposite directions in the absorption tower. The resulting TCS is further reduced with hydrogen in another fluidized bed reactor to form electronic-grade polycrystalline silicon:

$$2\mathrm{SiHCl_3} + 2\mathrm{H_2} \longrightarrow 2\mathrm{Si} + 6\mathrm{HCl}\,.$$

The newly obtained polycrystalline silicon meets the purity requirement of the electronic-grade silicon. It is then heated in a crucible, made of high-purity carbon, to a molten state for Czechralski crystal

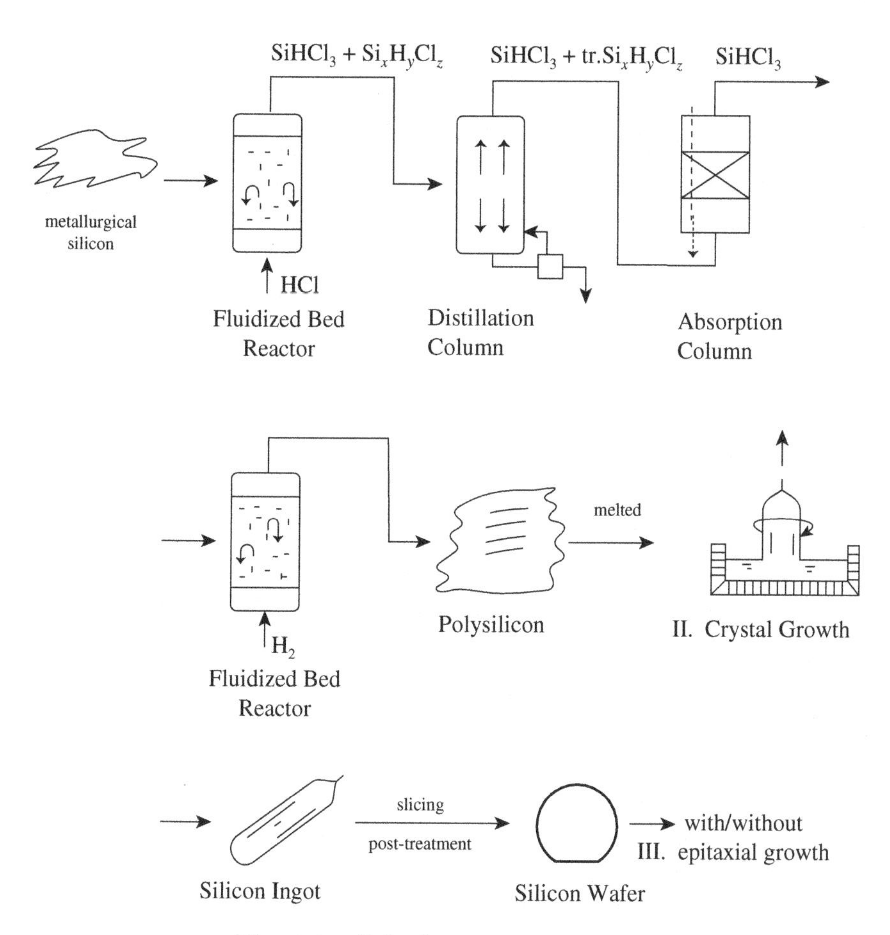

Fig. 1.8. Polysilicon refining process.

growth. A seed silicon crystal with a preferred crystal orientation is dipped into the surface of the molten silicon and pulled up, as illustrated in Fig. 1.9. Owing to the surface tension, some molten silicon is dragged out of the molten silicon surface. In the meantime, heat is transferred from this surface outward, resulting in crystallization. To maintain uniformity, the seed crystal is rotated while it is pulled up. The orientation of the grown silicon crystal follows that of the seed crystal. The seed crystal, together with the grown silicon, is pulled up very slowly, at a rate of several centimeters per day, allowing for heat transfer and to maintain the continuous silicon growth. In general,

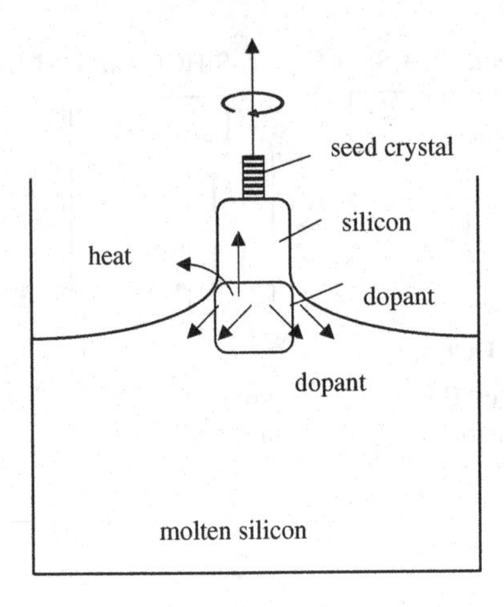

Fig. 1.9. Silicon crystallizes onto the seed crystal.

the substrate is lightly doped. Dopants, such as boron, phosphine, or arsenic, can be added to the molten silicon to obtain an n-type or p-type substrate.

During the crystal growth process, there is intricate heat and mass transport around the solidification zone. The solidification process that occurs at the molten and solid interface releases latent heat. The released latent heat is proportional to the amount of solid formed and must be conducted away via the conduction along the solid bulk. As a result, the maximum pulling rate is proportional to the thermal conductivity of the silicon bulk and the temperature gradient along the pulling-up direction. If the pull-up rate is faster than the growth rate, the crystal growth is discontinued. However, if the reverse situation is observed, the grown solid will remelt. This can result in defects and impurity striations. Furthermore, the maximum pulling rate is inversely proportional to the square root of the grown ingot radius. It is easy to comprehend that a larger ingot radius would take a longer time for the latent heat to transfer out of the solidification zone.

Mass transfer is primarily concerned with the dopant redistribution at the solid–liquid interface. Consistent dopant concentration along the ingot growing direction is very critical as it is related to device characteristics on each wafer. Theoretically, to obtain a uniform dopant concentration along the grown ingot length, it is essential to have a large bath of molten silicon such that the weight of the grown ingot is small compared to the bath.

For device manufacturing, a critical index of the silicon ingot is the crystal orientation. Flats or notches are made on the ingot to indicate the crystal orientation and for wafer alignment in subsequent process steps. The ingot is then ground to the desired diameter. After the desired diameter has been obtained, the ingot is sliced with a diamond saw into thin disks, called wafers. The slicing process is followed by a lapping step, in which wafers are mechanically lapped with counter rotating lapping machines, along with aluminum oxide slurry. The wafer then goes through a chemical–mechanical polishing process, in which the silicon surface material is removed both chemically and mechanically using an alkaline silica solution. This step is intended to obtain an extremely flat and smooth silicon surface. The final step of the substrate manufacturing process is wafer cleaning and drying. Wafers are often cleaned with a three-step cleaning procedure, starting with SC1 solution, ammonia, and hydrogen peroxide for removing organic surface contaminants and particles. Hydrofluoric acid is used to remove the native oxide and metal impurities. The cleaning step ends with SC2 solution, HCl, and H_2O_2 for a thin layer of super-clean native oxide formation. The state of the art wafer substrate size is 12 inches for production. However, eight- and six-inch wafer sizes are also being used for some production lines. The evolution trend for the size of the wafer substrate is to push for ever increasing wafer size. The reason for this trend is that the larger the wafer size, the more devices can be made simultaneously on the same wafer substrate. As a result, the manufacturing cost per die can be significantly reduced if the same level of wafer yield is achieved.

Example 1.1.

Assume that the housing facility and equipment required for the fabrication of a 12-inch wafer are twice more expensive than those required for the fabrication of an eight-inch wafer. If the eight-inch wafer fabrication is running at 80% wafer yield, what is the minimum 12-inch wafer yield to make the 12-inch line economically worthwhile?

To make the 12-inch line economically worthwhile, to the first-order approximation, one must have twice as many good dies on the 12-inch wafer compared to the eight-inch wafer:

$$A8 \times 80\% \times 2 = A12 \times Y\,.$$

Then, $Y = A8 \times 80\% \times 2/A12$. Therefore $Y = 71\%$.

In addition to the yield, there are other advantages with large wafer sizes. First, with a large wafer size, the cross-over wafer yield is lower. Therefore there is a higher potential to improve the yield. In addition, with the same real estate area, the 12-inch fabrication revenue can be doubled. Furthermore, the productivity or sales per head count can be significantly increased.

The silicon crystal orientation is of great importance in affecting the device performance. In a crystal, atoms have a unique arrangement. A unit cell is the smallest repeating unit in a crystal. In other words, by stacking up the unit cell, one can obtain the crystal structure. As shown in Fig. 1.10, more than one method of choosing a unit cell exists for a crystal. The unit cell constant refers to the

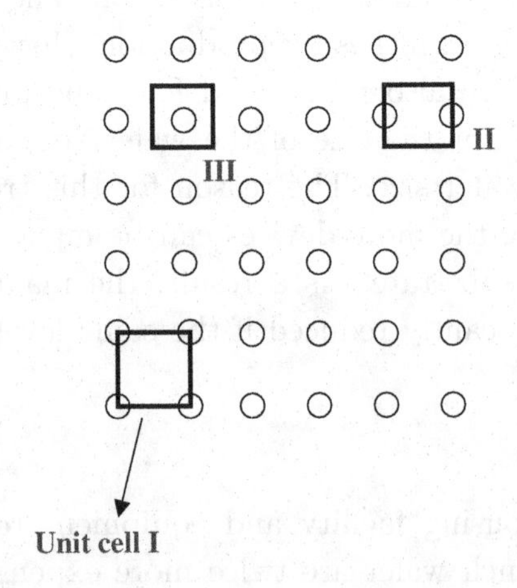

Fig. 1.10. More than one ways of choosing a unit cell.

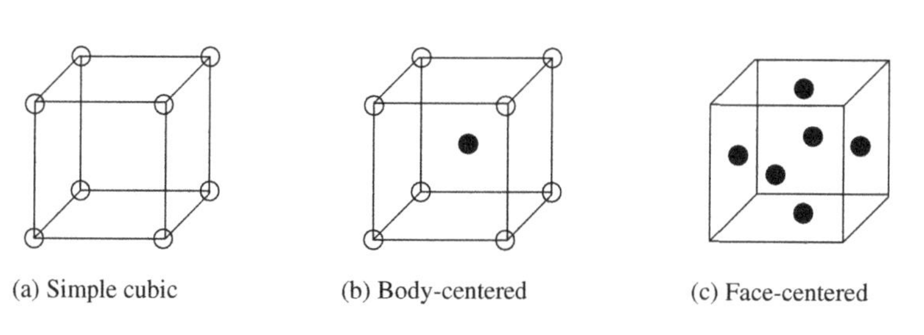

Fig. 1.11. Simple cubic crystals.

length of one side of the unit cell. If the unit cell constants happen to be the same along the three crystallography axes, the crystal is termed cubic. Figure 1.11 illustrates various crystal structures in the cubic class. Silicon crystal structure fits within the diamond structure, which is formed by two intercepting face-centered-cubic crystals. Each atom is surrounded by four other atoms through covalent bonds, forming a tetrahedron, as illustrated in Fig. 1.12. The Miller index is an index used to identify the crystal orientations or planes. The Miller index of a plane can bc obtained using the following steps:

(a) Take the intercepts of the plane with each axis, in terms of multiples of the unit cell length.
(b) Take the reciprocal of each intercept.
(c) Clear the fractional numbers.

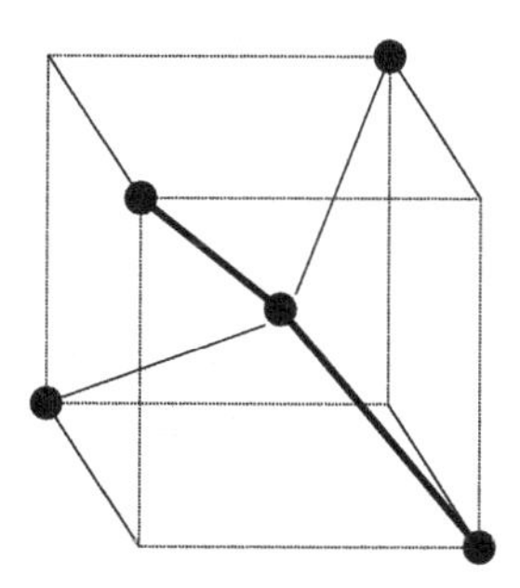

Fig. 1.12. The silicon tetrahedron in a diamond lattice cell.

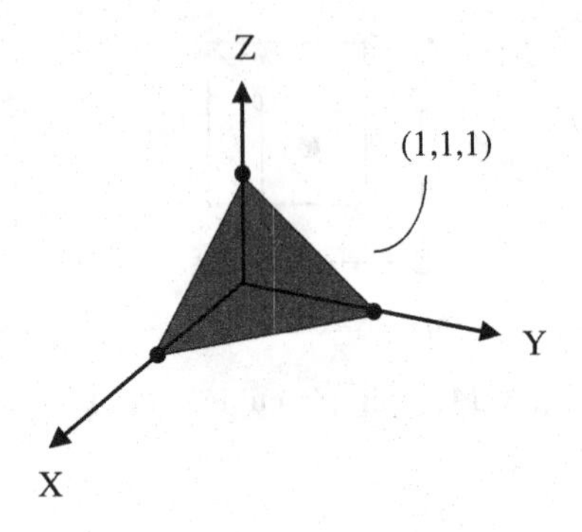

Fig. 1.13. The (111) plan.

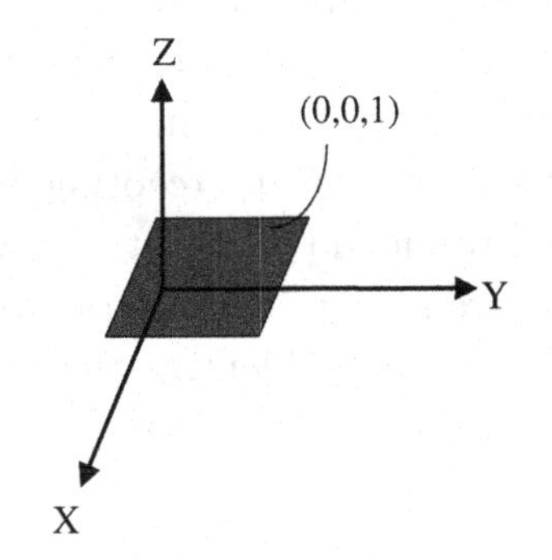

Fig. 1.14. The (0, 0, 1) plan.

For example, a plane intercepts the crystallographic axes at (1, 1, 1), as shown in Fig. 1.13. Therefore, the Miller index of this plane is (111).

Example 1.2.
Find the Miller index for the plane that parallels to the x–y plane, as illustrated in Fig. 1.14. The intercepts of this plane with the three axes are $(\infty, \infty, 1)$. Taking the reciprocal of each number results in (0, 0, 1). Therefore the Miller index of this plane is (0, 0, 1). The direction, normal to the (0, 0, 1), is written as [0, 0, 1]. In the case of cubic crystals, (0, 0, 1), (1, 0, 0), or $(0, 0, \bar{1})$ are crystallographically identical, and they are generalized as {001}.

One may imagine that the different crystallographic planes give different atomic packing densities. Substrate {1, 0, 0} is most commonly used for metal oxide semiconductor (MOS) device manufacturing owing to its low surface state density. Substrates {1, 1, 1} are most often used for bipolar device manufacturing. The {1, 1, 1} substrate has a higher oxidation rate as compared to {001} owing to its high atomic packing density. Furthermore, it is less prone to channeling during ion implantation.

1.5. Wafer Processing Flow

For the purpose of illustrating the wafer process module principles, a simplified process flow for making a MOS transistor is shown in Fig. 1.15. The wafer processing often starts with growing a silicon epitaxial layer. Using an epitaxial wafer substrate is optional for device manufacturing. An epitaxial growth process is to grow a single crystalline thin film on a single crystalline substrate. The reason for using this process is two fold. First, the epitaxial growth results in a purer silicon film, free of carbon and oxygen, for device manufacturing. Second, one can grow a lightly doped film on a heavily doped substrate to achieve a higher breakdown voltage, while minimizing the power dissipation in the substrate. The epitaxial growth is conducted in a cold-wall chemical vapor deposition reactor at atmospheric pressure.

For planar silicon process technology, it is essential to form local isolation (oxide islands) around each individual device so that each can function independently. To form the isolation islands, one must first form a patterned silicon oxide and silicon nitride stack. The silicon nitride has a very slow oxidation rate, and therefore it can be used as oxidation masks. However, the silicon nitride has such a high stress level that when directly deposited atop the substrate, it will damage or even crack the substrate. A stress buffer, such as silicon dioxide, must be deposited between the nitride and the substrate.

To form silicon oxide, the silicon wafers are placed in a furnace system in an oxygen or steam ambient:

$$\mathrm{Si} + \mathrm{O_2} \longrightarrow \mathrm{SiO_2}\,.$$

The oxidation rate is a function of temperature, silicon surface orientation, and oxidation time. On top of the silicon oxide, a chemical vapor deposition is conducted to form a thin silicon nitride film.

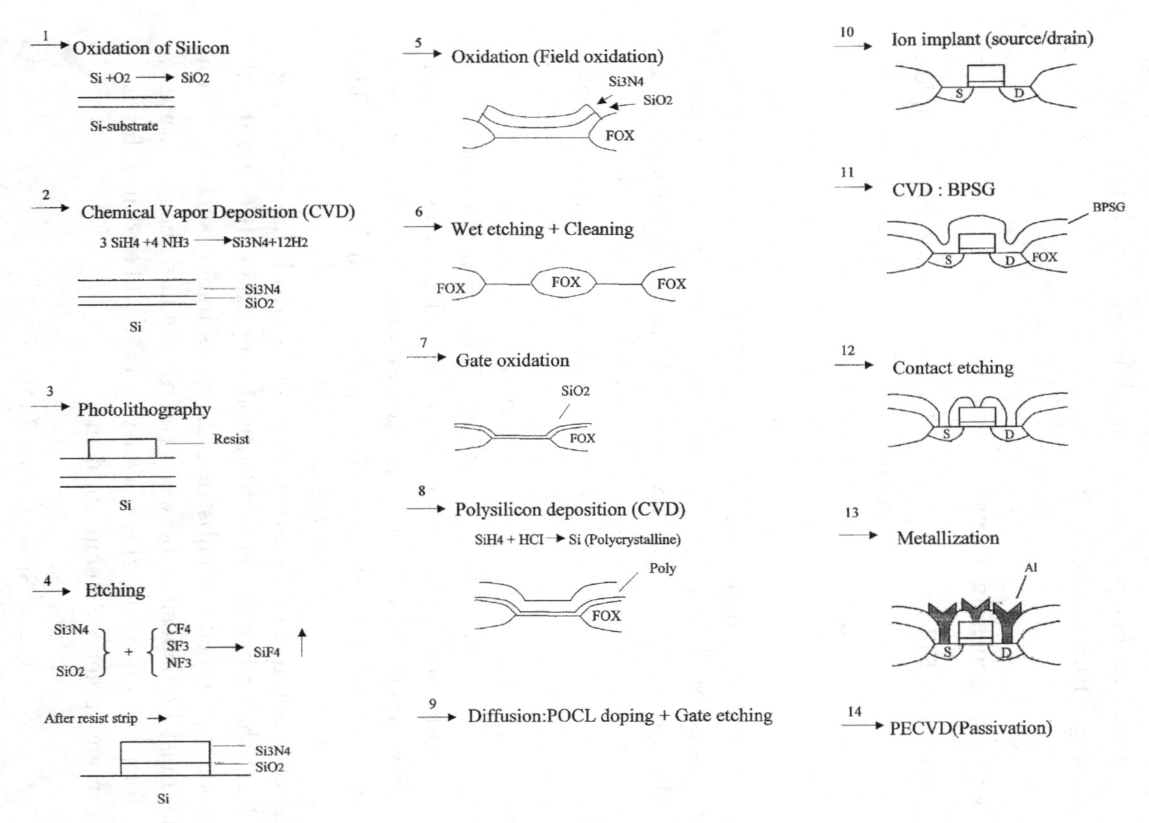

Fig. 1.15. Schematic MOS manufacturing process.

Chemical vapor deposition (CVD) is a process in which a precursor gas mixture passes over a heated substrate and thereby chemical reactions are initiated in the vicinity of the heated substrate, and a nonvolatile solid film is formed. CVD has been widely used in semiconductor manufacturing to deposit both conductive films, such as tungsten and tungsten silicide, and dielectric films, such as silicon nitride and oxide. For the silicon nitride, the reaction proceeds as follows:

$$Si + NH_3 \longrightarrow Si_3N_4 \,.$$

Now that the oxide-nitride stack is formed, the next step is to define the areas where the silicon oxide is to be formed. This is accomplished by using photolithography, which delineates the areas where the nitride has to be removed for the underlying silicon surface to be oxidized. The photolithography process is often followed by an etching process, in which the material in the undesired area is removed.

Photolithography involves complicated photochemical reactions. First, wafers are spin-coated with a photo-active polymer, which turns soluble in an alkaline solution on exposure to light. The exposure light illuminates on a photomask and incident on the polymer-coated wafer surface, generates the desired latent images. On developing with the alkaline solution, the mask pattern is transferred to the wafer surface, as illustrated in Fig. 1.16. Two important

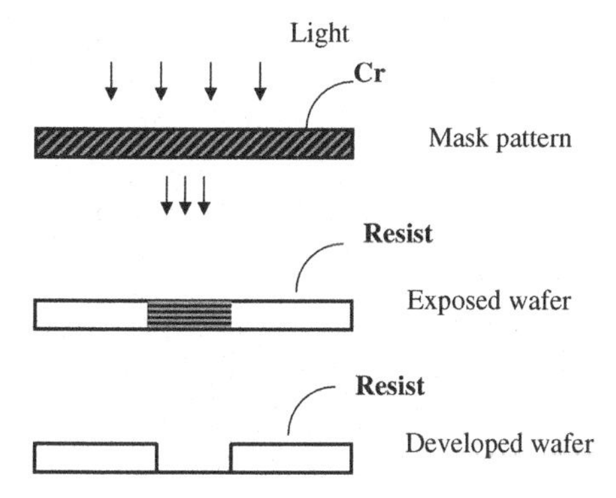

Fig. 1.16. Transferring mask patterns onto the wafer surface by using photolithography.

performance indices in the photolithography process are resolution and depth of focus. Both are related to the wavelength of the exposure light source and the sizes or numerical apertures of the lenses.

Photomasks are chrome patterns on the nearly optically impeccable quartz substrates. The chrome indicates the opaque area, and the quartz indicates the transparent areas. To generate a mask layer, a resist-coated chrome on a quartz blank is exposed to an electron beam, which is driven by the data file from the design layout information. Each local pattern is composed of a large number of electron beam shots. For advanced masks, the number of shots for one mask layer can be as large as billions. The exposed blank is then developed and etched to render the chrome patterns on the quartz. The etched pattern is then inspected using inspection machines, which compare the written pattern to the original data file or compare one die to the next die to assure that the pattern is perfect. Should any differences be detected, they are treated as defects, and they must be removed with a repair tool. Finally, a pellicle is mounted on the mask to protect the pattern from mechanical scratches and to make particles on the pellicle surface unprintable.

Resolution enhancement technology (RET) is required to enhance the photolithography performance owing to the fact that the operating wavelengths do not evolve as fast as the geometry shrinking pace. The most commonly seen RET includes phase shifting masks (PSMs) and optical proximity correction (OPC) techniques. Phase shifting primarily takes advantage of the fact that a partially transparent film with a certain refractive index can shift the phase of the light that comes out of the material. Consequently, by deliberately choosing a film thickness such that the shifted and nonshifted lights are of the opposite phases (180 degrees), the pattern image contrast can be significantly enhanced. OPC is an empirical technique in which the originally designed patterns are altered in such a manner that the printed patterns on the wafer can be matched to those of the originally designed ones, as shown in Fig. 1.17.

Up to this stage, a plasma etching is required to remove the silicon nitride that is not covered with the resist. A plasma etching is a material removal step. Energetic electrons in a plasma system (a partially

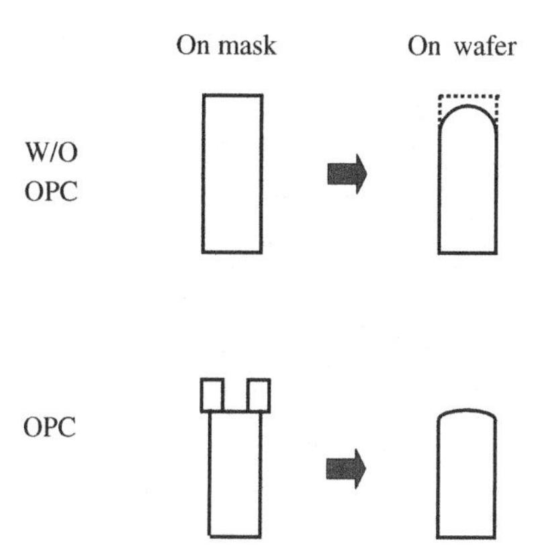

Fig. 1.17. Without OPC, severe line-end rounding is observed; with OPC, the rounding is improved.

ionized gas system) generate highly reactive free radicals, which react with solid surface and form volatile reaction products, thereby removing the surface material. Obviously, to remove the nitride and form the oxidation mask pattern, one must choose the proper chemical to react with the nitride but not with the resist material. After the etching, the resist is removed with a sulfuric acid solution. The wafers with the nitride-defined patterns are sent into a high-temperature furnace for field oxidation. The silicon volume expands as the oxidation reaction proceeds. Consequently, the nitride edge is bent, as illustrated in Fig. 1.18. The field oxide's thickness ranges from 3500 to 6000 Å. When the field oxidation is accomplished, the nitride and

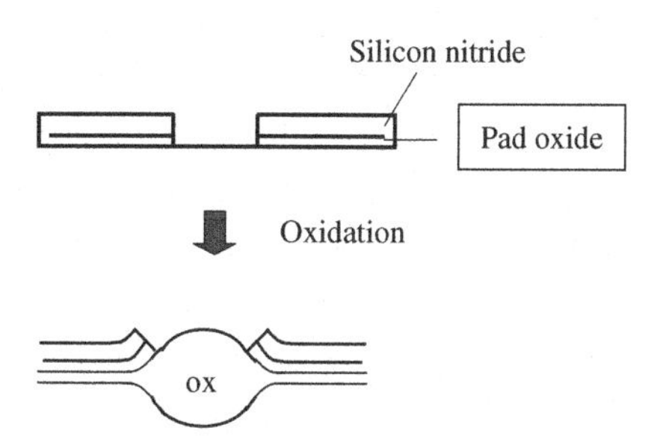

Fig. 1.18. After oxidation, the oxide volume expands, causing the nitride stack edge to bend.

oxide are removed with hot phosphoric acids and an HF solution, respectively, leaving the silicon surface with scattered oxide islands. To form devices, gate oxidation is first conducted to form a thin layer of high-quality oxide for the MOS transistors, followed by the CVD process of polysilicon deposition. The deposited polysilicon is highly resistive and must be doped to reduce its resistance.

The doping process is accomplished through dopant diffusion in a high-temperature furnace. Precursor gas is introduced into a high-temperature furnace, where it converts into dopant atoms, such as P or B, on the substrate surface. The atoms are then driven by thermal energy into the substrate to change the intrinsic polysilicon to doped polysilicon. The resistance change depends on the amount of introduced dopants. Applying photolithography and etching to selectively remove the polysilicon material, one can form gate and interconnect structures. The gate can be used for the ensuing self-aligned source and drain ion implantation.

Ion implantation is the process of doping the substrate in a precise manner. Dopants are ionized and extracted from an ionization chamber into an accelerator. The ion beams are accelerated to a certain high energy level and steered electromagnetically toward the substrate surface. With the accelerated energy, the ions penetrate into a substrate with certain depths. With a fixed ion–substrate pair, the higher the implant energy, the deeper the ion penetrating depth. After the source and the drain are implanted and annealed, the MOS device structure is pretty much complete. One has the source, drain, and gate. Each device is isolated from its surroundings with the field oxide islands and is capable of functioning individually. The process thus far is often called the front-end process. From this step, all the subsequent process steps are often called the back-end process and are supposed to connect these devices together to fulfill the circuit functions. Theoretically, one layer of metal can connect all the MOS transistors together and form a meaningful circuit. However, considering the packing density and line resistances, there is a tendency to use more than one layer of metals. The state-of-the-art 90-nm circuits are often composed of up to eight or ten layers of metals. To form the first-level connection, a thick dielectric layer is deposited. Holes on the dielectric layer are opened up by using photolithography and etching. The connecting material, usually aluminum for technologies of larger than 1.0 μm, is sputtered onto the substrate.

Sputtering is a process conducted in a plasma system. Heavy inert ions, such as argon, are accelerated toward a target material, knocking out the surface atoms, which then fall and land on a wafer surface. The deposited metal layer is then defined with photolithography and etching. Aluminum layers of 5000 to 15 000 Å are often required to form a proper metal interconnection. For a multilayer metal structure, it is very critical to maintain a proper planarization for the subsequent metal layers to form. Planarization plays a vital role in the back-end process.

Planarization is a technique to planarize the wafer surface and facilitate the ensuing process steps. A nonplanar surface is more prone to cause after-etching metal residues, high metal resistances, and poor photolithography performance. There is a wide variety of planarization techniques, ranging from high-temperature flow-, to spin-on-glass, etch-back, and CMP. For inter-metal isolation and planarization, a layer of silicon dioxide or its derivatives needs to be deposited. The final process step in making the device is to deposit a passivation layer, a silicon oxide on nitride stack. It prevents the device from mechanical scratches and mobile ion contamination. Because the deposition process occurs on top of an aluminum layer, the process temperature is limited to below 350°C to maintain the metal's integrity. As a result, the plasma-enhanced chemical vapor deposition (PECVD) technique is used.

PECVD is a film deposition process in which the precursor gas is dissociated into highly reactive radicals by the energetic electrons in plasma. The reactive radicals react with each other and deposit onto the substrate surface to form a nonvolatile film at low ambient temperature. The resulting films are often amorphous and nonstoichiometric in nature.

The above text elaborates the basics of process modules that are needed for making MOS devices. The actual manufacturing can be far more complicated, and it involves repetitive use of each module. The modules also evolve with technology migration. For example, as a device shrinks, the packing density increases, and more devices can be packed into a unit area of the silicon surface; long interconnects are inevitable. For long interconnects, the RC delay is doomed if no alternative approaches are taken. Copper has a far lower resistance than aluminum, and it is used as an alternative to aluminum. On the other hand, copper was not used until CMP became available because it is not plasma etchable. To further lower the RC delay, low dielectric constant materials are needed in addition to the lower interconnect resistance. Silicon dioxide is often doped with F or C to lower its dielectric constant.

1.6. What is the Semiconductor Industry Trying to Achieve?

The entire semiconductor industry pivots on one device, MOS transistor, as shown in Fig. 1.19. In other words, the industry is focusing on making circuits with MOS transistors and on shrinking the circuit feature sizes so as to give more powerful and faster circuits with lower cost.

There are two types of MOS transistors, the NMOS and PMOS. In the case of an NMOS device, the source and drain are commonly phosphorous-doped (n-type, rich in electrons) on the p-type substrate (rich in holes). If one sets both the source and the drain at grounded states, then by changing the applied voltage to the gate from positive to zero, for example, 5 V or state 1, to 0 V or state 0, one can turn the conducting channel on and off. When the conducting channel is

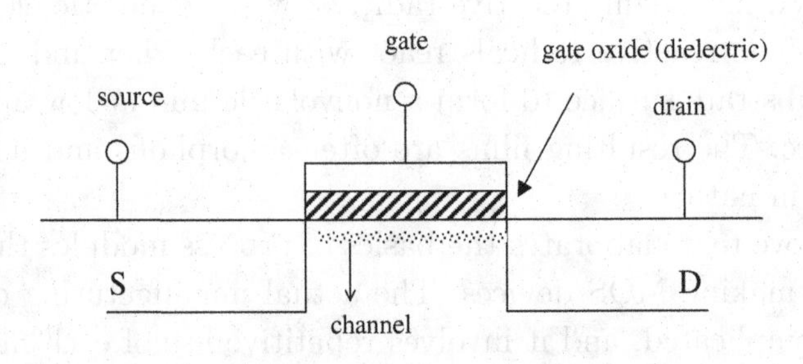

Fig. 1.19. An MOS transistor.

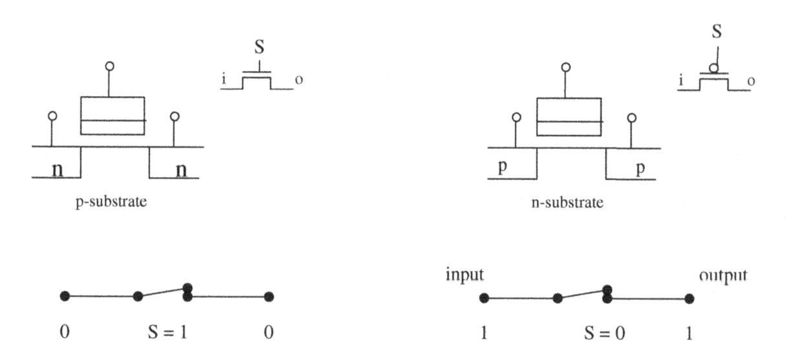

Fig. 1.20. The n-switch and p-switch.

on (at state 1), switching the drain from ground to any positive voltage (typically 5 V) would force electron flow from the source to the drain. Now, if we consider the source as the input and the drain as the output, then the NMOS device can function as an n-switch. If the gate is at 0, the channel is turned off. If the gate is at 1, the channel is turned on, and the state of the source (input) can be transferred to the drain (output). Analogously, one can explain the operation of a p-switch, as shown in Fig. 1.20. The switches are the fundamental building blocks for constructing complicated ULSI circuits. For example, by connecting the n- and p-switches end-to-end, and by tying the gate together, as shown in Fig. 1.21, one can obtain an inverter. Table 1.6 shows the inverter truth table. Various combinations of the switches can be made to represent various Boolean truth

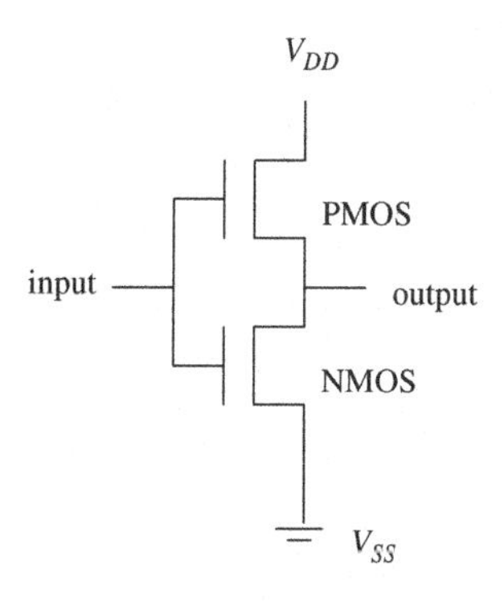

Fig. 1.21. A complementary switch.

Table 1.6. Inverter truth table.

Input	Output
0	1
1	0

tables that can develop into ULSI circuits such as logic, memory, imaging processor, and microprocessor chips.

Ever since it was learned how to put a large number of functional circuits on a silicon wafer by using planar processing technology, a real race began. The industry has been pushing for faster, cheaper, and higher-packing-density chips. Shrinking the chips both in horizontal and vertical dimensions is the only way to achieve all three goals. In the semiconductor industry, the feature sizes of an IC circuit are defined by a set of dimensions, called design rules. These rules are proposed by those who know the process technology capabilities and limitations. The technology nodes, or the manufacturing technology, as we will use in the ensuing contents, for example, 0.5 μm, 0.35 μm, or 0.18 μm, are referring to the transistor gate length. The gate length is the length of the gate in the current flow direction, as indicated by L in Fig. 1.22. Table 1.7 shows the critical feature sizes of 0.5-μm versus 0.35-μm production technology. Approximately, each feature is shrunk by a factor of 70%. The shrinkage in the device geometry calls for a change in process technology.

Figure 1.23 shows the technical evolution compared to the commonly known objects. Beginning in the early 1960s, the industry was making circuits with a characteristic length (most commonly referred to as the gate length) of 10 μm. In the year 2002, the worldwide

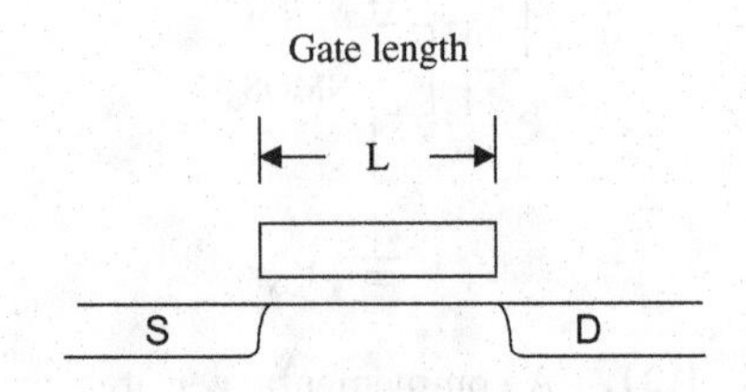

Fig. 1.22. The MOS transistor gate length.

Table 1.7. The critical feature sizes of 0.5-μm versus 0.35-μm technology. A 70% shrinking path (units: μm, line width/space).

	0.5 μm	0.35 μm
Gate length	0.5/0.5	0.35/0.35
1st contact hole	0.6/0.6	0.4/0.4
1st metal layer	0.7/0.65	0.45/0.45
2nd contact hole	0.7/0.7	0.5/0.5
2nd metal layer	0.7/0.65	0.45/0.45
3rd contact hole	None	0.6/0.6
3rd metal layer	None	0.7/0.7

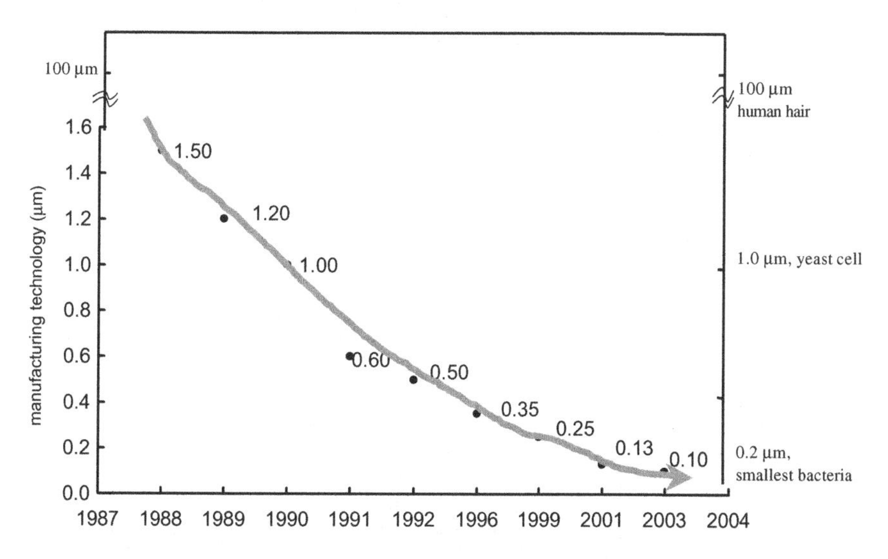

Fig. 1.23. Evolution of semiconductor manufacturing technology with respect to the known objects.

semiconductor industry was running a 0.13-μm technology for production. The 0.13-μm is much smaller than the size of a virus cell and measures only hundreds of atoms across. Figure 1.24 shows the gross number of dice that can be made per an eight-inch wafer for a 1-M (megabit) static random access memory (SRAM). By using 0.5-μm processing technology, one can have about 850 dies per eight-inch

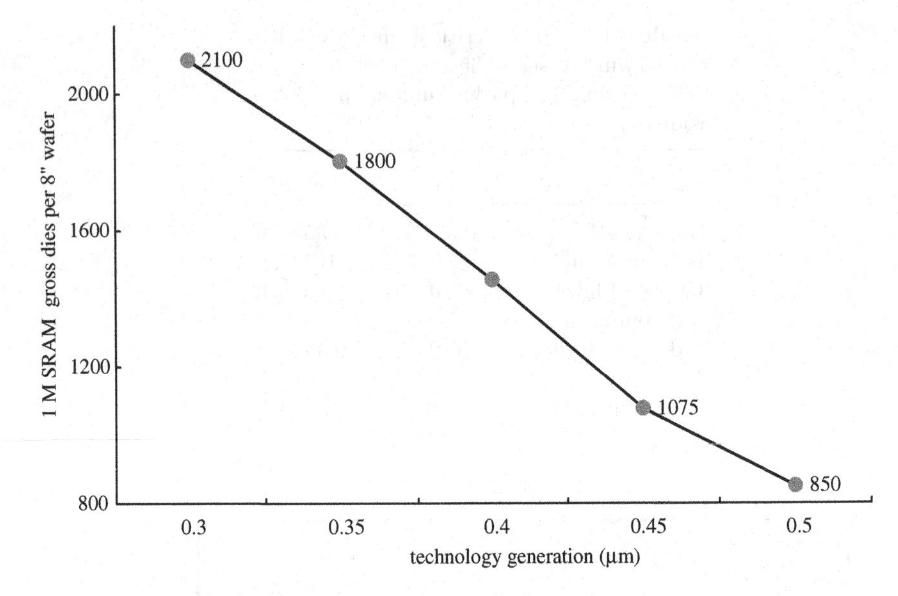

Fig. 1.24. The gross dies per wafer for 1-M SRAM made with different technology generations.

wafer. It will have about 1800 dies per eight-inch wafer when shrunk to 0.35-μm processing technology. The increase on dies/wafer is over 50%. Therefore the cost per die can be reduced by approximately 50%, assuming that the process cost does not increase significantly. Because the transistor gate length is scaled down by 70%, the circuit speed will significantly increase as well. Figure 1.25 indicates that the transistor saturation current increases as the gate length is scaled down. The device speed increases with the saturation current. Furthermore, one can clearly see from Fig. 1.3 that owing to the shrinkage from 0.8 μm to 0.4 μm, one can use a 0.4-μm technology to make a 64-M DRAM of 180 mm^2, as opposed to using 0.8 μm to make a 16-M DRAM of 130 mm^2. It is clear that the die area does not increase proportionally with the memory sizes due to technology migration. The shrinkage nearly triples the packing density. To shrink the device is much more involved than it appears. Sometimes, it extends the applications of existing module technologies, and other times, it requires new module technologies.

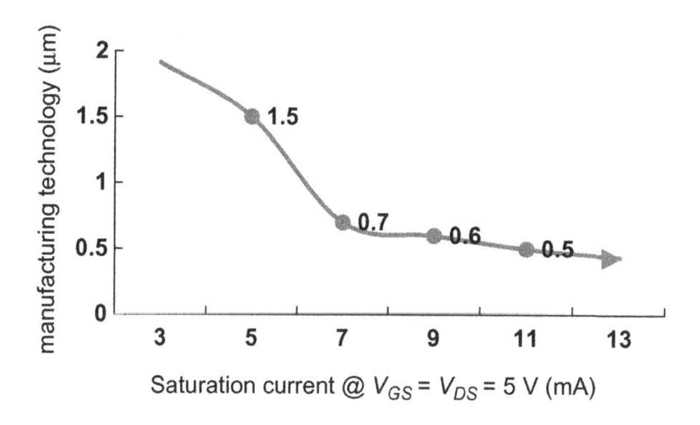

Fig. 1.25. Saturation current, directly related to device speed, increases as gate length shrinks.

As a result of the devices shrinking, more chip functions with faster speeds can be packed into a tiny chip at a lower cost. Consequently, applications of semiconductor devices have been multiplying over the last 50 years. The applications are so widespread that semiconductors affect almost all facets of our daily lives, ranging from computer information systems, communications, entertainment, and medical instruments to automotives. As the semiconductor technology continues to improve and advance, it will certainly result in wider and more powerful applications at lower prices.

1.7. The Never-Ending Effort — Yield Improvement

As the device geometry approaches the range of one-tenth of a micron, its functionality and yield strongly correlate to the cleanness of the manufacturing environment and its incoming materials. Any size of particles or contaminants that fall on a device during its manufacturing process can cause the device to become dysfunctional. Figures 1.26–1.28 illustrate that there are several cases in which particles can reduce yield:

(a) A particle falls on a mask, causing printable defects, and extra patterns show up on the wafer, as shown in Fig. 1.26.

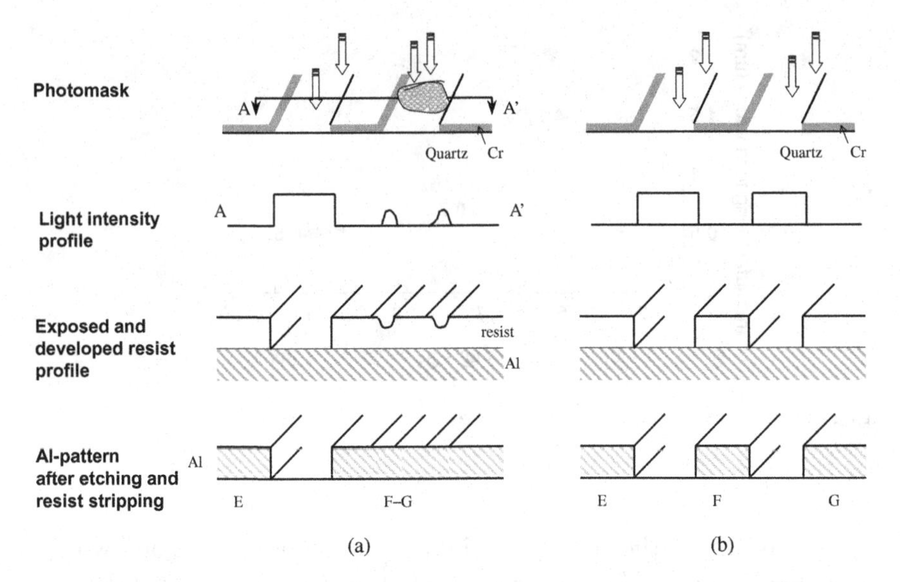

Fig. 1.26. (a) A particle on mask causing metal lines, F and G, to bridge, and (b) a clean mask gives three clear-cut metal lines: E, F, and G.

(b) A particle falls on a resist layer, preventing the resist from being exposed and hence causing the underlying conducting layer to be electrically shorted, as illustrated in Fig. 1.27.

(c) A particle falls on an oxide layer prior to the metal deposition process, causing a bump, which is etched away. This will result in a metal line opened electrically, as demonstrated in Fig. 1.28.

The defects on a wafer at any stage of the wafer processing can be detected by wafer defect inspection machines. It should be noted that not every defect would lead to wafer yield loss. However, to achieve high wafer yield, it is necessary to keep the defect count on a wafer as low as possible. The following equation is often used to correlate defect density and the wafer yield:

$$Y = \frac{1}{(1 + DA)^m},$$

where m represents the complexity of the manufacturing process and A represents the die area. In addition, D represents the defect density,

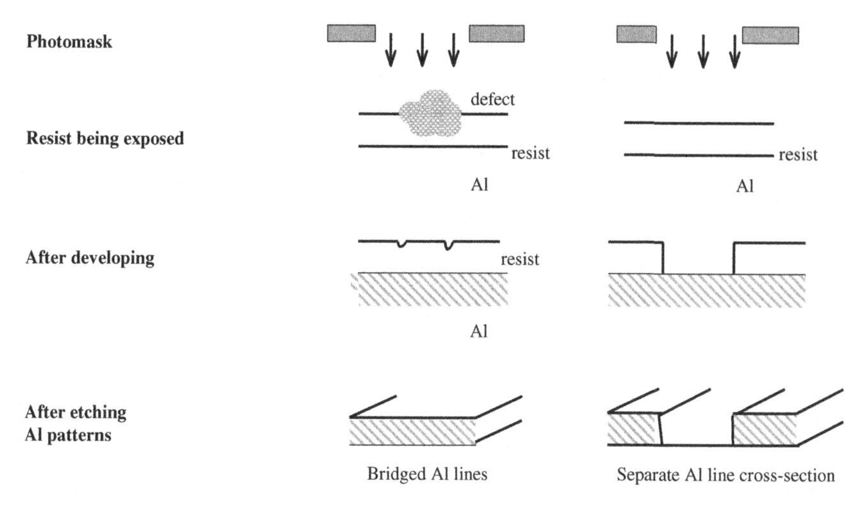

Fig. 1.27. A particle falls on a resist surface that is being exposed, causing Al lines to bridge under the particle.

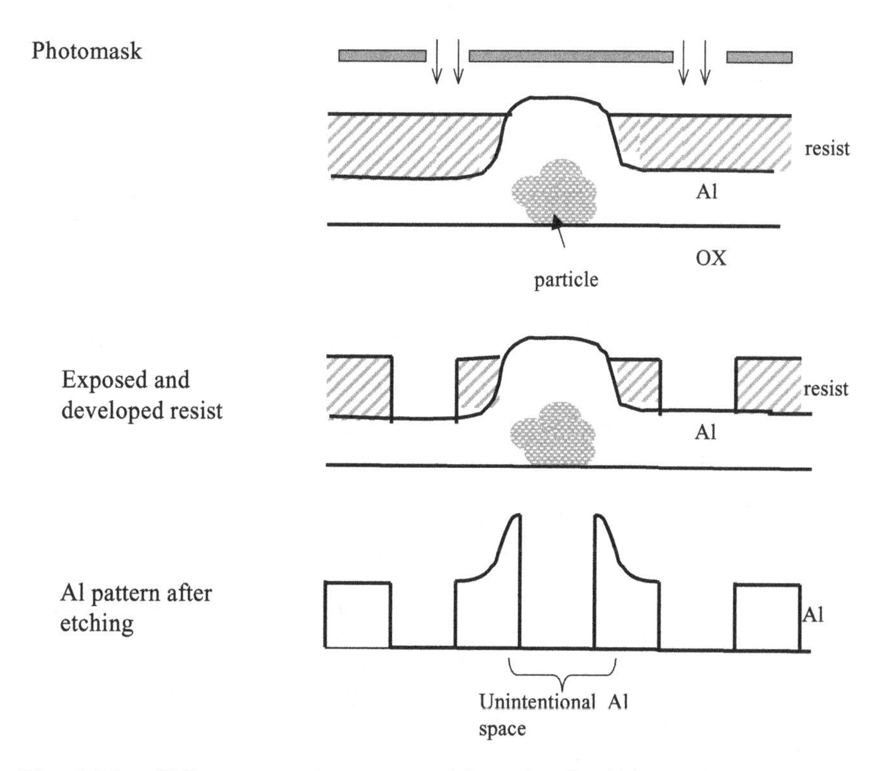

Fig. 1.28. Al line open owing to a particle under the Al line, causing a bump.

which is not necessarily equal to the defect count, as inspected by defect inspection machines. Rather, it is a number that characterizes the cleanness of a process line. The wafer yield, Y, equals the ratio of the final total number of good dies to the total number of dies made on the same wafer.

The defects come from various sources. To obtain a good wafer yield, the defect level needs to be kept as low as possible. The important aspects essential to achieving low defect levels are discussed as follows:

(a) The clean room environment. A clean room setup is a basic requirement for making semiconductor devices. Depending on the final device geometry, the cleanness (the class) requirements of the clean room could be different. For a class 1 clean room, the specification requires that the number of particles that are larger than 0.5 μm be less than one in a cubic foot space. Figure 1.29 shows a schematic of a conventional clean room. The air pressured by a fan filtering unit (FFU) goes through an ultra low penetrating air (ULPA). The ULPA has an efficiency of 99.999% in filtering 0.1-μm particles. In other words, only one out of 100 000 0.1-μm particles can penetrate through the ULPA. The filtered air flows downward in a laminar flow fashion so as not to disturb any more particles from the equipments and table surfaces or dead corners. The working tunnels stand on a raised grid floor to allow for continuous air circulation and filtration. In the air duct, there are air-conditioners that control the air temperature and humidity. Typically, a clean room is controlled at 45 ± 3°C and 45 ± 5% relative humidity. In addition, the particles, if generated in the working tunnels, can be carried down the grid floor and circulated toward the ULPA, and therefore fresh air (free of particles) is regenerated.

(b) Inappropriate process steps. For example, during a cleaning process, the wafer goes through acid and base solutions. If one or the other solution is not thoroughly rinsed, the reaction of the acid and base will result in salt formation, which often leads to particles. Particles can be generated in processing equipment as well. A moisture

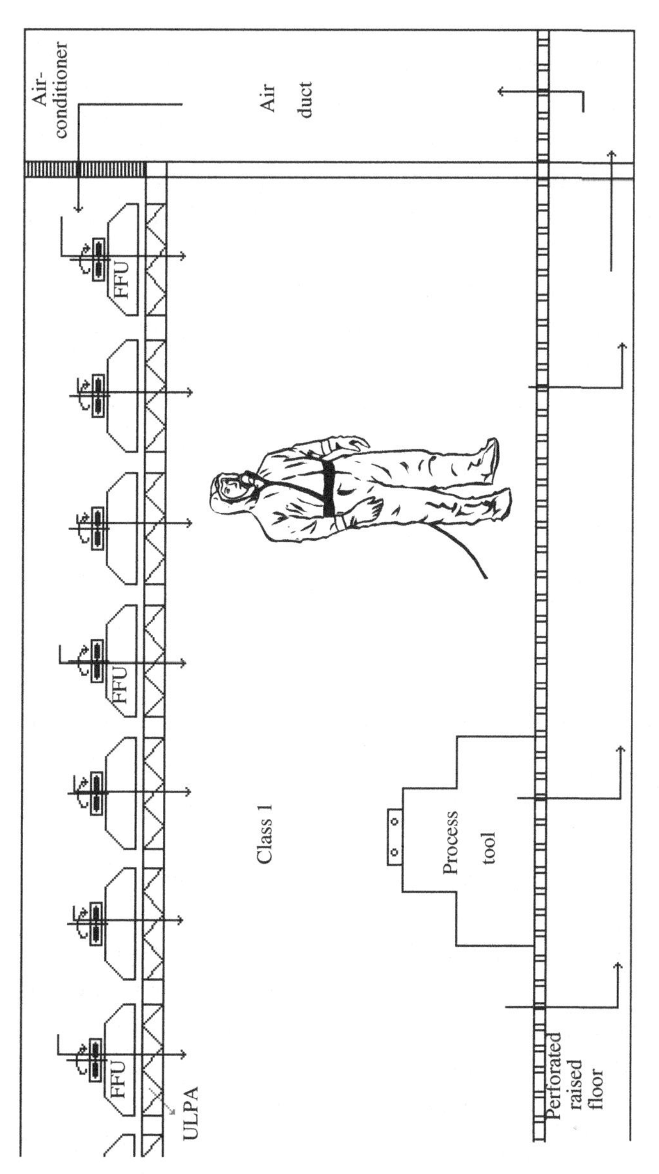

Fig. 1.29. A conventional clean room.

leak or an aging O-ring in a plasma chamber can result in particle formation. Process monitoring and equipment maintenance are critical in maintaining a clean semiconductor manufacturing line.

(c) Contamination from personnel. Particles can be generated from human bodies such as hair, cosmetics, perspiration, or saliva. There are two methods to keep an environment clean. One is to wrap the wafers, and the other is to wrap people. All personnel working in a clean room are required to wear clean gowns and breathing masks, as shown in Fig. 1.29. The air that is breathed out must be filtered before it is released to the ambient. This type of clean room gear is very uncomfortable. In a SMIF environment, each machine is equipped with a mini-environment chamber. People are not required to wear clean gowns similar to those shown in Fig. 1.29. Also, no out-breathing filtering mask is required. Therefore, it is far more comfortable for humans. Discipline in gowning is crucial in controlling the clean room environment.

(d) Chemicals and water quality. To manufacture IC of nanometer-range geometry, one has to be very stringent on the quality of the supply water and chemicals. For water, there are a few items to be noted, including organics, colloids, organisms and particulates. Humic and fulvic acids are often the organics that are seen in water, with molecular weights ranging from several hundreds to thousands g/mole. Colloids are aggregates of molecules with sizes ranging from 50 to 20 000 Å, and are invisible under microscopes. Organisms in water include bacteria, fungi, and viruses. The sizes are between 0.3 and 30 μm. For chemicals, particle counts and sizes are major concerns. The above-mentioned contaminants can be removed using various techniques. Ion exchangers can be used to remove ionic species and particulates that are larger than 5 μm. Powered ion exchangers can remove even smaller particulates. Some organics and colloids (aggregates of molecules) can be separated from the liquid using polymer adsorbents or activated carbons. Membranes with different pore sizes are often used to remove ionic and organic species. The pore sizes of the membranes are the key to their effectiveness in the removal process.

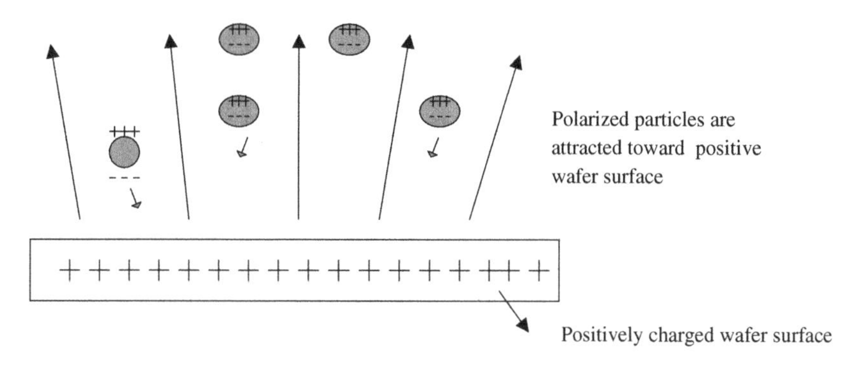

Fig. 1.30. A positively charged wafer surface can polarize particles, which are then attracted toward the wafer surface.

(e) Electrostatics in clean rooms. Electrostatic charges exist in clean rooms and can possibly result in wafer particle contamination or burnt-out circuits. A silicon wafer can charge up heat, evaporation, or its friction against the wafer container. It can also be inductively charged by its neighboring wafers or process equipment. The charged wafer surface can polarize and attracts floating particles, resulting in defects on wafers, as illustrated in Fig. 1.30. Movements or actions of workers in a clean room often generate static charges. These charges, if not properly discharged, can lead to damaging effects on the circuits or defects on the wafers. There are several common practices in a clean room that reduce the electrostatic charges: (1) air ionization is an approach that applies an intensified electrical field across a sharp point or edge to ionize the air; the resulting ions are carried away by the laminar flow in a clean room; (2) grounding all conducting surfaces and human bodies to eliminate the static charges; and (3) proper humidity control. High humidity tends to decrease the static charges formation. However, a clean room cannot be operated at a high humidity level since the equipment can rust. Normally, clean rooms are operated at about 45% relative humidity.

Chapter 2

BUILDING BLOCKS FOR INTEGRATED CIRCUITS

Following the semiconductor overview in Chapter 1, we will now look at the fundamental concepts of semiconductors, followed by the introduction of a few building blocks that comprise integrated circuits, and their underlying principles. These building blocks include resistors, *pn* junctions, and capacitors. The fundamentals of a metal oxide semiconductor (MOS) will also be introduced. Finally, a few categories of integrated circuits are demonstrated, with explanations of their major features.

2.1. Fundamental Semiconductor Concepts

There are myriads of publications on the theories of semiconductor devices. In this section, we will review the most fundamental equations required for understanding the basics of how such devices function. In addition, we will review important terms that are cited and used daily by those in the industry.

2.1.1. *Energy band theory and chemical bonding theory*

The number of free electrons in a material is the key to its electrical conductivity. Two theories (the energy band theory and the chemical bonding theory) are often used to interpret the mechanism of material conductivity.

An atom consists of a positively charged nucleus surrounded by negatively charged electrons, which revolve around the nucleus. The nucleus includes positively charged protons and neutral neutrons.

Owing to the attraction force between particles of opposite charges, each electron is acted on by the nucleus. According to Bohr's model, each electron takes a specific energy level:

$$E_n = \frac{-Z^2 m_0 q^4}{8\varepsilon_0^2 h^2 n^2} . \tag{2.1}$$

The energy level is related to a vacuum level, taken as zero; Z is the atomic number of the atom, m_0 is the free electron mass, q is the charge of the electron, ε_0 is the permitivity, h is the Planck's constant, and n is an index for the energy level.

Pauli's exclusion principle requires that each energy level accommodate only two electrons. When two atoms are brought together, the electrons of each atom are influenced by the two nuclei, and the original energy level is split into a so-called modified level to accommodate four electrons instead of two. As more atoms are brought together, such as in a solid material, the energy level is widened into a continuous energy band, which accommodates $2N$ electrons. N is the number of atoms in the solid material, as shown in Fig. 2.1. In a solid, the electrons that are farthest away from the nuclei, or the most loosely attracted by the nuclei, are called valence electrons. The energy band that accommodates these electrons is called the valence band. No valence electrons will conduct any electric current until they are freed from the valence band and pumped into the conduction band, where they are free to move around. At 0 K, the valence band is filled up with electrons, while the conduction band is empty. No electrons are allowed to take any energy levels between the conduction and valence bands; this gap is called the forbidden gap. The energy that is required to pump electrons from a valence band to a conduction band determines whether the material is a good conductor or not. For conductors, such as metals, the valence band and conduction band basically overlap, as illustrated in Fig. 2.2. On the other hand, for an insulator, the band gap is as large as several electronvolts. For example, the band gap of silicon dioxide is 8 eV. Meanwhile, the band gaps of semiconductors fall in between those of insulators and conductors.

The chemical bonding theory interprets this from another perspective. In a solid, atoms are chemically bonded to each other.

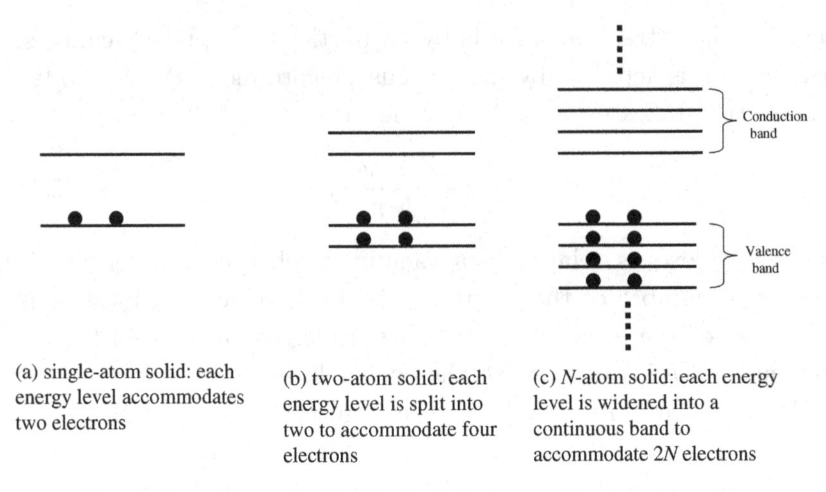

Fig. 2.1. Split of energy level as the number of atoms increases.

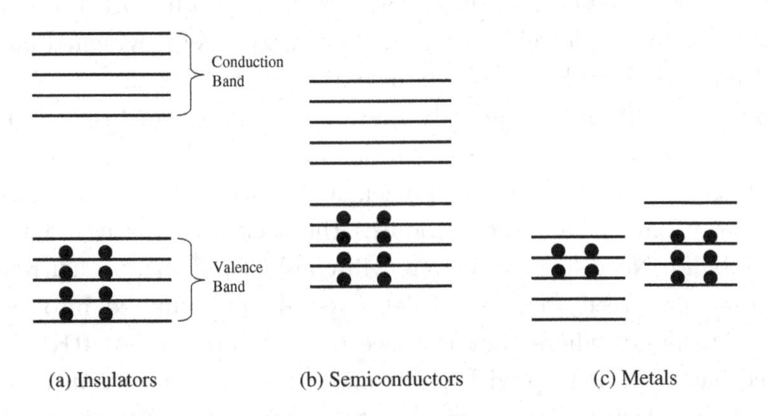

Fig. 2.2. Band diagrams of various materials.

Silicon or germanium has four valence electrons. When the atoms are chemically bonded, they form covalent bonds; each bond requires two electrons and is formed with one electron from each atom. No free electrons are available to conduct electric current until some of the bonds are broken by external forces. Each dissociated covalent bond generates two free electrons, which are subject to external electrical fields that force the electrons to move around and conduct current.

The amount of energy that is required to break a chemical bond is called bond dissociation energy. The band gap energy corresponds to the bond dissociation energy. The electrons resulting from the broken bonds can randomly move away from their original locations, leaving a vacancy, or a hole. As the electron moves from one vacancy to another, the hole moves in exactly the opposite direction. Each broken bond generates electron–hole pairs, and the concentration of electrons equals the concentration of holes in an intrinsic semiconductor material. The concentration is called the intrinsic carrier concentration:

$$n_e = n_p = n_i \,. \tag{2.2}$$

From the above discussion, it is apparent that n_i is a function of temperature and the band gap, as shown in Fig. 2.3. At the same temperature, a material that has a smaller band gap tends to have a higher intrinsic carrier concentration. The intrinsic carrier concentration of a material increases when the temperatures increase. Because

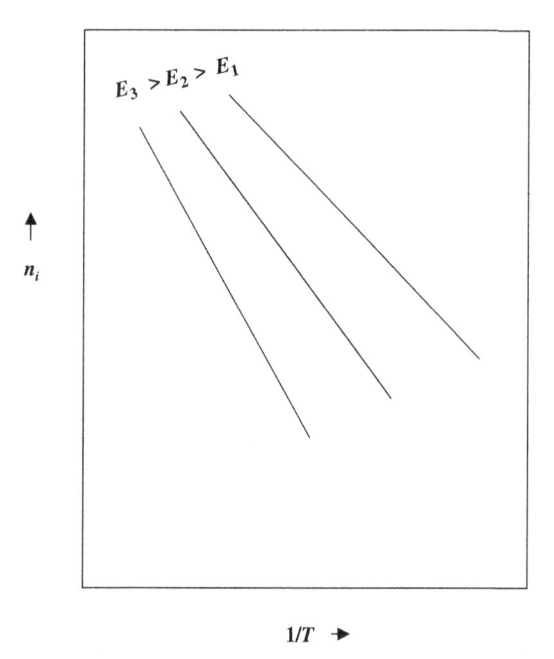

Fig. 2.3. The relationship between the intrinsic carrier concentration (n_i), temperature (T), and band gaps (E).

the lattice vibration is more rigorous at higher temperatures, the covalent bonds have a higher probability of breaking.

2.1.2. *Doping of a semiconductor element*

Doping is the process of adding foreign atoms to a semiconductor material. The foreign atoms are called dopants in the doping process. The most commonly used dopants are those elements in columns III and V of the periodic table. In a silicon crystal, each silicon atom has four valence electrons; it forms covalent bonds with its neighboring silicon atoms, as shown in Fig. 2.4. When the dopants are added into the crystal lattice, they take the place of silicon atoms at the crystal lattice site and are also covalently bonded to the neighboring silicon atoms. If the dopant has five valence electrons, such as P, As, or Sb, after forming the covalent bonding, there is one unbonded electron. This electron, when subjected to an external electrical field, becomes a charge carrier and conducts current. On the other hand, if the dopant happens to be Ga, B, or In, which have three valence electrons, it lacks one electron to form four covalent bonds with its neighboring silicon atoms. As a result, one vacancy (hole) is left, which can be easily filled by the neighboring moving electrons. In a

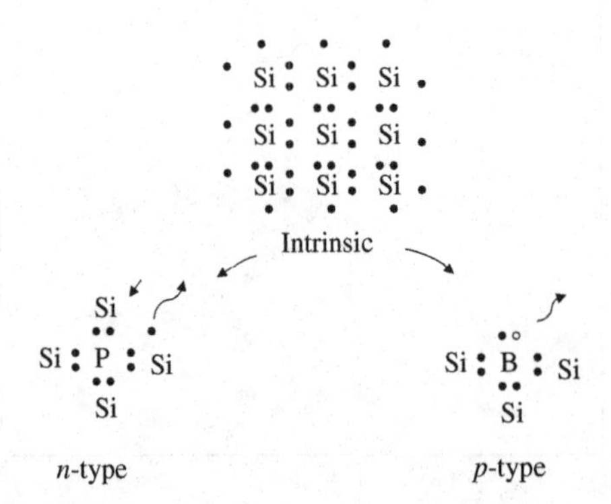

Fig. 2.4. A silicon atom forms covalent bonds with its neighboring silicon atoms. When doped with P, one electron is left unbounded; with B, one extra hole is formed.

silicon crystal, the motions of electrons and holes are equal and in opposite directions, and therefore the holes can be considered as positive charge carriers. Elements in column V of the periodic table are called n-type dopants or donors as they provide extra electrons when doped into silicon. On the other hand, elements in column III are called p-type dopants or acceptors when doped into silicon as they provide extra holes.

The extra electrons and holes resulting from the addition of dopants can be easily removed. At room temperature, the crystal lattice vibration energy is large enough to allow a complete ionization of these dopant atoms. In other words,

$$\begin{aligned} N_{\mathrm{D}} &= \text{donor-type dopant concentration} \\ &= \text{concentration of electrons} = n\,, \\ N_{\mathrm{A}} &= \text{acceptor-type dopant concentration} \\ &= \text{concentration of holes} = p\,. \end{aligned}$$

In both cases, n and p are much larger than n_i, which is the intrinsic carrier concentration. For complete ionization, all donor dopants become positive ions after they provide the extra electrons; the acceptor dopants become negative ions. Considering space charge neutrality,

$$\begin{aligned} q(p + N_{\mathrm{D}} - n - N_{\mathrm{A}}) &= 0\,, \\ \text{i.e.,} \quad p - n &= N_{\mathrm{A}} - N_{\mathrm{D}}\,. \end{aligned} \tag{2.3}$$

From the analysis of quantum mechanics, it indicates that

$$np = n_i n_i\,. \tag{2.4}$$

Therefore the following can be derived:

$$\text{if } N_{\mathrm{D}} - N_{\mathrm{A}} \gg n_i\,, \quad \text{then } n = N_{\mathrm{D}} - N_{\mathrm{A}}\,. \tag{2.5}$$

The corresponding hole concentration is

$$p = \frac{n_i^2}{N_{\mathrm{D}} - N_{\mathrm{A}}}\,. \tag{2.6}$$

Example 2.1.
Silicon is doped with 10^{16} As atoms/cm^3; find the concentrations of the electrons and holes.

Solution
Assuming complete ionization, $n = N_\text{D} = 10^{16}$ cm^{-3} and $p = n_i^2/n = (1.45 \times 10^{10})^2/10^{16} = 2.1 \times 10^4$ cm^{-3}.

2.1.3. *Mobility*

There are three major driving forces that cause electrons to move from one location to another in a solid, as shown in Fig. 2.5. One of them is the diffusion flux due to nonuniformity of the electron distribution; electrons flow from high to low concentration areas. The flux is proportional to the concentration gradient. The other two major driving forces, random thermal motion and drifting, occur regardless of the uniformity of the electron distribution. Random thermal motion provides no net current flow over a long period of time. The

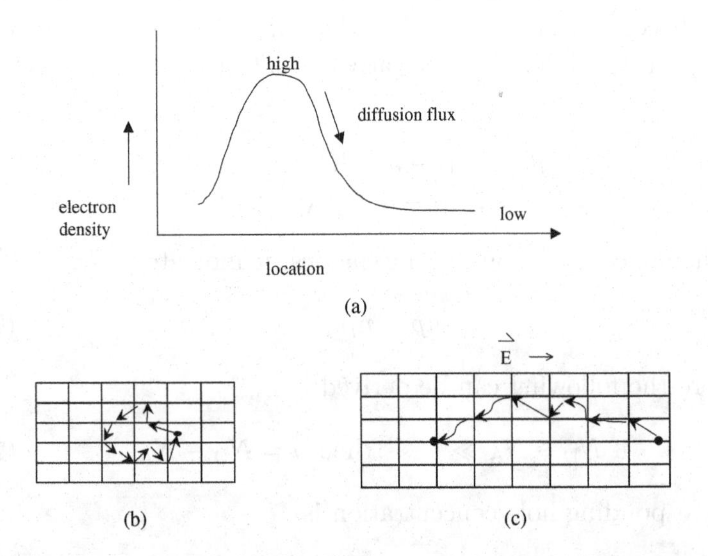

Fig. 2.5. Driving forces for electron flow in a solid. (a) Diffusion due to nonuniform spatial distribution of electrons; (b) random thermal motion; and (c) drifting due to external electrical field.

drifting component is due to an external electrical field. Under the electrical field, $\vec{E}$, the acceleration of an electron can be expressed as

$$\vec{a} = \frac{q\vec{E}}{m_e}, \tag{2.7}$$

where $\vec{a}$ is the acceleration of the electron, $\vec{E}$ is the electrical field, and m_e is the mass of the electron.

Along the direction of the movement, electrons can be scattered either by impurity ions or by a crystal lattice. On each scattering, electron velocity is reset to zero and reaccelerated. Imagine when the time interval between two consecutive collisions is t_{coll}: the average velocity, V_{drift}, is

$$V_{\text{drift}} = \frac{q\vec{E}t_{\text{coll}}}{2m_e} = \mu\vec{E}\,. \tag{2.8}$$

The mobility, μ, is the average drifting velocity of an electron at unit electrical field. Owing to impurity ion collision and crystal lattice scattering, the carrier's mobility decreases with total dopant concentration. Furthermore, Fig. 2.6 shows that at low temperatures, the mobility varies according to the dopant's concentration. Above room temperature, the mobility decreases with higher temperatures, and the dopant's concentration dependency tends to diminish. This is because at low temperatures, the impurity ion scattering dominates, while at high temperatures, the crystal lattice scattering tends to take over.

2.1.4. *Resistivity*

Considering a conducting bar with an electron density of n, and with an applied voltage of V, the current flow in this conductor can be expressed as

$$\text{Current flow, } I = qnV_{\text{drift}}A\,. \tag{2.9}$$

With Eq. (2.8), one can derive

$$\text{Resistivity, } \rho = \frac{1}{q\mu n}\,. \tag{2.10}$$

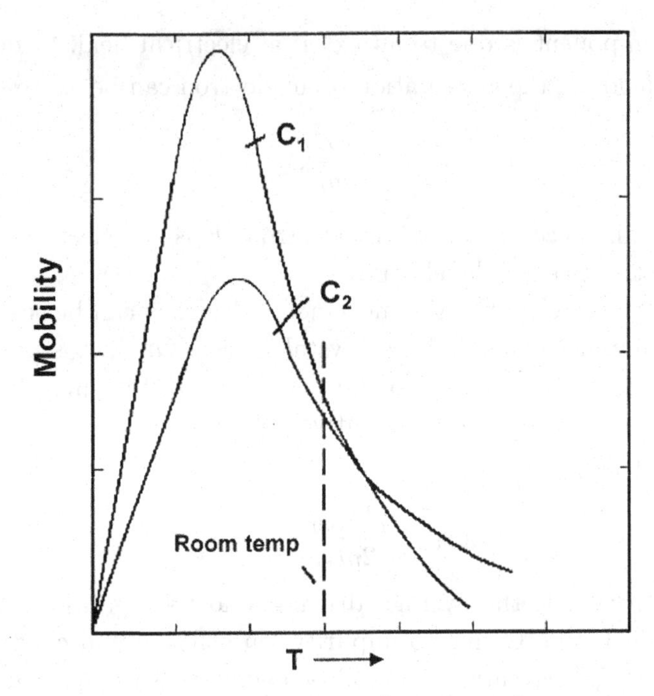

Fig. 2.6. At low temperatures, mobility depends on dopant concentrations, while at high temperatures, it is dominated by lattice scattering.

It shows that the resistivity of an n-doped or p-doped silicon decreases, as the n-type or p-type dopant concentrations increase, respectively. It can also be derived that when p- and n-type dopants coexist in a silicon material, the resistivity of the material equals

$$\frac{1}{q(\mu_n n + \mu_p p)} . \tag{2.11}$$

The mobility is a function of total impurity concentration.

Example 2.2.
At 300 K, a silicon material is doped with $1.3 \times 10^{16}\,\mathrm{P/cm^3}$ and $10^{16}\,\mathrm{B/cm^3}$.

Find, (a) the equilibrium concentration of the electrons and holes
(b) the mobility of the electrons and holes
(c) the resistivity of the doped material.

Solution

Assuming full ionization, $N_A \gg N_D$, which gives a p-type silicon,

(a) the concentration of the holes in silicon $= p = (1.3 - 1.0)\times 10^{16} = 3 \times 10^{15}\,\text{cm}^{-3}$ and the concentration of the electrons $= n = n_i^2/p = (1.45 \times 10^{10})^2/(3 \times 10^{15}) = 7 \times 10^{14}\,\text{cm}^{-3}$

(b) for total impurity concentration of $2.3 \times 10^{16}\,\text{cm}^{-3}$, one can find in reference literature (e.g., Grove, 1967) that the mobility of the electrons and holes are 9.5×10^2 and $3.5 \times 10^2\,\text{cm}^2/\text{Vs}$, respectively

(c) therefore, the resistivity of the doped silicon is

$$\begin{aligned}\rho &= \frac{1}{q(\mu_n n + \mu_p p)} \\ &= \frac{1}{1.6 \times 10^{-19}(7 \times 10^{14} \times 950 + 3 \times 10^{15} \times 350)} \\ &= 3.65\,\Omega\,\text{cm}\,.\end{aligned}$$

2.2. Resistors

In semiconductor processing, resistors can be formed by doping a deposited thin film or silicon substrate. The doping can be achieved either by an ion implantation or by dopant diffusion, and it is often accomplished in a selective manner. In particular, the silicon area intended to be doped is defined by a mask, which could be either silicon dioxide or silicon nitride. The mask can block the diffusing atoms or the implanting ions. As discussed earlier, the resistivity of a doped silicon equals

$$\frac{1}{\rho} = q(\mu_p(x)p(x) + \mu_n(x)n(x))\,. \tag{2.12}$$

Owing to the nature of the diffusion process, which will be discussed later, the dopant concentration in the silicon substrate is not uniform; instead, it is a function of the depth. The mobility is a function of the dopant concentration, and therefore the conductance of a diffused

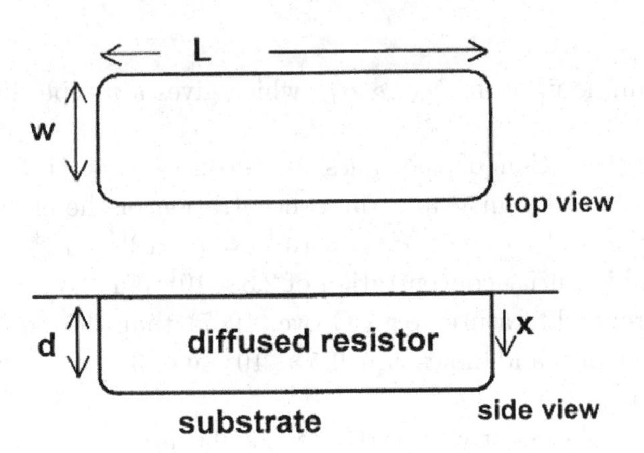

Fig. 2.7. Schematic of a diffused resistor.

resistor, as shown in Fig. 2.7, can be expressed as

$$U = \frac{1}{R} = \frac{A}{\rho L} = \int_0^d q(\mu_p(x)P(x) + \mu_n(x)n(x))wdx/L\,, \qquad (2.13)$$

where d is the depth and w is the width of the dopant profile. This value can be evaluated by numerical integration, or to approximate, one can take the bulk-averaged dopant concentration and mobility. Consequently, Eq. (2.13) becomes

$$U = q(\bar{\mu}_p\bar{p} + \bar{\mu}_n\bar{n})w/L = \frac{1}{R_s L/w}\,. \qquad (2.14)$$

In other words, the total resistance of the resistor is,

$$R = \frac{R_s}{\square}\,, \qquad (2.15)$$

in which $\square$ is the number of squares determined by circuit design (as shown in Fig. 2.8), while R_s is the resistance per square, determined by diffusion process. It includes temperature, time, and dopant concentration at the surface, as will be discussed later.

The other resistor type is formed by a thin film deposition via chemical vapor deposition. An example of this is the polysilicon resistor. Such an application is often observed in the static random access memory (SRAM) manufacturing process, in which the polyload is formed by about 550 Å of polysilicon deposition at temperatures

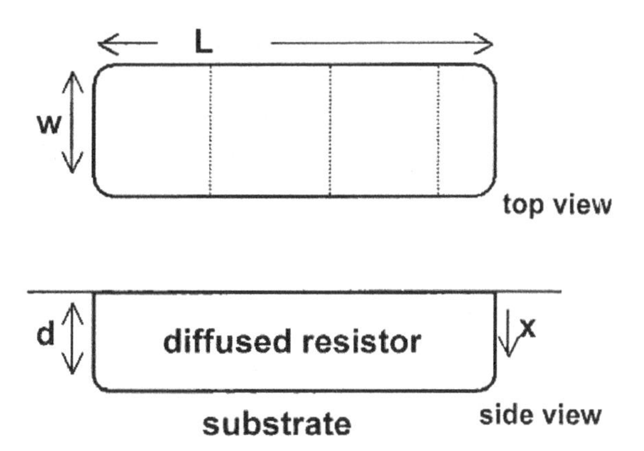

Fig. 2.8. The number of equivalent squares for a diffused resistor.

ranging between 600°C and 800°C. This is followed by an implantation process, using phosphorous, arsenic, or boron. After the implantation process, an annealing process is often required to activate the implanted dopants so that the desired resistances can be reached. The resistance is determined by choosing an appropriate implant dosage, annealing temperature, and resistor length. Figure 2.9 shows a typical polysilicon resistor. Both ends are heavily implanted with n-type dopants to achieve a good electrical connection with other

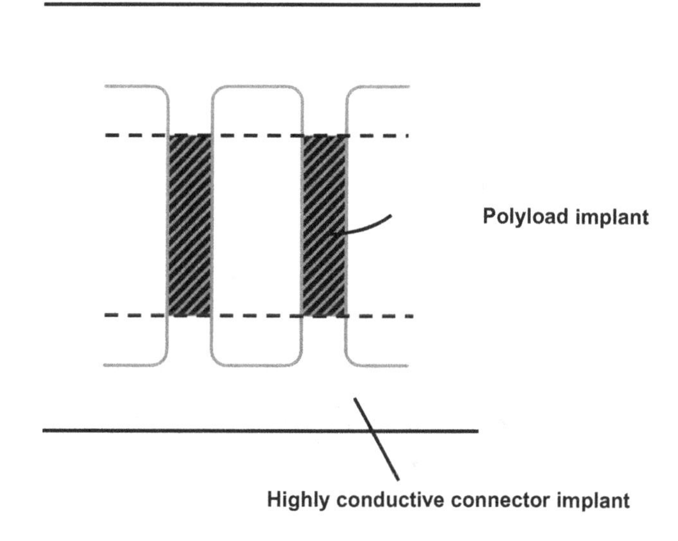

Fig. 2.9. A low-dosage polyload implant.

parts of the circuit. The lightly doped area between the heavily implanted areas is called the resistor load region. Typically, with a resistor length of 1.2 μm without any implant, the resistance value is about 5 TΩ; meanwhile, with 1.0×10^{13} As/cm^3 implant dosage, the resistance value is about 10 GΩ.

2.3. The *pn* Junctions

The *pn* junctions exist almost everywhere in a MOS circuit. Whenever *n*-type dopants are introduced into a *p*-type substrate, or vice versa, a junction is formed, as demonstrated in Fig. 2.10. By taking a closer look at the area near the junction, one can see that on one side, there is an *n*-type substrate, rich in electrons (the majority carriers); on the other side, there is a *p*-type substrate full of holes (the majority carriers). At the junction, electrons in the *n*-type substrate, and the holes in the *p*-type substrate, tend to swiftly combine, causing the junction region to be deficient in carriers. This area is called the depletion region, as shown in Fig. 2.11. The *n*-type dopants in

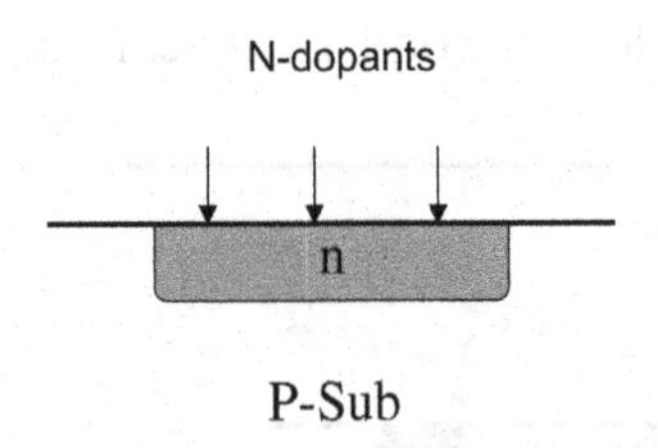

Fig. 2.10. A *pn* junction.

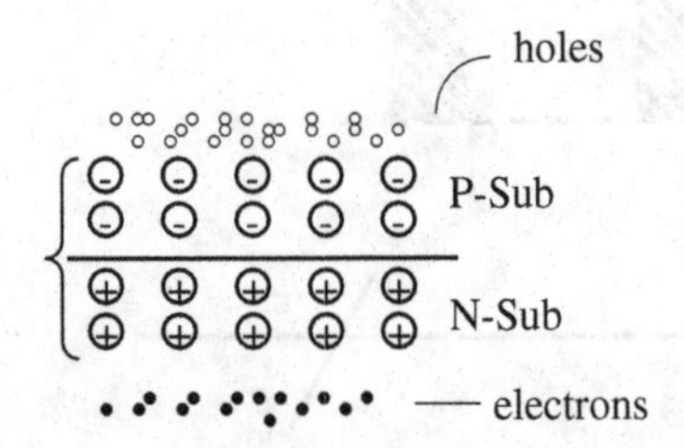

Fig. 2.11. The depletion region.

the n-type substrate near the depletion region become positive ions after losing electrons to the combination; on the other hand, the p-type dopants in the p-type region near the junction become negative ions after losing holes to the combination. The array of positive to negative ions in the depletion region hence builds up an electrical field across the region. The electrical field increases when more and more carriers flow to this region and recombine. An equilibrium state is finally achieved when the majority of the carrier flow is nearly stopped by the built-up field. On the other hand, for the minority carriers, holes in the n-type region have to overcome the barrier (repulsion of the positive ions) to flow into the depletion region; once they get into the region, they are swept across to the p-type region. This constitutes a minority carrier current. In contrast, the minority carriers, electrons in the p-type region, flow in the opposite direction. At equilibrium, the carriers' flows are equal and in opposite directions, namely, there is no net current flow.

Imagine that the dopant concentration in the p-type substrate is $N_{\rm A}$ and the dopant concentration in the n-type region is N_d. We also know that each recombination requires one electron and one hole; therefore, at equilibrium, it is clear that the higher dopant concentration substrate would have thinner depletion width, that is,

$$X_{\rm A}N_{\rm A} = X_d N_d \,. \tag{2.16}$$

Applying a reverse bias, as shown in Fig. 2.12, meaning applying a positive voltage to the n-type substrate with respect to the p-type substrate, causes the depletion width to increase because the majority of the carriers are attracted to the electrodes, and therefore more ionized dopants are observed. In contrast, by applying a forward bias, the depletion width decreases.

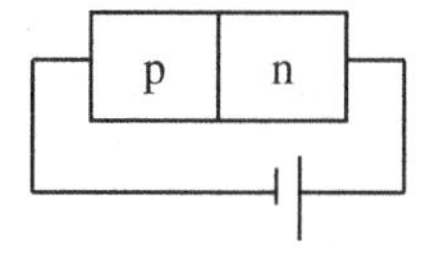

Fig. 2.12. A reverse bias across a junction.

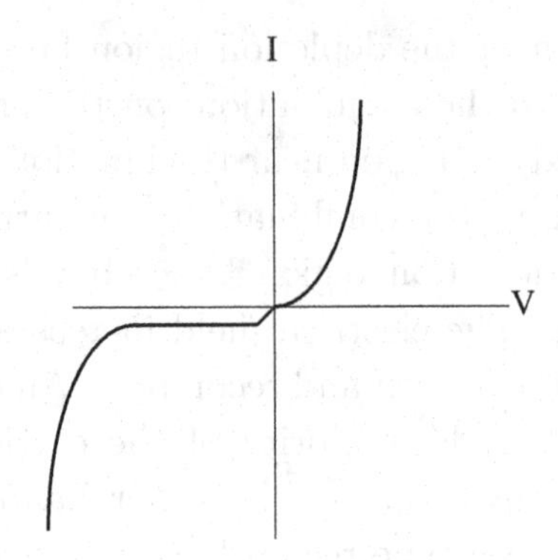

Fig. 2.13. *I–V* characteristics of a *pn* junction.

The reason for this is that the majority carriers gain more energy from the electrodes to cross the built-in junction barrier; therefore a large forward bias current is observed. Figure 2.13 shows the *I–V* characteristics of a *pn* junction.

At a constant temperature, the forward-to-reverse bias current ratio increases dramatically with the applied bias voltage. On the other hand, the ratio decreases with increasing temperature, namely, the off state deteriorates at higher temperatures. In circuit operation, to achieve a good on–off control, the forward-to-reverse bias current ratio should be larger than 10^3.

As one increases the reverse bias voltage, the depletion and the maximum electrical field, located at the *pn* interface, increase. If the increase continues, it will reach a point where the reverse current starts to increase rapidly. This voltage value is called junction breakdown. Junction breakdown voltage decreases with dopant concentration.

2.4. Capacitors

Capacitors are seen almost everywhere in an integrated circuit. Whenever two conducting lines or plates are separated by an insulator, a capacitor is formed. Capacitors can be built for certain purposes, such as the capacitors for dynamic random access memory (DRAM) or for analog devices. Capacitors such as junction depletion regions or two metal lines separated by a dielectric are generally considered parasitic capacitors.

Capacitors play key roles in DRAM cell operation. DRAM uses a capacitor to store the charges (the written messages). Owing to various leakage paths in the circuit, the stored charges need to be refreshed after a certain period of time. To have long refresh cycle times, the capacitor areas (values) have to be large enough to hold up the stored charges until the next refreshing signal comes in. Unfortunately, the DRAM cell area is largely determined by the capacitor area; to make the DRAM cell small, the capacitor area has to be kept small as well. As a result, three-dimensional capacitor structures are preferred. Figure 2.14 shows a stacked capacitor for DRAM applications. These capacitors can often achieve capacitances of larger than 35 fF/cell, for 0.25-μm manufacturing technology. Analog circuits often require a large capacitor to fulfill their signaling characteristics. Figure 2.15 shows a polyplate capacitor for an analog device.

A pair of conducting lines, such as polysilicon interconnects, or metal lines constitute a parasitic capacitor. As demonstrated in Fig. 2.16, a parasitic capacitor consists of parallel conducting lines.

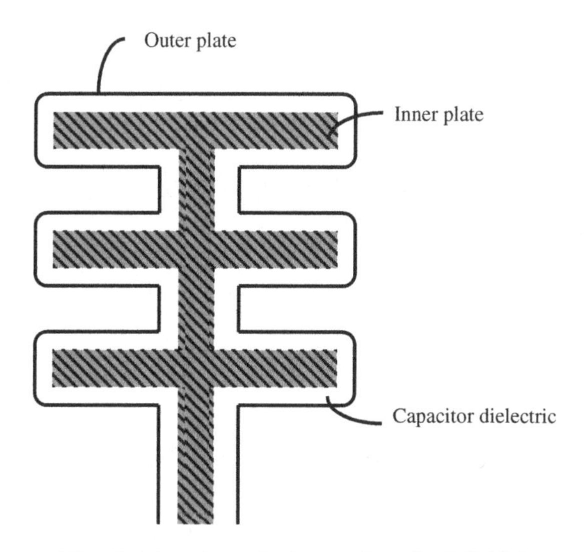

Fig. 2.14. A stacked capacitor for DRAM.

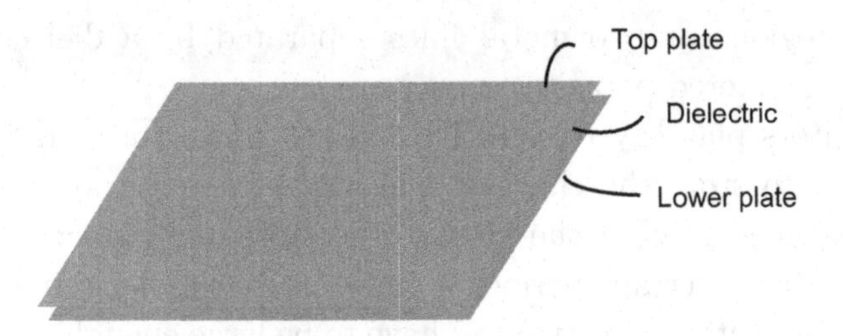

Fig. 2.15. A polyplate capacitor.

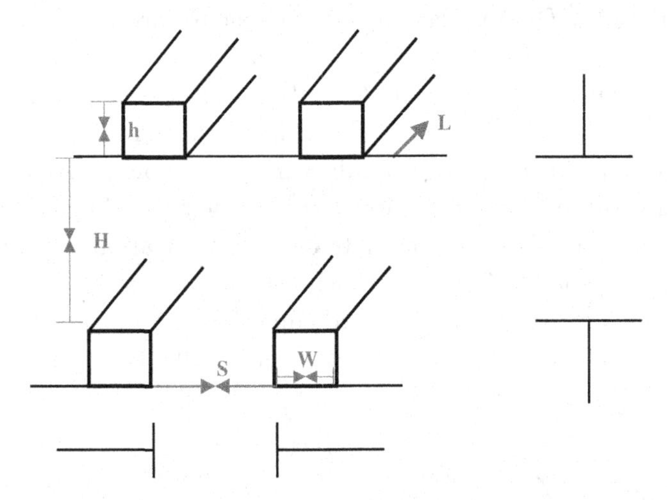

Fig. 2.16. The parasitic capacitors formed by neighboring conducting lines.

The capacitor between the conducting lines of the same level (C_1), or of different levels (C_2), can be expressed as

$$C_1 = \kappa\varepsilon \frac{Lh}{S}$$
$$C_2 = \kappa\varepsilon \frac{Lw}{H}. \tag{2.17}$$

Clearly, as technology shrinks, the conducting line spaces (S and H), widths (w), and thicknesses (H) also shrink. Therefore the parasitic capacitance and the conducting line resistance increase; so does the RC delay.

2.5. The MOS Transistor

The MOS transistors are commonly used as switches in an integrated circuit. The switches on and off correspond to true and false, respectively, in Boolean algebra. A MOS transistor, as shown in Fig. 2.17, consists of two heavily doped $n+$ regions on a p-type substrate. The two $n+$ regions are separated by a distance corresponding to the polysilicon gate length. An insulator (a thin oxide) exists between the doped polysilicon gate and the p-type substrate. Such configuration as shown in Fig. 2.17 is called NMOS. By using the opposite polarity of each portion, a PMOS can be obtained. The operational difference lies in the fact that electrons act as the majority carriers in an NMOS, while holes act as the majority carriers in a PMOS.

Normally, to operate the MOS transistor, voltages need to be applied to the gate, the source, and the drain terminals. For NMOS, the grounded end is the source, and the other end is the drain, which is often connected to a voltage, V_{dd}. The gate terminal is set at the gate voltage, V_g. Let us suppose that the source and drain are grounded, and V_g is set at a certain positive voltage, V_g. The depletion regions are formed at the junctions. The area under the gate is also depleted, as positive holes are expelled away from the silicon surface. The three depleted regions, which lack majority carriers, are connected together and form an isolation that separates the bulk silicon from the device itself electrically. If V_g is further increased to an even more positive voltage (greater than a threshold voltage, V_t), lots of electrons will be attracted toward the silicon surface to form an inverted channel. The inverted channel will be able to conduct electrons from source to drain, if the drain is raised to a positive voltage as well. As the drain voltage is increased, the depletion width widens

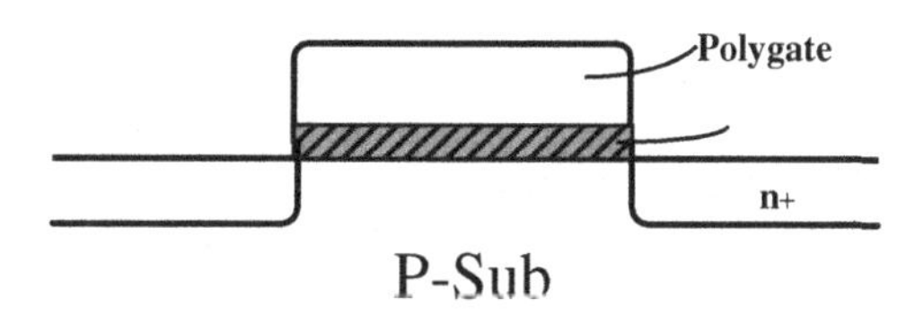

Fig. 2.17. An NMOS transistor.

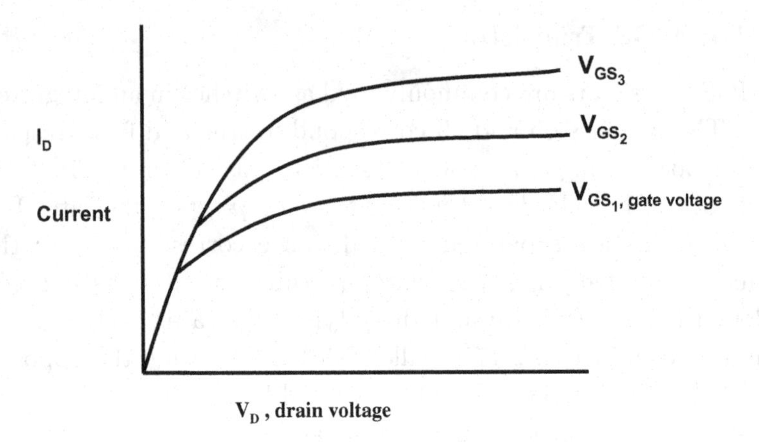

Fig. 2.18. A typical I–V curve of NMOS.

at the drain side, and the conducting channel tends to shrink at the drain, as long as $V_d < V_g - V_t$. When $V_d = V_g - V_t$, the conducting channel starts to pinch off at the drain. The pinch-off point is shifted to the left as V_d increases. Under such circumstances, the current flows from the source toward the pinch-off point and then is swept across the drain depletion region to the drain.

Figure 2.18 illustrates a typical current–voltage curve of a MOS transistor. At a constant V_g, the channel current increases linearly with V_d, until a pinch-off point is formed at the drain; this region is called the linear or resistive region. Afterward, the channel current levels off, regardless of the increase in V_d. This region is identified as the saturation region. On the other hand, if the V_d is held constant, while V_g is increased, the channel current will increase. This is because the channel becomes deeper at high V_g, allowing for more electrons to go through.

In the resistive region, the channel current, I, can be expressed as

$$I = \beta\left[(V_{\mathrm{G}} - V_{\mathrm{T}})V_{\mathrm{D}} - V_{\mathrm{D}}^2/2\right], \tag{2.18}$$

where $\beta = C\mu(W/L)$, C is the gate capacitance per unit area between the gate and the silicon surface, and μ is the mobility of electrons in the conducting channel.

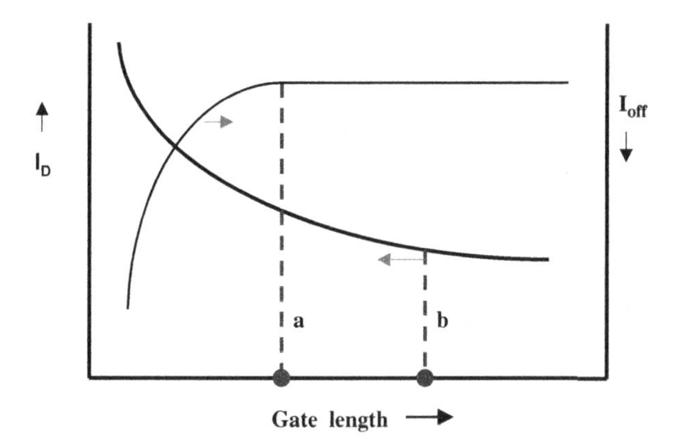

Fig. 2.19. The device window is defined by the difference between the gate length that gives the target transistor current and that at the I_{off} cliff point.

When $V_{\text{G}} - V_{\text{T}} = V_{\text{D}}$, the channel reaches the saturation region, and Eq. (2.21) is reduced to

$$I = \beta(V_{\text{G}} - V_{\text{T}})^2/2\,. \tag{2.19}$$

For a device to have good manufacturability, the device window has to be larger than the polysilicon gate dimension control. The device window is defined as the region between "a" and "b", as shown in Fig. 2.19. Namely, within this gate length range, the off current is constant. In addition, as the gate length decreases to a certain value, the threshold voltage of the device drops off significantly, or the device off current increases significantly, due to the short channel effect. This means that with the same gate voltage, the transistor with a shorter gate length will start to leak; that is, the transistor cannot be turned off when it is supposed to be turned off. This causes the whole chip to malfunction. For example, during a 0.25-μm chip manufacturing, if the device window is 0.02 μm, meaning that the shortest tolerable gate length is 0.23 μm in terms of short channel effect, then the gate dimension control has to be better than 0.25 μm $\pm$ 0.02 μm (3σ) to have the whole chip functioning properly. It can be observed that as the technology advances, the device window shrinks, and the gate dimension control becomes more and more difficult. For 0.1-μm technology, there is hardly any device window left, as defined traditionally. Therefore the gate dimension control is extremely critical for wafer yield.

2.6. Integrated Circuits

With the introduction of various building blocks for constructing an integrated circuit (IC) in the preceding sections, we are now ready to explore different ICs and their building blocks. In general, all ICs can be categorized into two broad groups: digital circuits and analog circuits. While most computers employ digital concepts, converting all the information into two states, 0 or 1, the operation of an analog circuit is closer to human nature. An analog circuit gathers and conveys information continuously. The two operation natures can be clearly differentiated by looking at two types of light switches. One is a digital light switch, which turns the light either on or off. The other light switch (the analog type) can manipulate the light brightness by varying the voltage applied to the lightbulb. In terms of design, one of the major differences between digital and analog circuits is that the latter uses capacitors and resistors more extensively than the former. There is a wide spectrum of digital circuits, including logic, memories, and microprocessors.

In terms of similarity in manufacturing characteristics, process technologies are divided into logic and memory products. Logic circuits include application-specific IC (ASIC), analog, and microprocessor chips; the memory circuits include a broad range of memory products such as DRAM, SRAM, and various types of nonvolatile memories.

ASIC consists of logic gates that are designed to fit a specific purpose, for example, the music chip embedded in a Christmas card or in a baby doll. Analog chips are applied in situations where continuous, instead of discrete, information is needed. The microprocessor, or central processor unit (CPU), is the brain of all computers. It manipulates and conveys information and, if necessary, stores the information. It also makes decisions and issues instructions to related functional parts of a circuit. Over the last 18 years, CPU has almost become an acronym for Intel, which invented the first single-chip

CPU, 8080, in 1974; this CPU contains 6000 MOS transistors. Since then, Intel has dominated the CPU market. In the year 2000, Pentium 4 made its debut with 42 million MOS transistors on a single chip.

Memory products are used for data storage in a computer system. Prevailing memories in the market include SRAM, DRAM, and nonvolatile memory (NVM). Each distinguishes itself by its own unique way of storing information. Random access memory (RAM) means that the access to a specific part of the information does not have to be accomplished in a sequential manner, as is the case in a conventional magnetic tape. As long as the column and row addresses are identified in the RAM, the information to be written or read can be accessed accordingly. Volatility is the term used to broadly categorize memory products. If the stored information in a memory circuit remains after the power is turned off, it is called a NVM such as erasable programmable read only memory (EPROM) or ROM. On the other hand, if the stored information goes away with the power, it is called a volatile memory, such as SRAM or DRAM. NVMs can be used in situations where the stored information must be retained when the power goes off. Such applications as computer rebooting systems, peripheral configuration and setup, or music toys all need NVMs. DRAM is often used in a large memory system such as computers' main memory. Compared with other memories, DRAM provides a cost advantage as it is produced on a very large scale worldwide. SRAM, on the other hand, is used when a memory system requires fast speed and low standby power, such as for the memory for a PC or PDA.

SRAM stores data with a six-transistor flip-flop, as indicated in Fig. 2.20. The six transistors include two PMOS load connected (L_1 and L_2 to V_{dd}) and two NMOS pull-down (P_1 and P_2), connected to ground. When writing, the two bit-line transistors, B_1 and B_2, are charged at high and low settings, respectively; while turning on both word-line transistors, W_1 and W_2, the states of C_1 and C_2 are set accordingly. By doing so, it gives the memory cell the state 1, and the opposite operation gives the 0 state. This information is firmly latched by the flip-flop circuit. During reading, the cell address is

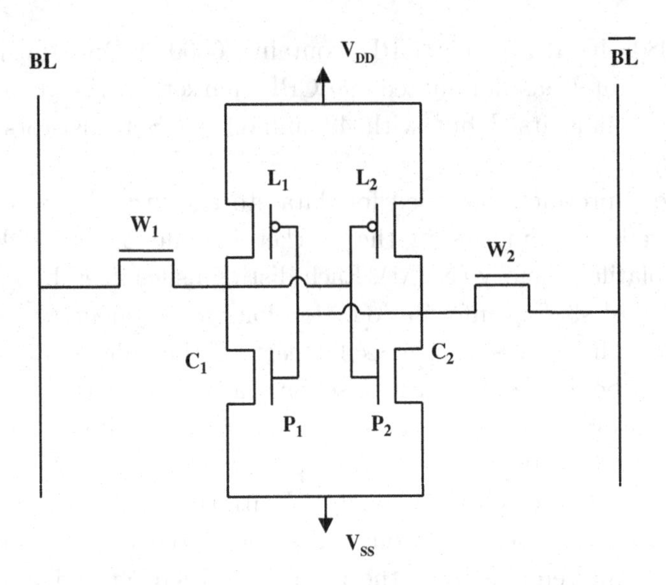

Fig. 2.20. A six-transistor CMOS SRAM cell.

selected by row and column decoders that address the specific row and column, or the cell location. The word-line transistors W_1 and W_2 are turned on.

Bit-lines, B and $\overline{B}$, are precharged to high states. Reading "1" causes a current flow through W_2 and C_2. It results in a difference in voltages between the two bit-lines. The resulting differential voltage is then amplified by a sense amplifier to determine whether the cell is at state 1 or 0.

DRAM has a simpler cell structure and therefore smaller cell sizes than SRAM. As shown in Fig. 2.21, it consists of an NMOS word-line transistor and a capacitor. The logic state of the stored data is determined by the amount of charges in the capacitor. Writing data into a cell, the column and row addresses select a specific cell and turn on the NMOS word-line transistor so that the voltage in bit-line can be charged onto the capacitor. In reading "1," the NMOS word-line transistor is on, and the current flows toward the bit-line. The voltage is then compared to a reference level to distinguish state 1 from state 0. The data or charges stored in the capacitors may leak through various paths such as junction leakage

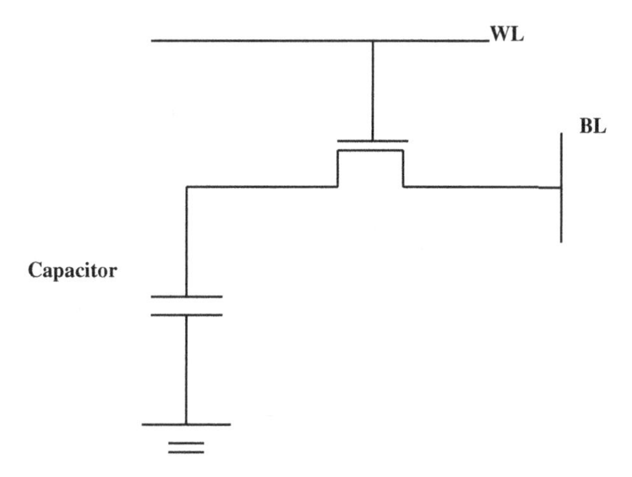

Fig. 2.21. A typical DRAM cell.

or device subthreshold leakage. As a result, the state 1 stored in the capacitor may decay to a weak 1 and then to 0. Therefore the stored charges need to be refreshed periodically. Furthermore, one can compensate for the leakages, and therefore avoid data loss, by enlarging the cell capacitances. However, this would increase the DRAM cell size and contradict the cost advantages of using DRAM.

An EPROM cell is a single-transistor cell with a stacked gate, as shown in Fig. 2.22. The upper gate is the control gate, while the lower gate is the floating gate. During writing, the drain and control gates are raised to high voltages, around 18–20 V. The high drain voltage creates a significant electrical field that generates high-energy electrons. They are attracted by the control gate to inject ion through silicon oxide toward the floating gate, which is completely surrounded by silicon oxide. Once the electrons are collected into the floating gate, they are trapped there. As the number of electrons reaches a certain level, they will saturate and repel further electrons from coming into the floating gate. The floating charges resulting from the above cause the threshold voltage of the NMOS transistor to shift toward the positive direction by about 5–10 V. During reading, both the gate and drain are connected to 5 V; the written cell does not give a transistor current, while the unwritten cell does. This distinguishes the 1 from 0 in a cell.

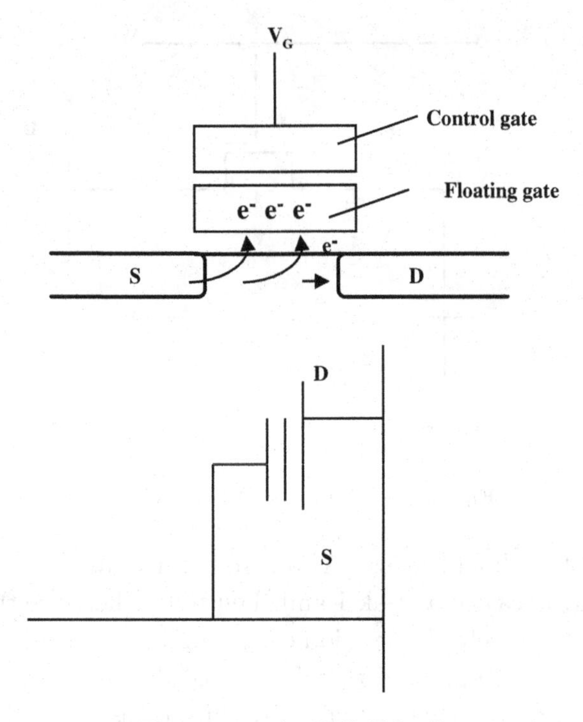

Fig. 2.22. An EPROM cell.

Recently, there has been more and more interest in making embedded chips in which a chip includes logic circuits and memory arrays, or even analog circuits. One can observe that as technology continues to shrink, more functional blocks can be integrated into one single chip, driving the size smaller and the cost lower.

Chapter 3

THERMAL OXIDATION

Wafer manufacturing normally starts out with a thermal oxidation, followed by a device isolation process. Thermal oxidation is the module that makes the so-called silicon planar process possible. The first section provides an overview of oxidation approaches, chemicals used, and oxidation systems. It is then followed by an explanation of oxidation modeling to gain additional insight into the species transport mechanism during oxidation. Such important applications of oxidation in field isolation as well as gate oxides will be explained in detail. The last section is devoted to illustrating different preoxidation cleaning approaches that play critical roles in determining the quality of the resulting oxide.

3.1. Introduction

Thermal oxidation is a process that employs oxidants to oxidize a bare silicon surface to silicon dioxide at elevated temperatures. Silicon dioxide is an excellent insulator, with a resistivity larger than $10^{16}\ \Omega\,\text{cm}$. While it is not soluble in water, it readily dissolves in HF solutions. Although silicon oxide films can be formed with chemical vapor deposition (CVD) or thermal oxidation, thermal oxidation of bare silicon provides the best oxide quality in terms of purity, density, and insulation. However, thermal oxidation has some application limitations. It requires the presence of a silicon surface, and it must be conducted at relatively high temperatures, that is, higher than 800°C.

Thermal oxidation can be used for various purposes in device manufacturing flow. First, thermal oxidation has been the only reliable method of growing the gate oxide for a transistor. The transistor current is inversely proportional to the gate oxide thickness. Furthermore, the quality of the gate oxide dictates the quality of the device and therefore the quality of a product chip. Second, thermal oxidation has been used to form a stress buffer between the nitride and the silicon for field oxidation. Third, thermal oxidation has been used as the only approach for field oxide (FOX) formation. Fourth, thermal oxidation has been used as a means of removing a damaged layer on the silicon surface. For example, after the field oxidation, the nitride and oxide dual layers are stripped, and the underlying silicon surface might have been damaged due to the excessive stress resulting from nitride bending. The damaged silicon surface is oxidized first; this process is called sacrificial oxidation. It is then wet-etched to remove the damaged layer on the surface. Finally, thermal oxidation is conducted on the silicon surface to form a masking layer to block dopant ion implants or dopant diffusion. The oxide bulk can mechanically trap the implanting ions and slow down the dopant diffusion.

Thermal oxidation can be conducted in a dry or wet ambient, with oxygen and water, respectively:

$$\mathrm{Si} + \mathrm{O_2} \longrightarrow \mathrm{SiO_2}$$
$$\mathrm{Si} + 2\mathrm{H_2O} \longrightarrow \mathrm{SiO_2} + 2\mathrm{H_2}\,.$$

Dry oxidation provides a better quality thermal oxide, but it also has a slower oxidation rate than wet oxidation. In general, when an oxide layer has an influence on the final device structure, or has intimate contact with the device silicon surface, or the required thickness is too thin to have good control in a wet ambient, it will be grown with a dry oxidation method. For example, gate oxides and pad oxides are formed using the dry oxidation method, while FOXs and sacrificial oxides are grown using wet oxidation. Furthermore, an oxidation procedure can be designed with the two oxidation approaches to achieve a specific purpose. For example, a dry-wet-dry oxidation method has been used in growing thick field isolation oxide to achieve good oxide quality at the interface with the silicon surface and fast throughput for the majority of the bulk oxide. Oxidation can also be conducted in a high-pressure ambient. The oxidation rate increases with the partial pressure of the oxidant. However, high-pressure oxidation has never been popular in device manufacturing. One reason for this is that the oxidation seldom becomes a bottleneck for a device manufacturing flow. Also, the high-pressure systems are hard to control in case of leakage. Remember that almost all of the systems that operate in semiconductor fabrication are under high vacuum. Should any leakage occur, the processing gases (noxious or non-noxious) will not leak out to ambient and harm people. Finally, having a high-pressure system could be very dangerous as it can explode if anything goes wrong.

Some additive gases can be added to improve the oxide quality; for example, gate oxide is often grown with chlorine, which is added during the oxidation process. Chlorine can getter the mobile ions, and therefore reduce the trapped mobile ions, such as sodium or potassium, in the interstitial sites. It can also bind with the dangling silicon bonds at the silicon–silicon oxide interface, and thereby reduce the interface state charges. Chlorine can be introduced by adding hydrogen chloride (HCl), tri-chloro ethylene (TCE), or tri-chloro ethane (TCA) to the oxidation process. The added chlorine increases the oxidation rates. However, excessive incorporation of chlorine in the grown oxide may degrade the oxide quality and device stability.

Oxidation is often conducted in a furnace, as illustrated in Fig. 3.1. The furnace is equipped with a few elements, such as the heating chamber, the quartz tube, the quartz boat, and the gas delivery system. Wafers are placed on a quartz boat with a narrow spacing between them, as dictated by the process design, to insure a uniform film growth across the furnace. The quartz tube forms the process chamber, which also separates the wafers from the heating elements. The resistive heating coils are placed around the quartz tube. In this arrangement, the throughput of the process depends on the length of the tube and the wafer spacing. The wafer spacing is critical as it has to be large enough to allow enough oxidants to be transported into the spacing to achieve uniform film thickness across the wafer. On the contrary, it is intended to be small to increase the throughput. The

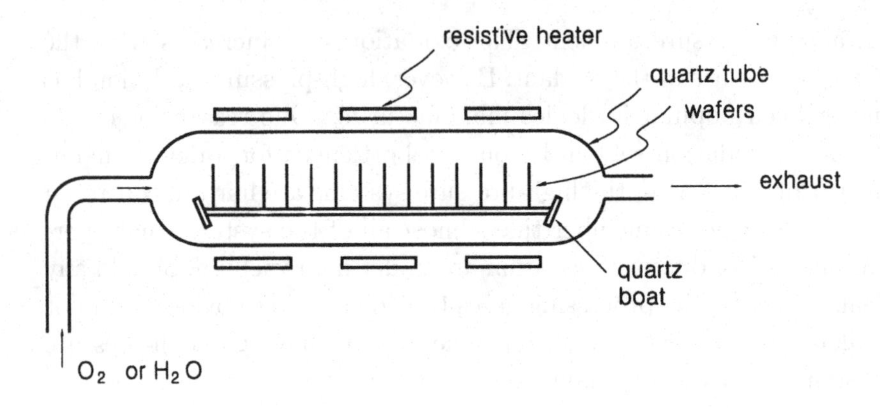

Fig. 3.1. A typical furnace system, used for oxidation.

quartz tube cannot be too long since normal processes demand high temperatures of up to 1150°C without sagging. In general, the size varies with different designs. Normally, the tube is about 2 m long, and the wafer spacing is about a few millimeters apart. The chamber is heated with a resistive heating system. The temperature profile in a furnace is divided into three zones: the ramping up, the flat, and the ramping down zones. The purpose is to prevent wafer damage due to abrupt temperature changes, which can lead to low yields. The gas delivery system consists of a few components. The gas tanks are usually installed in a safe gas cabinet, at some distance away from the furnace system. The regulators and control valve monitors can be used to adjust the gas pressure as it travels into the process chamber; mass flow controllers control the mass flow rates.

During wet oxidation, water is delivered by bubbling nitrogen through a water tank to bring out the water vapor, as shown in Fig. 3.2. To achieve a more accurate control on the water flow rates, a pyrogenic steam system is often used. Hydrogen and oxygen are introduced with a mole ratio of slightly less than 2:1 at the furnace's entrance. At around 400°C, with oxygen, hydrogen bursts into the water vapor. During operation, a batch of wafers (around 100–200) can be placed on a quartz boat either manually or automatically by using a robot arm from wafer boxes. Normally, a few dummy

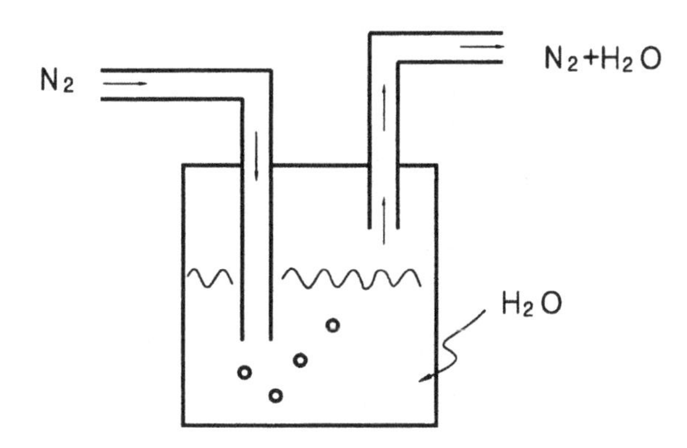

Fig. 3.2. An N_2 bubbling system for wet oxidation.

wafers are placed at the front and rear ends of the quartz boat in an attempt to avoid the nonuniform film thicknesses due to entrance and exit effects. After the wafers are loaded on the quartz boat, the boats are slowly pushed into the furnace. It takes about 60–90 min to complete the wafer loading. As the wafers move in, they experience a gradual temperature rise in lieu of a temperature shock, which leads to high wafer stresses, and therefore the wafers warp or break. The same approach is applied to the wafer unloading when the process is completed. During the process, the required time depends on the final oxide thickness. Roughly 125 Å of gate oxide takes about 30 min at 900°C and about 120 min for loading and unloading.

As the wafer diameter increases, it becomes more and more difficult to sustain a uniform temperature and oxidant concentration profiles across the wafer due to gravity in a horizontal furnace. A vertical furnace, as illustrated in Fig. 3.3, can solve these issues. A vertical furnace is very much the same as a horizontal furnace, except that the system stands upright, and therefore the gas flows upward. The gravity slows down the gas flow, but the flow uniformity is better along the wafer perimeter. The result is better uniformity of heat and mass transfer across the wafer diameter, and therefore a better film thickness uniformity can be obtained. For 8-in and 12-in wafer production, the vertical furnaces have gained much popularity.

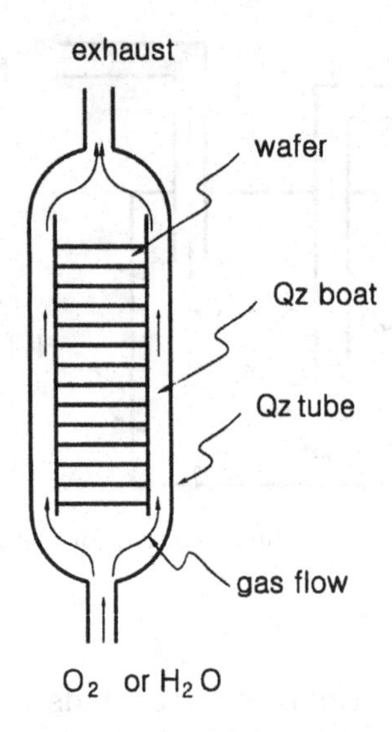

Fig. 3.3. A vertical furnace system.

3.2. Oxidation Mechanism and Modeling

During the silicon oxidation process, the oxidants have to reach the silicon surface to initiate the oxidation reactions. As the oxidation proceeds, more and more oxide is built up. The oxidants then have to diffuse through the bulk of the oxide to reach the silicon surface. The oxide bulk acts as a diffusion barrier for the oxidants' diffusion. Therefore, as time increases, the oxidation slows down, as illustrated in Fig. 3.4. In the case of pure silicon, one silicon atom is tetrahedrally bonded to four neighboring silicon atoms. Once converted to silicon dioxide, the structure survives, but the four corner silicon atoms are replaced with bridging oxygen atoms. As the oxidation proceeds, the newly formed silicon dioxide at the silicon oxide interface tends to push out the previously formed silicon dioxide. Furthermore, the number of molecules per unit volume for silicon is 5×10^{22}, and for silicon dioxide it is 2.2×10^{22}. As shown in Fig. 3.5, the volume expands after oxidation. As a result, the formed oxide film continuously experiences a compressive stress. As long as the oxidation process continues, this push and expand mechanism results in bond breakages, and hence an amorphous silicon oxide film is formed. Some

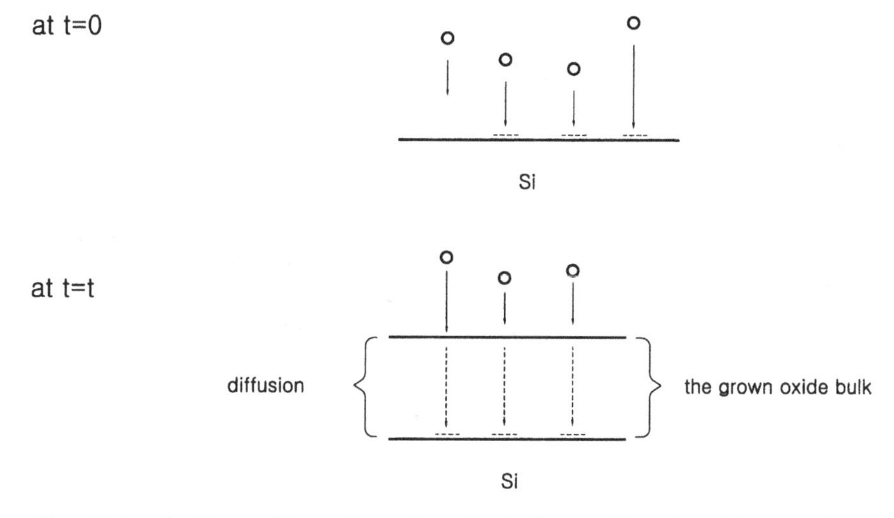

Fig. 3.4. Oxidants diffuse through oxide bulk, reaching the silicon surface, where oxidation takes place.

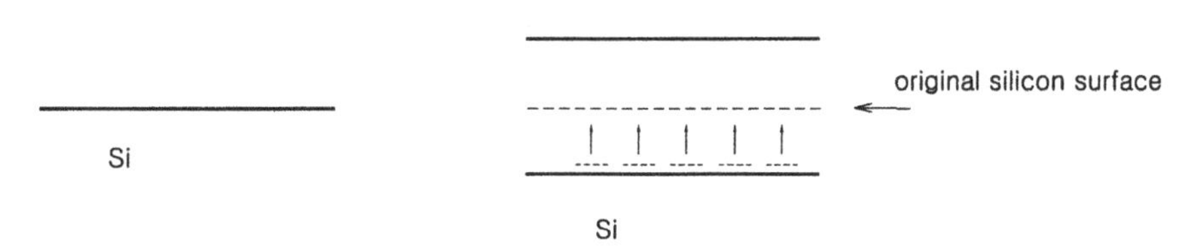

Fig. 3.5. Volume expands as oxidation proceeds.

stress can be relaxed via viscous flow of silicon dioxide at 960°C and above.

3.2.1. *Macroscopic modeling*

In a semiconductor manufacturing line, engineers are often required to estimate the time needed for a desired oxide thickness. For this, modeling work is essential. Assuming that at any instant in time, t, the oxide thickness formed is x_0, as indicated in Fig. 3.6, and considering the oxide bulk, the one-dimensional governing equation for the oxidant diffusion can be described as

$$D\frac{\partial^2 C}{\partial x^2} = 0\,. \tag{3.1}$$

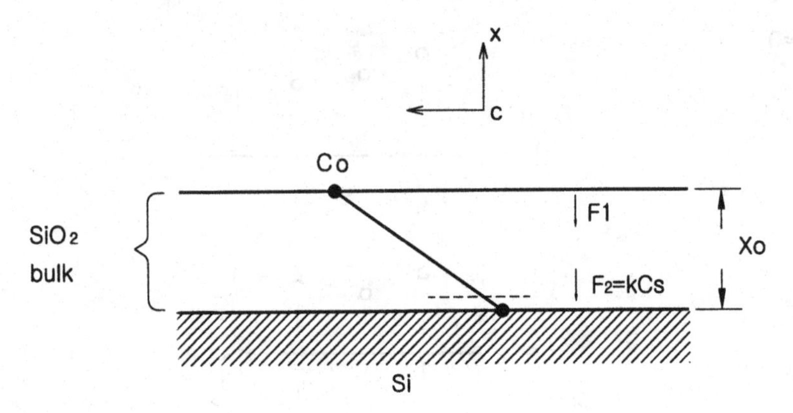

Fig. 3.6. A linear concentration profile of oxidant is assumed across the silicon oxide bulk.

With boundary conditions, $C = C_0$ at $x = x_0$ and $C = C_s$ at $x = 0$, where C_0 and C_s are the concentrations of the oxidant at the oxide surface and the silicon–silicon dioxide interface, respectively. The solution to Eq. (3.1) is

$$C = \frac{(C_0 - C_s)x}{x_0} + C_s\,. \tag{3.2}$$

The deposition flux equals the diffusion flux, evaluated at the silicon surface:

$$F = D\frac{\partial C}{\partial x}_{x=0} = D(C_0 - C_s)/x_0\,. \tag{3.3}$$

Now, two scenarios can be discussed here.

3.2.1.1. *Case 1*

Assume that the surface oxidation rate is very fast such that the oxidants are consumed on reaching the silicon surface; in other words, $C = C_s = 0$ at $x = 0$. This scenario corresponds to the oxidation via a thick oxide, where the oxidation rate is determined by the oxidant diffusion through oxide bulk. With Eq. (3.3), and assuming a quasi-steady state, the following can be concluded:

$$\frac{dx_0}{dt} = \frac{F}{M} = \frac{DC_0}{Mx_0}, \tag{3.4}$$

where M is the number of oxidant molecules needed to form a silicon dioxide molecule. For example, M equals 2.2×10^{22} cm^{-3} for dry oxidation and $2 \times 2.2 \times 10^{22}$ cm^{-3} for wet oxidation.

Integrating Eq. (3.4), one concludes that

$$t = \frac{M}{2DC_0}x_0^2 - \frac{M}{DC_0}\left[\frac{x_i^2}{2}\right]. \tag{3.5}$$

By grouping the constants, Eq. (3.5) can be rewritten as

$$t = \alpha x_0^2 - \beta\,. \tag{3.6}$$

The grown oxide thickness is proportional to the square root of time. In other words, the oxidation rate slows down with increasing oxidation time.

3.2.1.2. *Case 2*

For a more general case, assume that at the silicon surface the reaction is first order with respect to the oxidant concentration:

$$F_2 = k_s C_s\,. \tag{3.7}$$

One concludes that

$$\frac{dx_0}{dt} = \frac{F}{M} = \frac{DC_0}{M(x_0 + D/k_s)}. \tag{3.8}$$

Integrating Eq. (3.8), one obtains the following:

$$t = \frac{M}{2DC_0}x_0^2 + \frac{M}{k_s C_0}x_0 - \frac{M}{DC_0}\left[\frac{x_i^2}{2} + \frac{D}{k_s}x_i\right]. \tag{3.9}$$

Grouping the constants, Eq. (3.9) leads to

$$\frac{x_0^2}{B} + \frac{x_0}{B/A} - (\tau + t) = 0\,. \tag{3.10}$$

The equation can now be solved for the thickness as a function of time. The constants can be obtained by fitting the actual growth rates of an oxidation system to Eqs. (3.6) or (3.10). Once the constants are estimated at different temperatures, they can be expressed in Arrhenius form, $k = k_0 \exp(-E/kT)$.

The equation can then be used to predict the oxidation rates at different temperatures and times. Owing to the fact that there are assumptions made in deriving the model, some constants may not have exact physical meanings; instead, they are a set of fitting parameters. For example, the initial oxide thickness at time zero (x_i) is 250 Å for dry oxidation and 0 Å for wet oxidation.

The oxidation rates are found to be dependent on several factors. The silicon surface crystal orientation dictates the atomic packing density, and therefore it affects the oxidation rates. Wet oxidation provides higher oxidation rates than dry oxidation due to the more porous nature of the wet oxide, which facilitates oxidant diffusion. Addition of chlorine in the oxidation process not only improves the oxide quality, but also accelerates the oxidation rates. It is believed that the chlorine can accelerate the oxidant diffusion and catalyze the surface oxidation reaction. On the other hand, pressure and temperature affect the oxidation, as dictated by chemical reaction kinetics. All of these different factors affect the oxidation rates and therefore fit to different sets of parameters.

Figure 3.7 illustrates the growth rates versus oxidation times using Eq. (3.10). It can be observed that the oxidation process can be divided into three different regimes. The first regime occurs when the oxidation time is short such that

$$x_0 \gg \frac{x_0^2}{A} .$$

Consequently, this leads to

$$x_0 = \frac{B}{A}(t + \tau) . \tag{3.11}$$

In this regime, the thickness is linear with time, which indicates that the oxidation process is limited by the surface reaction rate.

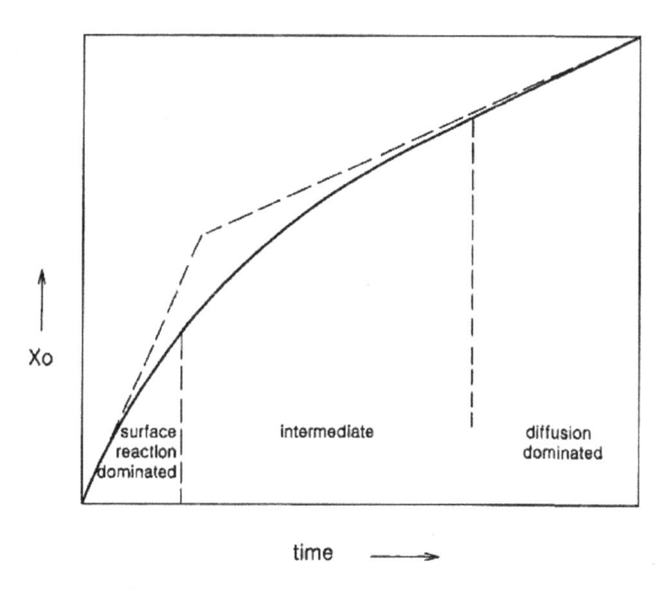

Fig. 3.7. An oxidation process can be divided into three different regimes.

Therefore the constant B/A is dependent on the crystal orientation, with the value of $\langle 111\rangle$ being larger than that of $\langle 100\rangle$.

Second, for the long oxidation time when a thick oxide layer is built up on the silicon surface, the oxide thickness can be approximated as

$$x_0 = \sqrt{Bt}\,. \tag{3.12}$$

B is the parabolic rate constant. In this regime, the oxidation rate is limited by the oxidant diffusion through the oxide bulk.

Third, for the intermediate regime, where the oxidation rates are dictated nearly equally by both diffusion and oxidation,

$$x_0 = \frac{-A + \sqrt{A^2 + 4B(t+\tau)}}{2}\,. \tag{3.13}$$

The above model, Eq. (3.10), is called the Deal-Grove model. It applies to the normal oxidation process, and it predicts the oxide thickness as a function of time. However, there are a couple of cases for which the Deal-Grove model does not seem to work well. One is initial oxidation or the oxidation of ultrathin oxide, while the

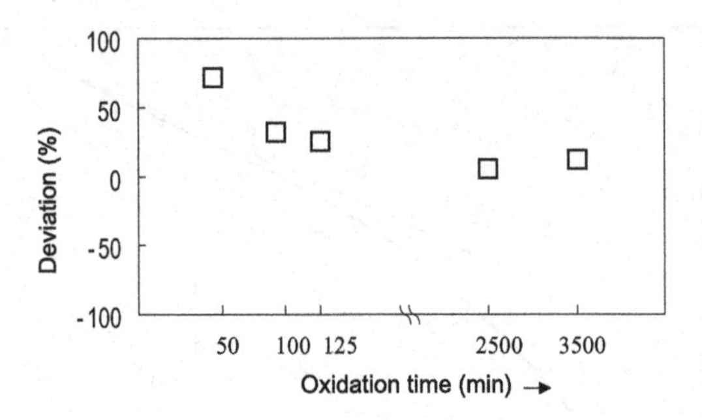

Fig. 3.8. The deviation between the wet oxidation data of (110) oriented Si and Deal-Grove model. The model tends to underestimate the oxidation rates for short oxidation times.

other is the geometry-dependent oxidation rate. For long wet oxidation times, the model fits the experimental data quite well, as illustrated in Fig. 3.8. However, the model significantly underestimates the actual growth rates in early oxidation regimes (less than 1000 Å). Similar phenomena have been observed with dry oxidation as well. On the other hand, if the oxidation rates of a field area versus the surface of a nanometer column (or a thin line), as indicated in Fig. 3.9, are compared, as in Fig. 3.10, it can be seen that as the oxidation proceeds, the core diameters of the nanometer columns shrink. After a certain period of time, the shrinkage seems to level off, indicating that the oxidation is self-limiting. The onset of the leveling off point seems to be temperature-dependent. The data in Fig. 3.11 explain that the oxidation process progresses faster on surfaces with larger-diameter cores. This could possibly be due to the fact that on smaller cores, the existing oxide has to push out more for the newly grown oxide. This results in higher stresses, and therefore it retards further oxidation. This unique characteristic may be very helpful in

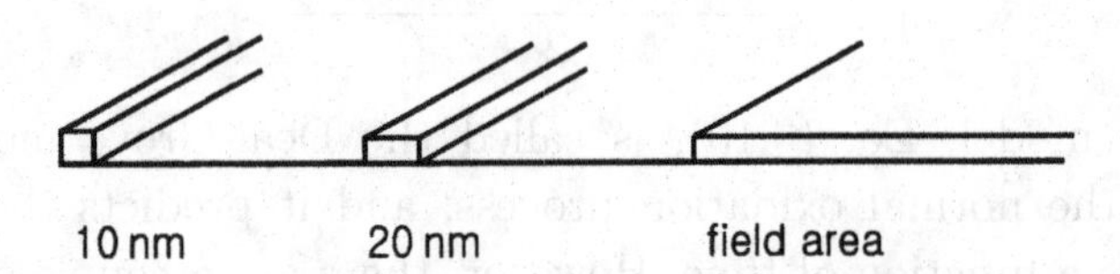

Fig. 3.9. Structures of nanometer columns and field area for oxidation experiments.

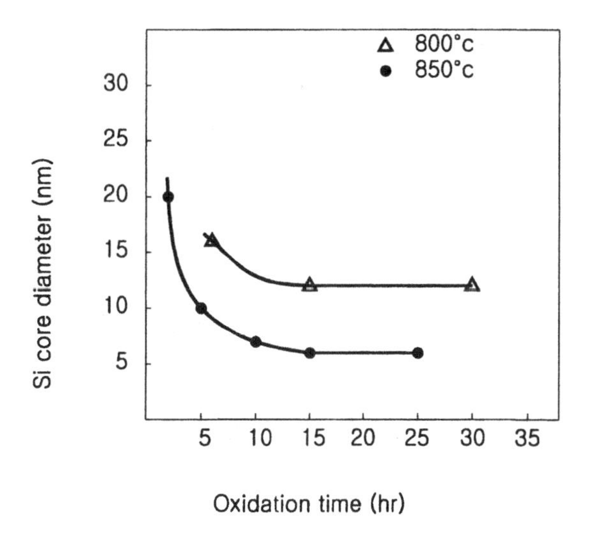

Fig. 3.10. Silicon core diameter shrinks with the dry oxidation time.

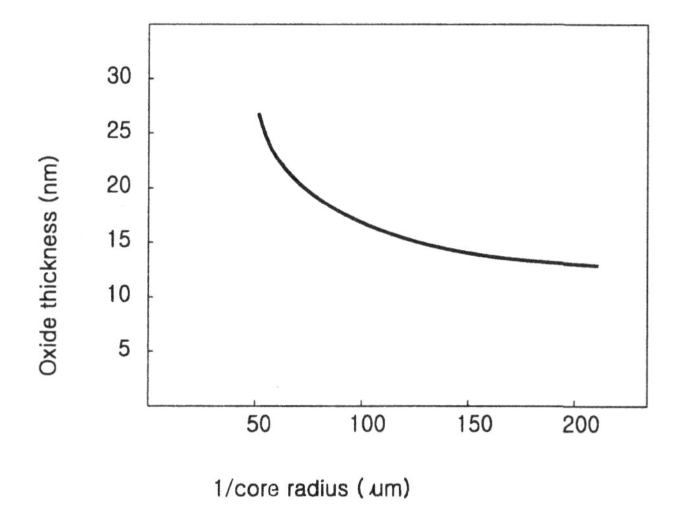

Fig. 3.11. For a given dry oxidation time at 800°C, the grown oxide thickness decreases with decreasing core radius.

micromachining, in which uniform sizes of silicon lines can be grown, even when starting out with different sizes.

The above macroscopic model provides oxide thicknesses in terms of oxidation time and conditions but sheds no light on the spatial dependency of the oxidation rates in a furnace.

3.2.2. *Microscopic modeling*

By reviewing the schematic of the oxidation furnace, shown in Fig. 3.1, it is apparent that the wafers stand vertically, with a narrow space between each other. The reactant gas mixture, composed of oxidant and nitrogen, flows along the annular area formed by the wafer edges and the reactor tube. In this region, the mass transfer of the reactant relies primarily on convection and partly on diffusion. The reactant then diffuses into the wafer-to-wafer space, driven mainly by the concentration gradient between the wafer edge and the wafer center. The concentration gradient is created by the consumption of the reactant during oxidation. A model can be formulated to describe the oxidation reactor, as shown in Fig. 3.12. A complete model should consist of continuity, momentum, mass, and energy balance equations. However, the model can be greatly simplified with the following assumptions:

(a) Since the reactor temperature profile is basically flat for the wafers to be oxidized, as indicated previously, and assuming that the heat of oxidation reaction is negligible, the reactor can be considered isothermal.
(b) There are no homogeneous reactions; volume expansion and contraction due to oxidation are negligible.
(c) Mass transfer due to radial convection in the annular region is negligible. Furthermore, the mass transfer due to convection in the region between wafer spacing can be ignored.
(d) The system operates at steady state (no time variation).

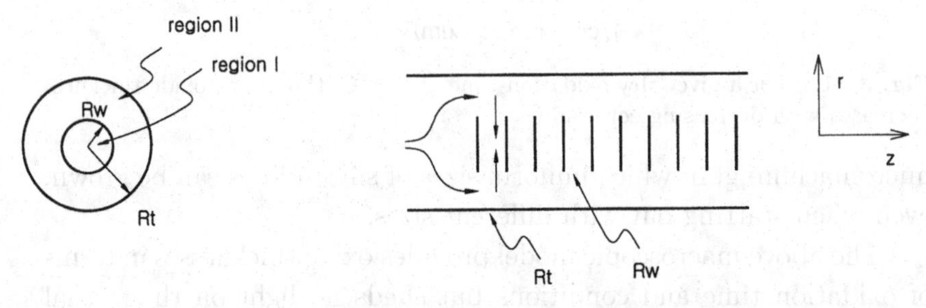

Fig. 3.12. Schematic of the oxidation reactor (furnace) for model formulation.

With the above assumptions, the model reduces to

$$V_r \frac{\partial C}{\partial r} + V_z \frac{\partial C}{\partial z} = \frac{1}{r} \frac{\partial}{\partial r} \left(rD \frac{\partial C}{\partial r} \right) + D \left(\frac{\partial^2 C}{\partial z^2} \right), \tag{3.14}$$

where C is the oxidant concentration. The left hand side of the equation represents the convective mass transfer; the right hand side represents the diffusive part. Taking the average of the equation over r for the annular region and over z for the regions between wafers, one concludes the following:

(a) For the space between wafer,

$$\frac{\delta}{r} \frac{\partial}{\partial r} \left(rD \frac{\partial C}{\partial r} \right) = 2\kappa C ,$$

where δ is the spacing between wafers and κ is the oxidation rate constant as related to the gas phase oxidant concentration. The boundary conditions are as follows:

$\partial C / \partial r = 0$, at $r = 0$ (center of a wafer); this states the symmetry at the wafer center
$C_{\mathrm{I}} = C_{\mathrm{II}}$, at $r = R_w$; at the wafer edge, the oxidant concentration maintains continuity.

(b) For the annular flow region,

$$V_z \frac{\partial C}{\partial z} = D \frac{\partial^2 C}{\partial z^2} + \frac{1}{R_t^2 - R_w^2} \left(-DR_w \frac{\partial C}{\partial r} \right)_{r=R_w}, \tag{3.15}$$

where R_t and R_w are the radius of the reactor tube and wafer, respectively. The associated boundary conditions are as follows:

$$C = C_0 \text{ (at the inlet)}$$

$$\frac{\partial C}{\partial z} = 0 \text{ (at the outlet)} .$$

These two equations form a model that describes the oxidation rate distribution along the axial and radial directions. The second term in the right-hand side of Eq. (3.15) accounts for the oxidant consumption due to the oxidation reaction. The model can be solved numerically to obtain the concentration profile and hence the oxidation rates along the axial direction. It can be predicted that the axial oxidant concentration decreases along the length of the tube due to the consumption of the oxidants along the way. Maintaining a relatively constant axial concentration is key to ensuring oxide film uniformity from wafer to wafer. For example, by increasing the inlet gas velocity on the left-hand side of the equation and bringing in more reactants, a more uniform film growth will be achieved. Increasing the wafer spacing can increase the oxidant supply to the wafer surface and improve the growth uniformity across the wafer film. In addition, the temperature profile along the axial direction can be adjusted to compensate for the depletion effects. In this case, a nonisothermal model must be used.

3.3. Isolation Technology

In silicon planar technology, a large number of devices are made on the same silicon surface. The isolation among the devices is essential, and it is made of oxide. In this section, we will introduce the three most used isolation technologies, local oxidation (LOCOS), poly-buffered LOCOS (PBLOCOS), and shallow trench isolation (STI). LOCOS or its variations have been serving the purpose quite well; its lifetime spans from device technology of larger than 2 μm down to 0.35 μm. Beyond the 0.35-μm technology, it was gradually replaced by STI. In general, a successful isolation technology should satisfy the following requirements:

(a) It should provide good isolation among devices. Figure 3.13 shows the parasitic devices that need to be disabled by the isolation.
(b) It should not create a severe topography. LOCOS, for example, tends to result in humps above the silicon surface, as illustrated in Fig. 3.14. The ensuing polysilicon gate layer must then lay over a topography, which leads to polysilicon necking or notching during patterning. A good isolation technology should create as little topography as possible.
(c) It should give good gate oxide quality. With an inappropriate nitride-to-pad oxide ratio, the underlying silicon surface may be

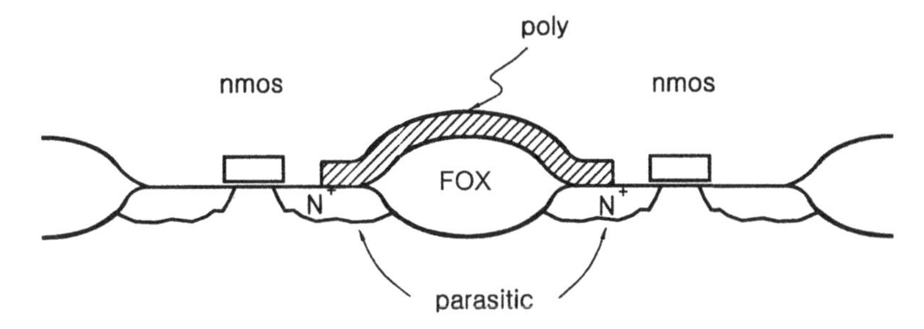

Fig. 3.13. A parasitic NMOS field device.

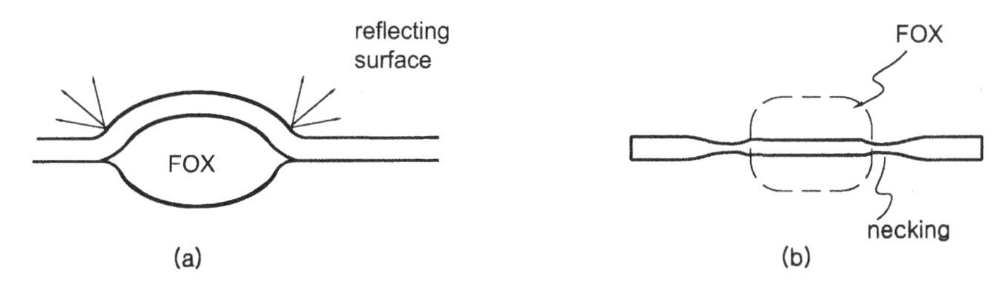

Fig. 3.14. (a) Polysilicon across the field oxide (FOX) topograph may result in reflecting surfaces. (b) Polynotching or necking may result.

damaged. This could lead to poor gate oxide quality. For STI, the sharp top corners may degrade the gate oxide quality or device performance. An oxidation step is often carried out to round off the sharp corners, as demonstrated in Fig. 3.15.

(d) It should take as small an area as possible to increase the device packing density.

(e) The FOX formation should be easy to implement and control.

The field oxidation for isolation was mentioned earlier in Sec. 3.1. To be more specific, this isolation technique in the industry is called LOCOS. In the early years, it was never thought to be compatible for technologies below 0.8 μm, but today, it does extend beyond the

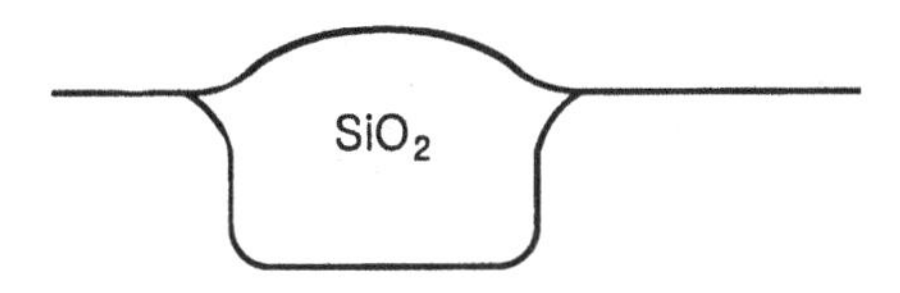

Fig. 3.15. For STI, rounded top corners tend to produce better gate oxide integrity and device characteristics.

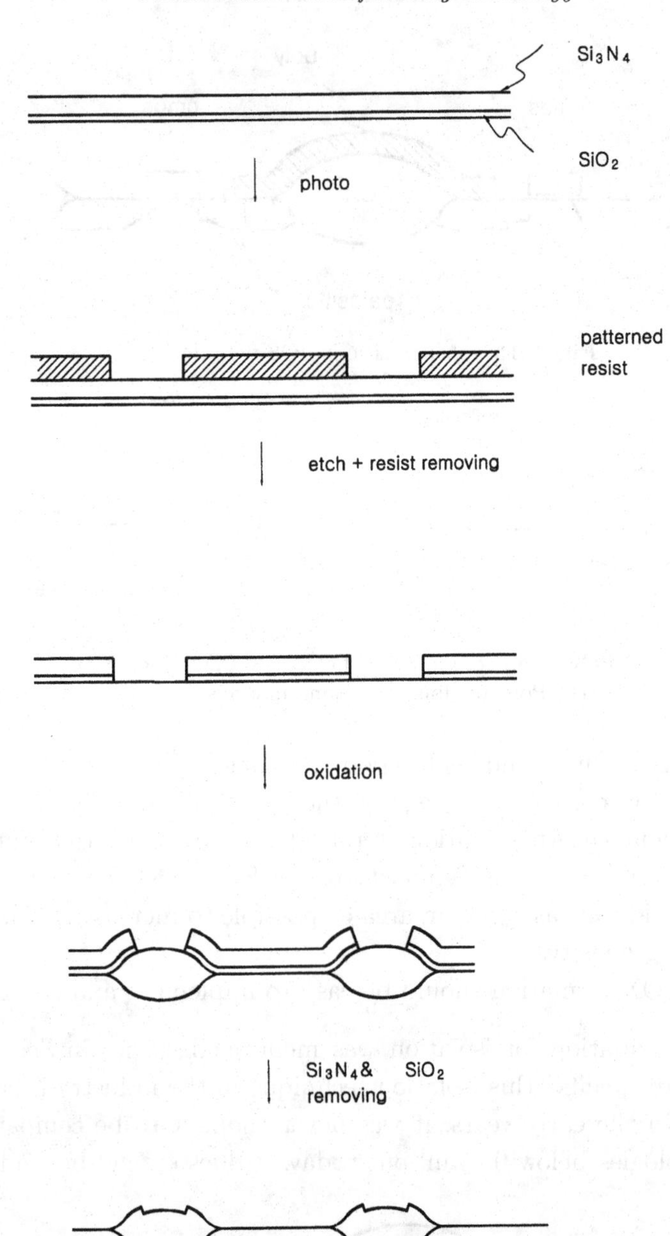

Fig. 3.16. The process flow of forming field isolation.

original expectations with slight modification. A process flow of forming the LOCOS is shown in Fig. 3.16, where the nitride-oxide stack is patterned. The wafers are then oxidized to obtain a thick oxide in the nonprotected area. Because the silicon volume nearly doubles after oxidation, the silicon nitride edge is pushed up. The oxidants diffuse much faster in oxide than they do in nitride; in addition, oxidants diffuse laterally. Therefore the oxide thickness decreases as it moves away from the nitride edge, as indicated in Fig. 3.17. After the field oxidation process, the nitride layer is first etched with an HF solution to remove a thin oxide from the top of the nitride, which is formed due to slight nitride oxidation. Then the nitride layer is removed by using a hot phosphoric acid bath. The HF dip removes more oxides at the nitride edge; this is possibly due to its high stress, which results in a bird's head shape, as demonstrated in Fig. 3.16. Therefore it is called a bird's beak. The bird's beak length (BBL) is a characteristic of the process technology. The FOX area takes up real estate on the silicon surface, and it does not contribute to the active components of a circuit; therefore the BBL is intended to be as short as possible. BBL associates with the photolithography limits. For example, the typical design rule for 0.5-μm technology is 1.4 μm, with a BBL of 0.1 μm and 0.2 μm. The longer the BBL, the smaller the space that the photolithography has to resolve, as demonstrated in Table 3.1. The reason for this is that the BBL must be biased on the mask to

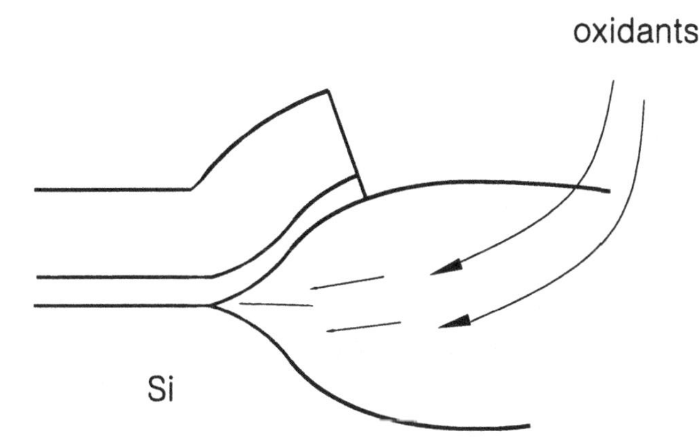

Fig. 3.17. Oxidants diffuse toward silicon surface.

Table 3.1. With a fixed pitch rule for active area, the larger the bird's beak length (BBL), the smaller the spacing that has to be resolved. Pitch rule of 1.4 μm.

BBL (μm/side)	line/space (μm)
0	0.7/0.7
0.1	0.9/0.5
0.2	1.1/0.3

achieve the desired width and space on the wafer, after the oxidation. With a BBL of larger than 0.2 μm the photolithography has to resolve a width of smaller than 0.3 μm, which is challenging for I-line photolithography technology.

Shortening the BBL is one of the main obstacles to overcome in shrinking a technology. A straightforward approach is to decrease the stress buffer, which is the underlying pad oxide. Figure 3.18 illustrates the effects of decreasing the pad oxide thicknesses on the BBL. The BBL can be reduced to less than 0.2 μm if the pad oxide thickness is less than 100 Å. The risks involved are that the excessive

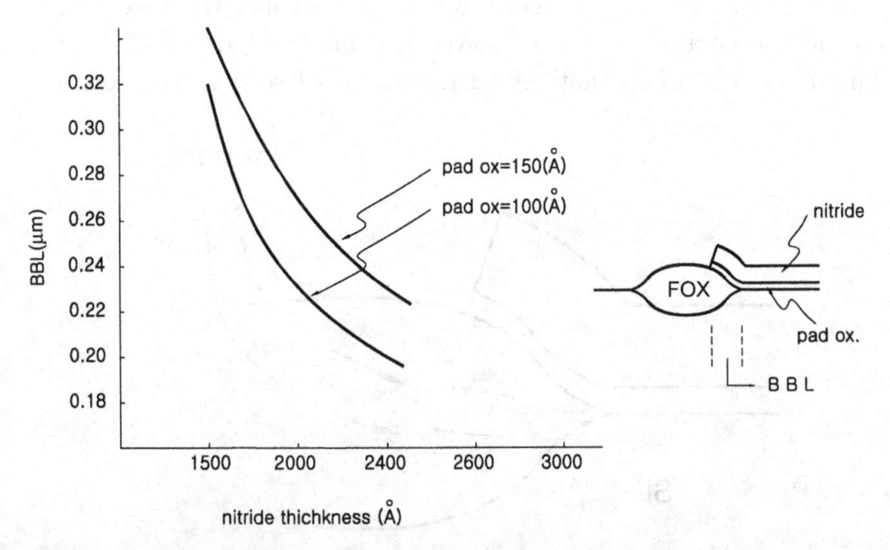

Fig. 3.18. The bird's beak length (BBL) versus pad oxide and nitride thickness.

Fig. 3.19. The defects (damages) observed on silicon surface due to excessive stress.

stress exerted by the nitride could damage the underlying silicon surface. Figure 3.19 demonstrates an example of stress-induced defects around field oxide. The defects cause device leakage or gate oxide quality concerns. Another approach is to thin down the field oxide's thickness. By doing so, the effective isolation for the parasitic devices are thinned and shortened accordingly. Device engineering must take this into account.

In an attempt to decrease the BBL without creating excessive stress, another method, PBLOCOS, has been proposed. The architecture of a PBLOCOS is illustrated in Fig. 3.20; a polysilicon of a few hundred angstroms is used between the nitride and the pad oxide to absorb the stress of the nitride. The polysilicon etching is done halfway through the polysilicon layer, followed by an oxidation step. This structure allows for a higher nitride-to-pad thickness ratio to reduce the BBL without inducing excessive stress on the silicon surface. The defects that are observed in conventional LOCOS

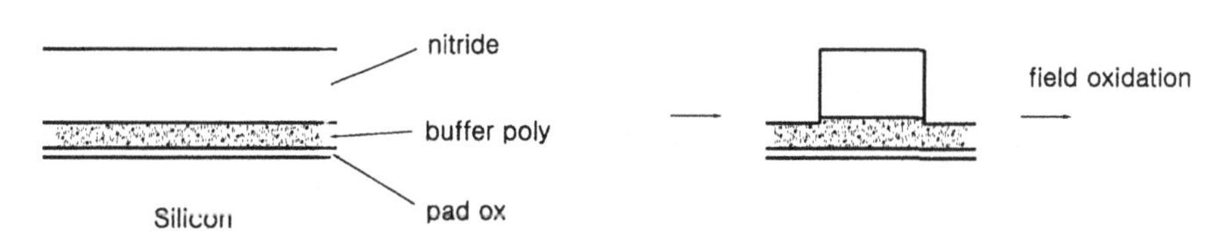

Fig. 3.20. The architecture of PBLOCOS process.

with a nitride/pad oxide ratio of 2000 Å/150 Å are not seen with a PBLOCOS structure.

A striking phenomenon associated with the wet field oxidation is the white ribbon effect or the Kooi effect. There is a ribbon around the bird's beak edge that is composed of silicon oxide and silicon nitride. During the wet oxidation, water reacts with silicon nitride and forms ammonia:

$$H_2O + Si_3N_4 \longrightarrow SiO_2 + NH_3\,.$$

The ammonia can further react with the silicon surface to form silicon nitride:

$$Si + NH_3 \longrightarrow Si_3N_4\,.$$

This nitridation can compete with the oxidation, or even overtake the oxidation, when the oxidant concentration is relatively low (e.g., at the bird's beak edge).

The white ribbon must be removed by performing a sacrificial oxidation to ensure a uniform and good-quality gate oxidation at a later stage. In the case of PBLOCOS, there are two sources of silicon for the oxidation and nitridation, giving rise to a twin white ribbon effect. One source is on the silicon surface, while the other source is on the polysilicon surface, as shown in Fig. 3.21. The white ribbon effect can be eliminated in a dry oxidation ambient.

The STI technique has been widely accepted for 0.25 μm and beyond. It essentially eliminates the drawbacks of conventional LOCOS such as BBL and topographs. The process flow of a typical STI process is illustrated in Fig. 3.22. About 1000 Å of nitride on 100 Å of pad oxide are grown on the silicon surface; this is followed

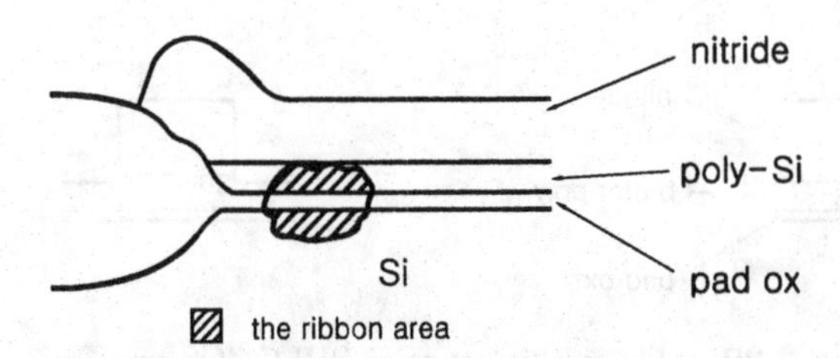

Fig. 3.21. The twin white ribbon effect in the PBLOCOS process.

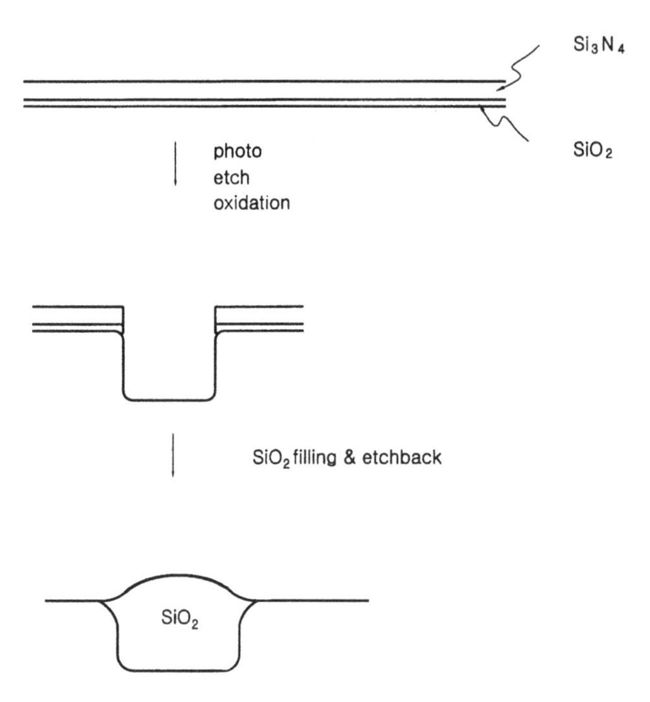

Fig. 3.22. A representitive process flow of forming shallow trench isolation.

by patterning and etching. The trench depth is about 3500–4500 Å, and it varies with technology migration. An oxidation process is carried out to smoothen the sharp corners. CVD oxide is then employed to fill the trenches. Chemical mechanical polishing is carried out to planarize the topography and stop at the nitride layer. The nitride layer is then removed using a dry etching step. A sacrificial oxidation is carried out, and the oxide is removed to expose the refreshed and clean silicon surface for gate oxide growth. The resulting isolation structure is nearly BBL- and topograph-free.

Dopant segregations are often observed after oxidation, especially in the case of field oxidation. This is due to the fact that dopant solubilities in silicon are different from those in oxide. Dopants of the p-type, such as boron, have a lower solubility in silicon than they do in oxide. Consequently, after oxidation, the silicon surface is depleted of the p-type dopants. Dopants of the n-type, such as As or P, have the opposite characteristics, as indicated in Fig. 3.23. The dopant segregations need to be taken into account during device

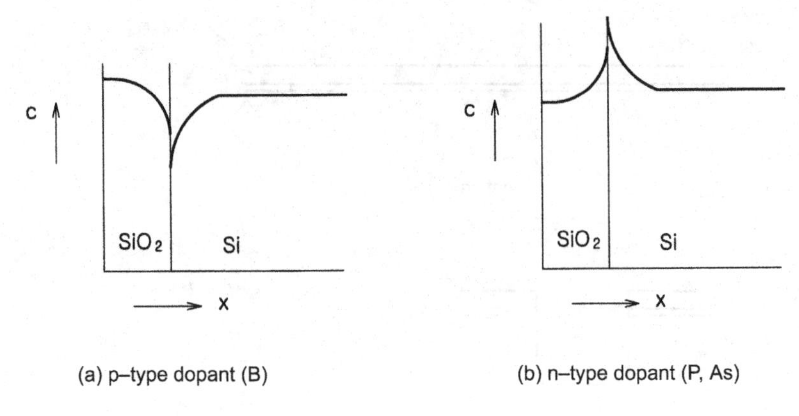

Fig. 3.23. Dopant redistribution during silicon oxidation.

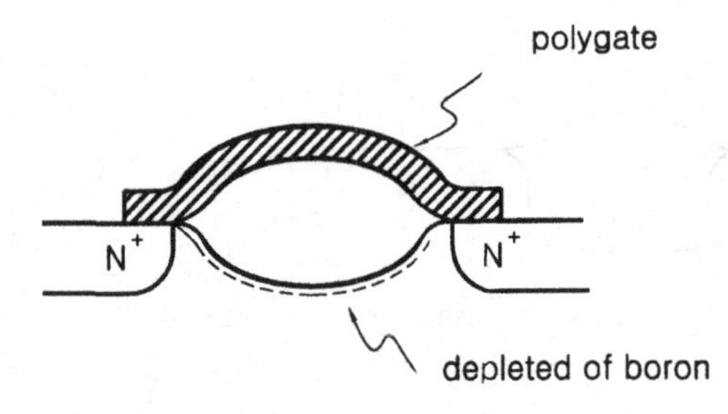

Fig. 3.24. Silicon surface under FOX is depleted of boron, lowering the threshold voltage of the parasitic field device.

tuning. For example, as illustrated in Fig. 3.24, the silicon surface is depleted of boron, which can lead to N-field device failure and low threshold voltages. Therefore a field implant is often implemented through the field oxide bulk to make up for the boron loss at the silicon surface and to maintain a proper parasitic device breakdown voltage, as demonstrated in Fig. 3.25.

3.4. Gate Oxide

Gate oxide is a thin layer of silicon dioxide sandwiched between the silicon substrate and the gate. When a voltage is applied on the gate layer (across the gate oxide), equal amounts of opposite charges are induced at the surface of the silicon. Thereby the device channel can be turned off and on. In addition, the gate oxide prevents

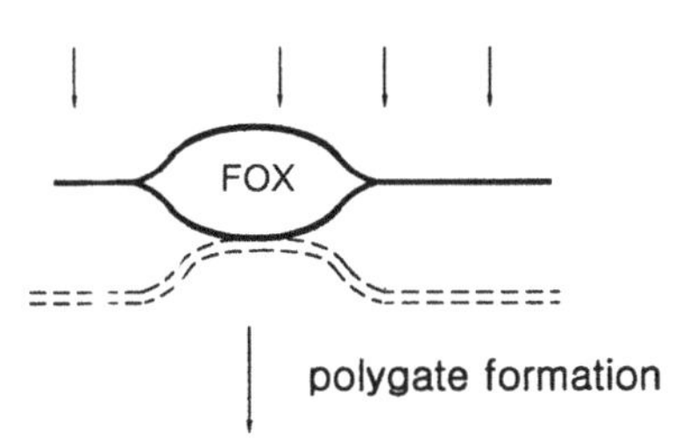

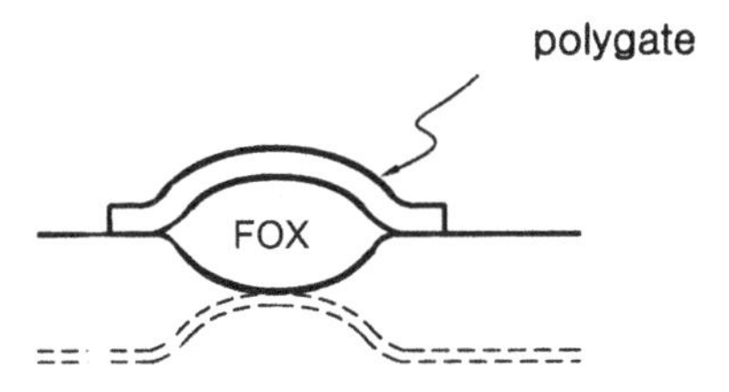

Fig. 3.25. P-field implant is implemented before gate formation to compensate for the boron loss.

the dopants of both sides from interdiffusing. The performance and reliability of a MOS device is largely determined by the quality and integrity of the gate oxide. With respect to the device performance, as illustrated in Fig. 3.26, the gate oxide's thickness decreases monotonically as the technology evolves. Despite decades of semiconductor

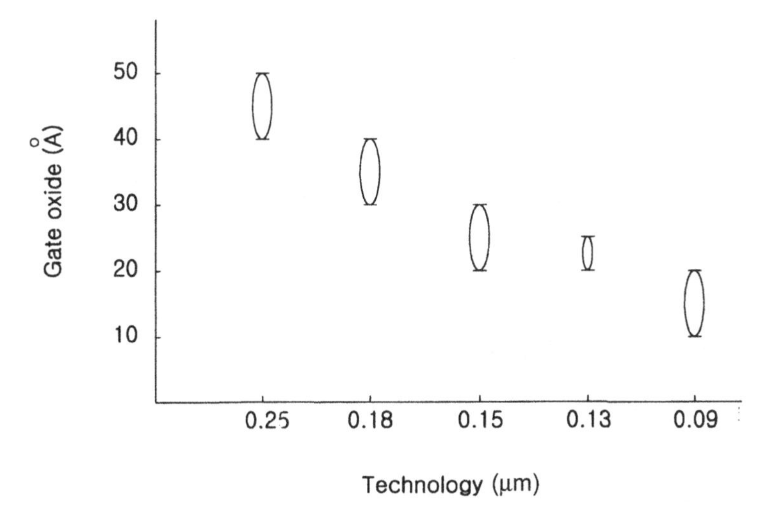

Fig. 3.26. The gate oxide thicknesses for advanced technologies.

technology development, thermal oxidation still remains the most reliable method for growing gate oxides. There are several factors that may affect the grown gate oxide integrity. The first factor is organic residues due to incomplete cleaning. During oxidation, the organic residues burn off and leave carbon contaminants buried in the oxide bulk, causing local weak spots. The second factor involves local thinning of the gate oxide due to defects such as white ribbon debris. The white ribbon is mainly composed of nitride, which is difficult to oxidize and leaves a local thinning area. The third factor is that metallic or mobile ions are left on the silicon surface from the wet-clean solution. These ions are so tiny that they can contribute to oxide pin holes. Finally, crystal defects, such as the presence of silicon dioxide near the silicon surface, can lead to local weak spots for the gate oxide. These local weak spots, or thinning areas, give rise to oxide breakdown during electrical stresses with either constant currents or constant voltages.

Three common electrical stresses for gate oxide quality screening are electrical field to breakdown (EBD), charge to breakdown (QBD), and time-dependent dielectric breakdown (TDDB). The EBD involves testing to find out what electrical field the oxide can stand up to before it breaks down. A large sample number of gate oxides in the form of capacitor structures are required for this purpose. Each is stressed with a constant voltage across the oxide bulk. The voltage is ramped up with a constant slope, as demonstrated

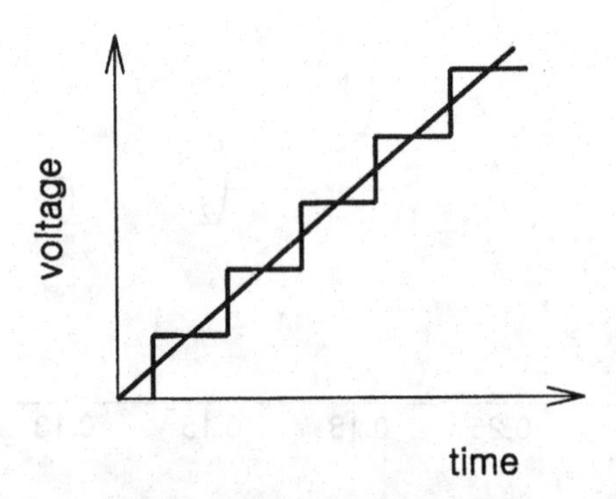

Fig. 3.27. A voltage ramp method to test gate oxide quality. The breakdown voltage is expressed as millivotts per centimeter.

in Fig. 3.27, until a point is reached at which the voltage across the oxide suddenly drops. The cumulative failure percentage is then plotted against the electrical field. For such a plot, the fewer the early failures and the higher the intrinsic breakdown, the better the gate oxide quality. Similar to EBD, instead of ramping up in voltage, QBD ramps up in current with a constant slope. The current ramps up until the oxide breaks down. The total charges introduced through the oxide can be calculated. The more charges that an oxide can take up before breaking down, the better the oxide quality will be. The above-mentioned EBD or QBD give the index of the dielectric strength of the oxide or flag the major oxide quality issue but show less correlation to the reliability of a device over time. In particular, how long can the gate oxide can survive under certain voltage operating conditions? A TDDB can partly answer this question. TDDB stresses the oxide with a constant voltage or constant current until the oxide breaks down. The stressing time before breakdown indicates the oxide quality. The voltage is often higher than the normal device operating conditions.

As device geometry shrinks, the gate oxide thickness decreases. For subquarter micron devices, pure silicon dioxide can no longer satisfy the device's performance requirements, largely due to two factors. First, the boron in the dual-gate structure can penetrate through the gate oxide and reach the silicon substrate, and therefore alter the device's characteristics. Second, for gate oxides thinner than 20 nm, electrons can directly tunnel through it. To meet the requirements with less than 20 nm, a few alternatives have been proposed. Among them, silicon oxynitride seems to be the most promising. Silicon oxynitride (SiN_xO_y) inhibits the boron penetration and improves the hot carrier performance. However, it may reduce electron mobility. The desired silicon oxynitride films can be grown using various approaches. They can be grown in a furnace with NO or N_2O. Alternatively, they can be formed from silicon dioxide annealed in NO or N_2O. The overall reactions involve the following:

$$2N_2O \longleftrightarrow 2NO + N_2\,,$$

$$N_2O \longleftrightarrow 2N_2 + O_2\,,$$

$$\mathrm{Si} + \mathrm{O_2} \longleftrightarrow \mathrm{SiO_2}\,,$$

$$\mathrm{Si} + x\mathrm{NO} \longleftrightarrow \mathrm{SiO}_x\mathrm{N}_y + \frac{(x-y)}{2}\mathrm{N_2}\,.$$

The overall nitrogen incorporation and the nitrogen profiles are functions of the process conditions and the involved reactants. In general, for furnace oxidation with nitric oxide or nitrous oxide, the nitrogen incorporation percentage is less than 10. The optimal nitrogen profile is determined by the ultimate desired purpose, for example, some desire to have nitrogen piled up at the silicon–silicon oxide interface to improve the hot carrier performance. Others may like to have nitrogen piled up at the polysilicon gate–silicon oxide interface to prevent boron penetration. It has been demonstrated that the silicon oxynitride film grown in NO has uniform nitrogen incorporation across the bulk. Furthermore, the film growth seems to be self-limiting since the grown oxynitride acts as a strong diffusion barrier. The oxynitride films grown with N_2O or silicon oxide annealed in NO or N_2O all result in nitrogen being piled up at the silicon–oxide interface. On the other hand, reoxidation of an oxyntride film gives rise to the nitrogen content being pushed up toward the oxynitride surface or the oxynitride–polysilicon interface.

The silicon oxidation can also be carried out in a rapid thermal process (RTP) system. The advantages of the RTP system as compared to the conventional furnace are short cycle time and low thermal budget. The RTP system is a single-wafer process unit, while the furnace is a batch system, which processes a few hundred wafers at a time. Oftentimes, when the number of wafers is not sufficient to run a batch process, the wafers must be queued for more to come to the stage. Furthermore, a furnace process often takes about 3 h to finish, even though the real processing time is only about 30 min. This is due to slow loading and unloading to avoid thermal shock. The thermal cycle experience during this time period can be large enough to drive dopant diffusion, which could be too significant to be acceptable for 100-nm device engineering. In an RTP system, the cycle time per wafer can be significantly reduced. It ramps up to the process temperature at a rate of about 100°C/s, which results

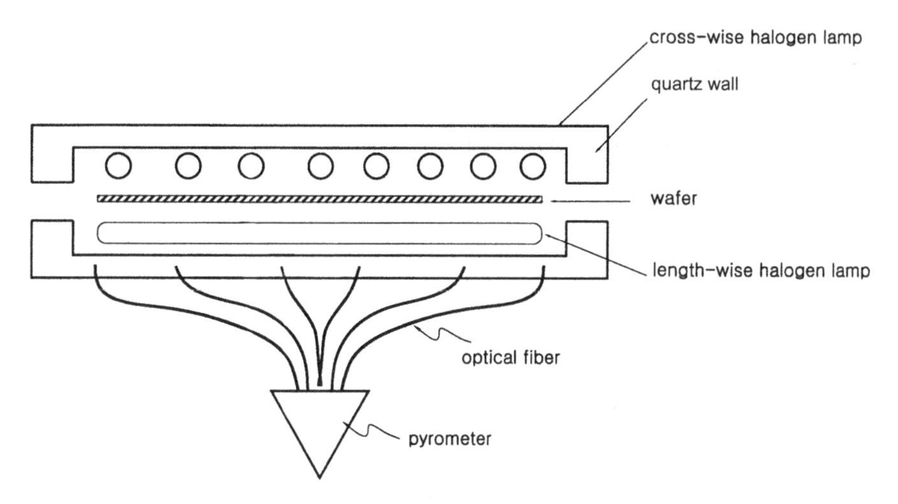

Fig. 3.28. An RTP system with halogen lamp heating.

in finishing the process, including load and unload, in about a couple of minutes. Therefore the unnecessary thermal budget is greatly reduced. The RTP can be heated up with either a halogen lamp or with a resistive heating element, as demonstrated in Fig. 3.28.

3.5. Cleaning before Thermal Oxidation

Typically, gate oxides are grown in a dry oxidation ambient with the addition of chlorine to getter the mobile ions and metal contaminants. Gate oxidation requires dedicated furnaces to maintain proper furnace cleanliness for producing good-quality gate oxide. Furthermore, to avoid contamination, gate oxide precleaning is very critical in determining the gate oxide quality, and it must be kept separate from the wet benches that are used for other purposes.

Cleaning is the most frequently repeated step in a semiconductor manufacturing line. It often ends a process module such as photoresist strip and polysilicon doping. It sometimes precedes a process module such as film deposition and thermal oxidation or annealing. The purpose of the cleaning process is to remove the particles; organic contamination such as HMDS or photoresist residues; and

Table 3.2. Often used wafer cleaning recipes and chemicals involved.

Recipe name	Chemicals	Removal
SC_1	$NH_4OH/H_2O_2/H_2O$	Particles or metal impurity
SC_2	$HCl/H_2O_2/H_2O$	Heavy metals, metal hydroxides
SPM	H_2SO_4/H_2O_2	Heavy organics (resist residues, particles)
HF	HF/H_2O	Silicon oxide
BOE	$HF/NH_4F/H_2O$	Silicon oxide

inorganic contamination such as metal ions or mobiles ions. There are two types of cleaning — wet-cleaning and dry-cleaning.

Table 3.2 indicates the wet-cleaning procedures that are commonly used in semiconductor manufacturing. The wet-cleaning procedures originate from the RCA cleaning methods, developed in 1965, and they are composed of SPM, SC_1, and SC_2. SPM is the mixture of sulfuric acid, hydrogen peroxide, and ultrapure water (deionized water, or DI water) with temperatures around 100–130°C. SPM is used for removing heavy organic contaminants. SC_1 is the mixture of ammonium hydroxide, hydrogen peroxide, and DI water. It is used for removing particles. SC_2, a mixture of hydrogen chloride, hydrogen peroxide, and DI water, is used to remove metallic particles. Furthermore, hydrofluoric acid or diluted hydrofluoric acid is often used to remove oxides. Alternatively, buffered oxide etch (BOE) can be used for the purpose of oxide removal. One has to be cautious with the wet-cleaning approaches because a long immersion time of silicon surface in hot water, alkaline solution, or BHF can lead to a roughened silicon surface, which can be detrimental to subsequently grown gate oxide integrity.

A conventional wet bench consists of a series of chemical tanks. Each tank contains the desired acid or base solution; the final step is a quick down rinse (QDR) tank with DI water. The operating parameters include the chemical concentrations and ratios, operating temperatures, residence times, flow rates, rinse times, and the purity of the chemicals and water. A proper procedure must be optimized in temperature, concentration, and elapse time to obtain the maximum

Table 3.3. The approach, mechanism, and nature of various methods of dry-cleaning.

Dry-cleaning	Mechanism	Nature of cleaning
Ar sputtering	Momentum transfer	Physical
Hydrogen plasma	Plasma reaction	Chemical
NF_3 plasma	Plasma reaction	Chemical
O_3/UV	Free radical reaction	Chemical
HF vapor	Chemical reaction	Chemical

cleanliness and minimum surface damage. Finally, wafers must be completely dried before being unloaded from the bench.

A major drawback of the wet-cleaning method is its consumption of a huge amount of water and chemicals. Some of these chemicals are not environmentally friendly. As indicated in Table 3.3, some dry-cleaning methods have been proposed to avoid the drawbacks of wet-cleaning. Some of them remain in the development stage, and some are being used widely in manufacturing such as Ar sputtering, hydrogen anneal, and remote plasma cleaning. Dry-cleaning methods can be chemical or physical in nature. The physical dry-cleaning approach removes the contaminants by particle bombardment. An example of this is Ar sputtering for precontact metal sputtering. The Ar ions bombard the native oxide at the bottom of the contact holes. The momentum transfers from the bombarding ions to the substrate surface atoms or molecules and sputters them off the substrate surface. The chemical dry-cleaning approach removes contaminants by means of chemical reactions.

After the acid and base solution treatment, the wafers must be thoroughly rinsed with deionized water. In general, several rinse cycles are required, and each rinse cycle is assisted with megasonic power for the removal of tiny particles. A crucial part, and the last step, of the cleaning is the wafer drying. A proper drying should not leave any particles or water stains on the wafer surfaces. The water stains are essentially contaminants that are aggregated by surface tension as water dries out; this is demonstrated in Fig. 3.29. There are different drying approaches such as spin drying, IPA drying, and

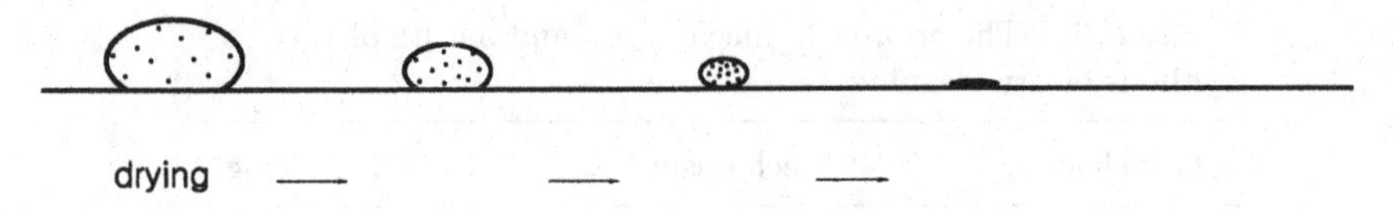

Fig. 3.29. The concentration of contaminants in a drying water droplet increases as drying proceeds. The droplet volume shrinks, and finally, water stains form.

Marangoni drying. In a spin drying process, wafers are loaded in cassettes that are fixed on a rotating stage. A clean inert gas, such as nitrogen, flows downward or upward through the wafer surface, while the stage rotates. The water on the wafer surface is removed by both the centrifugal force from the rotation and the dragging force due to the gas flow, as illustrated in Fig. 3.30. During the IPA drying process, the wet wafer surfaces are exposed to hot IPA vapor, which is carried by nitrogen, as shown in Fig. 3.31. The IPA vapor condenses on the cold wet wafer surface and displaces the water droplets to sheet off the wafer surface. The highly volatile IPA dries out the

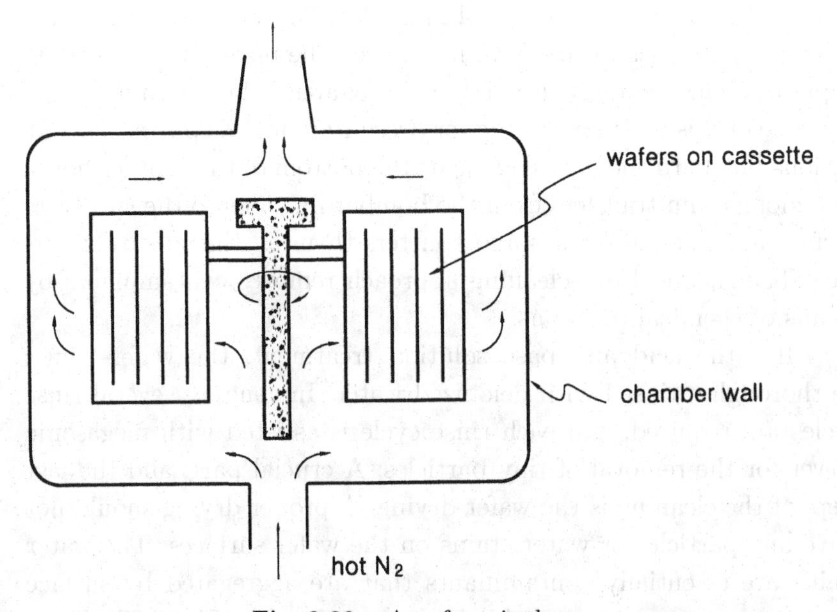

Fig. 3.30. A wafer spin dryer.

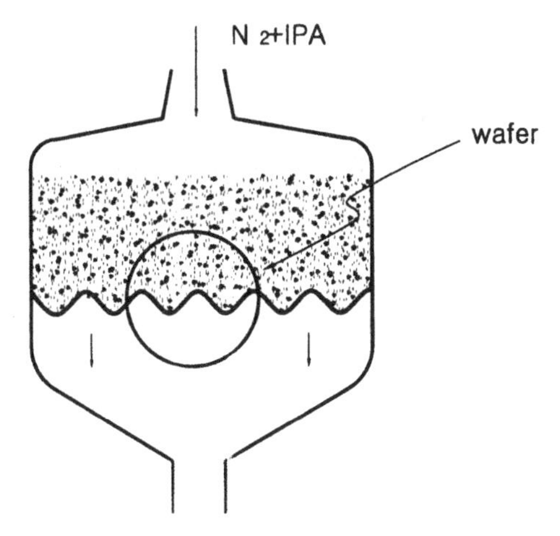

Fig. 3.31. In an IPA drying process, a wafer is exposed to hot IPA which dries out the residual water on wafer surface.

wafer surface. The Marangoni drying process dries a wafer surface by means of the Marangoni effect. The Marangoni effect, which is named after the nineteenth-century Italian scientist Marangoni, occurs when a fluid flows along liquid–liquid or liquid–gas interfaces, driven by a surface tension gradient. During the Marangoni drying process, water

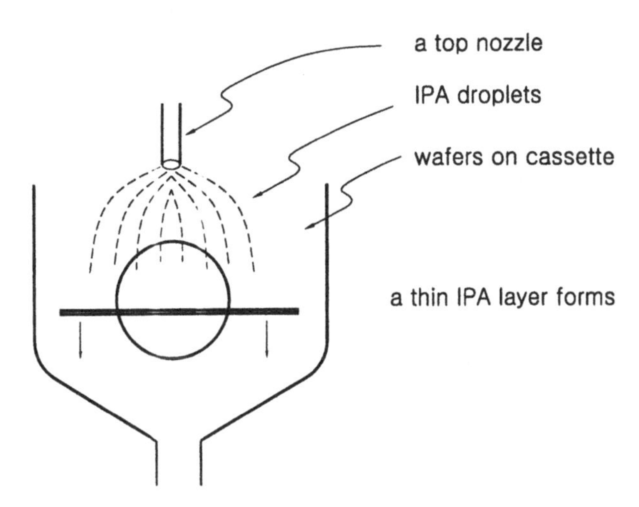

Fig. 3.32. A Marangoni drying process: IPA droplets form a thin layer at water surface as water drains out.

level is slowly drained out, while the IPA droplets continuously fall from the top nozzle and form a thin layer at the water surface, as illustrated in Fig. 3.32. There is a surface tension gradient at the IPA–water interface. The surface tension of water drags the water together with contaminants back to the tank and leaves the wafer surface above the water dry and clean.

Chapter 4

GAS KINETICS AND PLASMA PHYSICS

As discussed in Chapter 3, after the silicon surface oxidation, the wafer moves to the silicon nitride deposition and etching step. This chapter is intended to lay out the foundation for chemical vapor deposition (CVD) with or without plasma enhancement and plasma-enhanced etching. The first section provides an overview of gas kinetics, including particle velocity, mean free path, collision rate, and collision cross-section. These elements are essential in understanding the mechanism for the deposition and etching processes. The second section illustrates the plasma state and how plasma is formed. It then moves on to explain the electron impact phenomena, such as excitation, dissociation, and ionization, which sustain the plasma state and enable the plasma-enhanced reactions. This chapter concludes with the introduction of various plasma reactor configurations.

4.1. Gas Kinetics and Ideal Gases

Gas kinetic theory helps in understanding the deposition and etching processes from the viewpoints of particle interactions and bulk-averaged behavior of a gas system. It sheds light on how gas phase reactions are initiated and how they proceed.

4.1.1. *Ideal gas law*

When dealing with gas kinetics, real gas behavior is complicated to predict. Therefore the gas behavior is often derived based on some assumptions such as the ideal gas law. On the basis of these assumptions, the real gas behavior can be approximated. Under certain circumstances, the ideal gas approximation can be very close to reality.

Ideal gases are hypothetical and have common characteristics:

(a) *There are no intermolecular interactions.* In the case of a real gas, the molecules move individually, but they are attracted to each other. But when two molecules are getting too close, they can be repulsive to each other. For an ideal gas, there are no intermolecular interactions, meaning that the molecules are free from each other.

(b) *The gas particles have zero volume when at rest.* Any real gas molecule has a finite volume and mass. When the system temperature is lowered to 0 K, the gas molecules condense and reach a certain volume, as shown in Fig. 4.1. The ideal gas has zero volume.

(c) *An ideal gas satisfies a governing equation, also known as the ideal gas law:*

$$PV = nRT \tag{4.1}$$

or

$$P\overline{V} = RT\,, \tag{4.2}$$

where P is the system pressure (pascals, $\mathrm{N/m^2}$), V is the volume of the system ($\mathrm{m^3}$), $\overline{V}$ is the molar volume ($\mathrm{m^3/mol}$), n is the number of moles of gas in the system (moles), R is the ideal

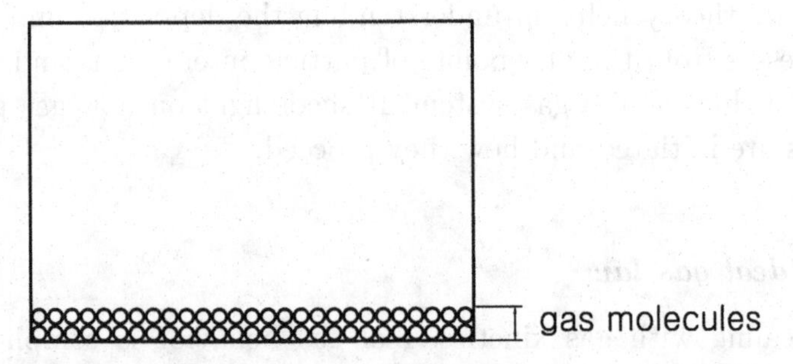

Fig. 4.1. The volume of a real gas after condensation at 0 K.

gas constant (8.314 J/mol K or Nm/mol K), and T is the absolute temperature (K). In other words, in a closed system, the state of an ideal gas system can be determined when two of the system's intensive variables are fixed. For example, if the system temperature and pressure are known, one can derive the system molar volume, and vice versa. With a constant molar volume, the system pressure increases with temperature. Since the molecules have a higher kinetic energy at higher temperatures, they impart more energy to the container wall on colliding with it. On the other hand, in a constant temperature system, a higher pressure tends to squeeze the gas volume and results in lower molar volume. If the system pressure is kept constant, the volume expands when the temperature increases.

In reality, there is no ideal gas; however, some conditions exist under which the real gas behaves nearly ideally. A real gas system at high temperature and low pressure will approach the ideal behavior. This is because at such a condition, molecules are farther apart and more energetic. Experience shows that the gas behavior in a system at pressures under 10 atm and above room temperature can be approximated with the ideal gas law. In semiconductor manufacturing processes, all systems, such as annealing, oxidation, etching, or deposition, are operated under high vacuum, lower than in the mTorr range, and above room temperature. Therefore the ideal gas law prevails.

Example 4.1.

Assume a polysilicon CVD system with SiH_4 at 1000 K. What is the molar volume at 1 atm? What is the molar volume when the pressure decreases to 0.5 atm?

Solution

At such a high temperature and low pressure, the system can be assumed as an ideal gas. Therefore

$$\overline{V} = \frac{RT}{P} = \frac{0.082\ l\cdot\text{atm/k}\cdot\text{mol} \times 1000\,\text{K}}{1\,\text{atm}} = 82.05\,l/\text{mol}\,.$$

At 0.5 atm, the molar volume is

$$\overline{V} = \frac{RT}{P} = \frac{0.082\, l\cdot\text{atm/k}\cdot\text{mol} \times 1000\,\text{K}}{0.5\,\text{atm}}$$
$$= 164.1\, l/\text{mol}\,.$$

It can be observed that as the pressure is halved, the volume doubles.

4.1.2. *The mean traveling speed of a gas molecule*

Let us assume a closed ideal gas system with a number of molecules N. All molecules travel without preferential direction, that is, randomly, with an average kinetic energy proportional to the system temperature, $(1/2)mv^2 = (3/2)kT$. There is an energy distribution among all the molecules. At any instant in time, molecules undergo elastic collisions with each other. It can be derived from Maxwellian molecule energy distribution that all molecules travel at an arithmetic mean speed, $\langle v \rangle$, and that

$$\langle V \rangle = \frac{1}{N}\sum_{1}^{N} V_i = \sqrt{\frac{8RT}{\pi M}}\,, \tag{4.3}$$

where M is the molecular weight of the gas molecule.

Example 4.2.
A tungsten deposition system operates at 200 mTorr and 500 K. What is the average speed of the molecules that are traveling?

Solution

$$\langle V \rangle = \frac{1}{N}\sum_{1}^{N} V_i = \sqrt{\frac{8RT}{\pi M}}$$
$$= \sqrt{8 \times (8.314\,\text{J/mol/k})(500\,\text{K})/(3.1416 \times 0.297\ \text{kg/mol})}$$
$$= 188.8\,\text{m/s}\,.$$

4.1.3. *Collision frequency and mean free path*

In a CVD process, the desired final product is a nonvolatile film on the substrate surface. As a prerequisite to the deposition reaction, the molecule has to collide with the substrate surface. Therefore the collision rate between the precursor gas and the substrate surface plays a determining role in film growth rate. As shown in Fig. 4.2, the gas molecule undergoes random motions in a closed system. It can be derived from the Maxwellian velocity distribution function that the collision rate per unit area, F, is

$$F = n_p \sqrt{\frac{RT}{2\pi M}}, \tag{4.4}$$

where n_p is the number of molecules per unit volume and M is the molecular weight of the molecules. One approach for estimating the film deposition rate is to assume that the deposition rate equals the collision rate times a sticking coefficient, as shown in Fig. 4.3. The sticking coefficient is defined as the percentage of the impinging molecules that are stuck on the substrate or film surface and become part of the deposited film. The sticking coefficients are determined experimentally.

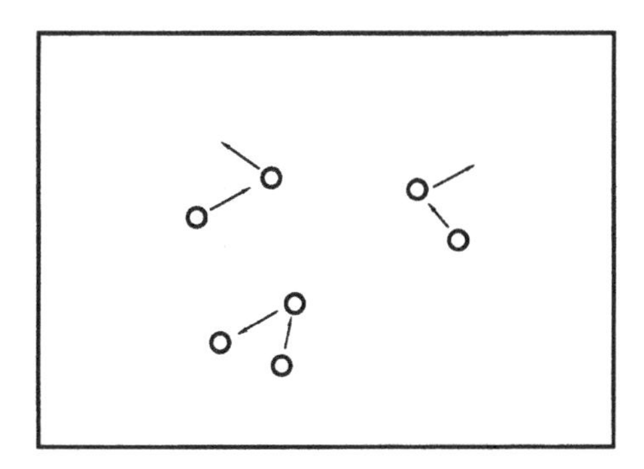

Fig. 4.2. Gas molecules undergo random motion at any instant in time.

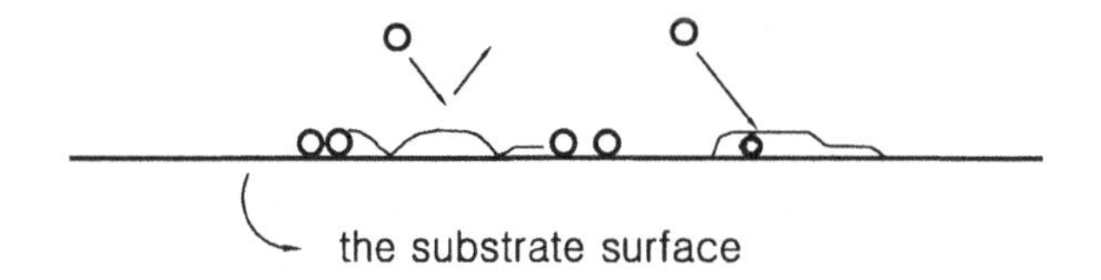

Fig. 4.3. Depositing molecules impinge on substrate surface; some bounce back, while the others are incorporated into the growing film.

Example 4.3.
A low-pressure CVD system for polysilicon, operated at 1 mTorr and 1000 K, contains SiH_4. Calculate the flux of SiH_4 that is colliding onto the substrate surface.

Solution
The molecular density for such a system can be calculated as

$$n_p = \frac{PN_A}{RT} = \frac{0.001/760\,\text{atm} \times (6.02 \times 10^{23}/\text{mol})}{(0.082\,\text{l}\cdot\text{atm/mol}\cdot\text{K}) \times 1000\,\text{K} \times 0.001\,\text{m}^3/\text{l}}$$
$$= 9.6 \times 10^{18}\text{m}^{-3}\,.$$

The collision frequency per unit area is estimated as

$$F = n_p\sqrt{\frac{RT}{2\pi M}} = 9.9 \times 10^{18} \times \sqrt{\frac{(8.314\,\text{J/mol}\cdot\text{K})(1000\,\text{K})}{2 \times 3.14 \times 32 \times 10^{-3}\,\text{kg/mol}}}$$
$$= 2.0 \times 10^{21}\,\text{s}^{-1}\,\text{m}^{-2}\,.$$

To account for the homogeneous reaction, it is important to understand the collision between molecules. Molecules undergo collisions at any instant in time with other molecules. Considering each individual gas molecule as a rigid body, as shown in Fig. 4.4, any other molecules that come into the circle, with an area of πd^2, will collide with the molecule. The collision cross-section of the two molecules

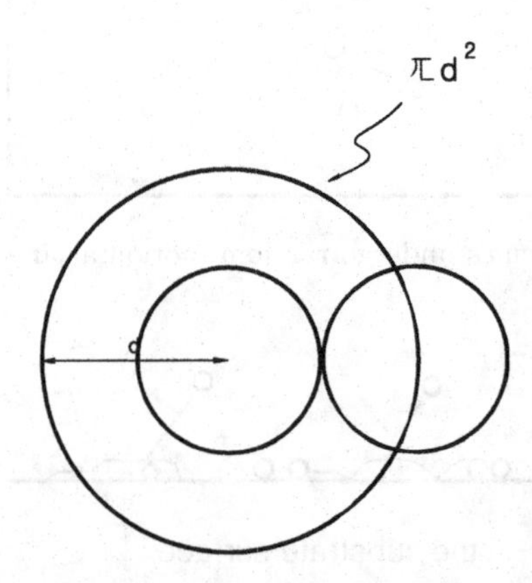

Fig. 4.4. The collision cross-section, πd^2, of particles with a diameter of d.

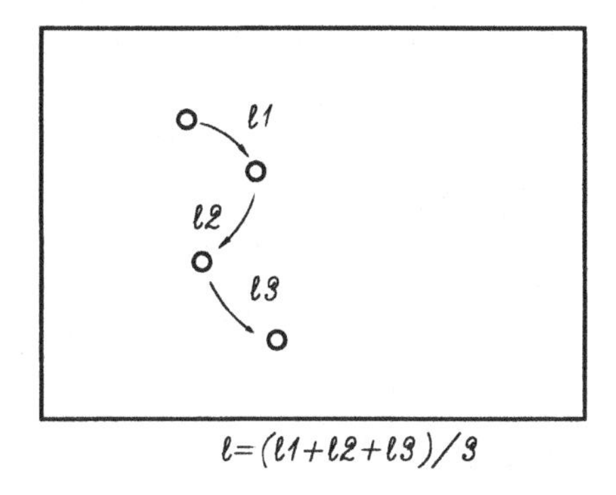

Fig. 4.5. Mean free path is the average distance that a molecule can travel between collisions.

is πd^2. Assuming that each molecule travels at an arithmetic mean speed of $\langle v \rangle$ in a system with gas molecular density n, then the number of collisions that occur in the system is q and can be derived as

$$q = \sqrt{2}\langle v \rangle n\pi d^2 \,. \tag{4.5}$$

The mean free path, λ, as shown in Fig. 4.5, is the average distance that a molecule travels between two consecutive collisions. Therefore

$$\lambda = \frac{\langle v \rangle}{q} = \frac{1}{\sqrt{2}\pi n d^2} = \frac{kT}{\sqrt{2}\pi d^2 P}\,, \tag{4.6}$$

where k is the Boltzman constant, $3.3 \times 10^{-24}\,\text{cal/K}$, or $1.38 \times 10^{-23}\,\text{J/K}$.

Example 4.4.

Calculate and compare the mean free path of a CVD precursor with a diameter of 10^{-10} m at 600°C and 1 Torr and 0.1 mTorr.

Solution

$$\lambda = \frac{kT}{\sqrt{2}\pi d^2 P} = \frac{(0.082/6.02 \times 10^{23}\,\text{atm}\cdot\text{l/K}) \times 873\,\text{K} \times 10^{-3}\,\text{m}^3/\text{l}}{\sqrt{2} \times 3.14 \times 10^{-20}\,\text{m}^2 \times (1/760\,\text{atm})}$$
$$= 0.00203\,\text{m}\,.$$

At 0.1 mTorr, the mean free path becomes 20.3 m.

This explains why low-pressure CVD provides better thickness uniformity than atmospheric-pressure CVD. The reason is that the mean free path of a molecule in a low-pressure CVD system is longer than that in an atmospheric-pressure system. Therefore there is a higher probability that the molecules travel a longer distance without being diffracted by collisions, which allows them to reach the corners of the substrate surface. This will result in better uniformity.

4.2. What is Plasma? How is it Formed?

Plasma is a state of matter that occurs at temperatures above 10 000 K. Consider a solid matter enclosed in a perfect container, which is assumed to have an infinitely high melting point. As the temperature of the container increases, the matter will be heated and will undergo phase changes, from solid to liquid and then to vapor. In the vapor phase, the molecules will undergo random motion with a kinetic energy proportional to the system temperatures. Energy gets transferred from one to the other, as molecules collide with each other. When the temperature reaches a certain level, the received energy from the collision can be enough to cause excitation, and the ensuing relaxation causes the gas system to glow. As the received energy gets higher, it can dissociate some of the molecules into individual atoms. As the temperature increases further, the transferred energy will be large enough to knock out the electrons from the outer energy shells of the molecules, and the gas starts to ionize. Most gases start to ionize at temperatures around 10^4–10^5 K. A partially ionized gas containing ions, electrons, and neutrals is referred to as plasma, which is the most common state of matters in nature. For example, the sun and stars are nothing but gigantic natural plasmas.

Plasma can also be created using electrical fields. Electrons and ions gain energy from the electrical field and then pass it on to neutrals via collisions. The collisions could lead to elastic scatter, excitation, dissociation, or ionization, depending on the electron energy and the threshold energy of each event. Figure 4.6 shows the I–V characteristics of a DC discharge system. As one applies a voltage

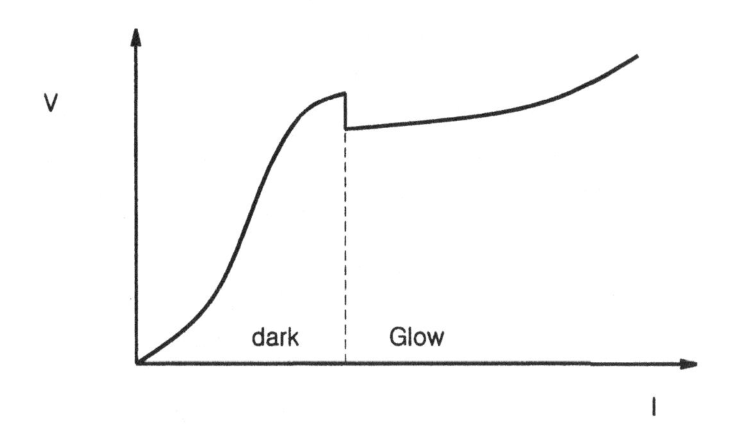

Fig. 4.6. I–V characteristics of a DC glow discharge.

across two parallel plates, the gas essentially becomes an insulator at low applied voltages. A small current results from external agents such as cosmic rays. If the voltage continues to increase, at some point, the gas system will break down and start to conduct current. The voltage at which the current is increased by a large factor is called the breakdown voltage of the gas system. After the transition, the system starts to glow. As the system current increases, a second transition occurs, in which the glow changes over to an arc discharge. The fact that the system conductivity varies is the result of varying extents of ionization. Table 4.1 shows interesting milestones

Table 4.1. Discoveries in gaseous electronics.

Year	Concept	Originator
1600	Electricity	Gilbert
1808	Arc	Davy
1834	Cathode and anode	Faraday
1860	Mean free path	Maxwell
1879	Fourth state of matter	Crookes
1898	Ionization	Crookes
1906	Electron	Lorentz
1928	Plasma	Langmuir
1935	Velocity distribution function	Allis

of gaseous electronics in human history, from the discovery of electricity in 1600 and the establishment of related theories to today's applications of plasma in heating, material processing, and chemical reactions.

In a plasma system, there are electrons, ions, and neutrals. A plasma system can be well characterized by both electron temperature and electron density. Figure 4.7 shows the electron temperatures and densities for some naturally formed and man-made plasma systems. Today, applications of plasma seen in microelectronic processing include sputtering, plasma etching, and plasma-enhanced chemical vapor deposition (PECVD). Most of the plasmas

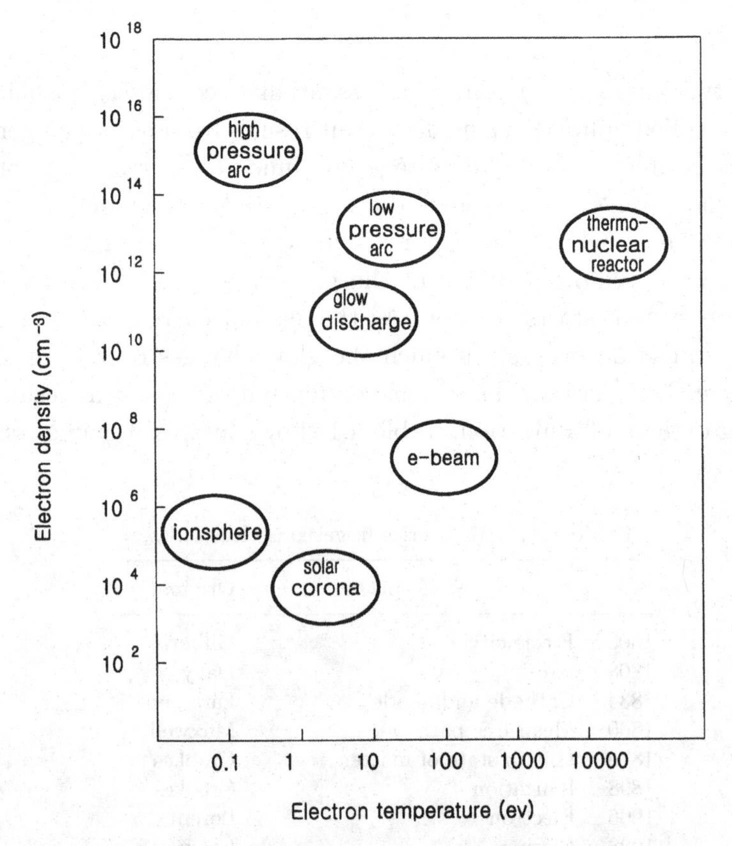

Fig. 4.7. The electron temperature and density ranges for some natural and man-made plasma systems.

Table 4.2. Characteristics of glow discharge systems in semiconductor processing.

Properties	Ranges
Pressure	0.001–1 Torr
Electron temperature	1–10 eV
Electron density	10^8–10^{12} cm^{-3}
Neutral and ion energy	0.025–0.035 eV
Ionization fraction	0.001%–0.00001%
Free radical density	< 30%

are RF-excited under low pressure. Table 4.2 shows some characteristics of a plasma system used in microelectronic device fabrication.

Now, let us consider the quantitative aspect of the energy transfer in plasma. Consider a gas system that is contained in closed parallel electrode plates, as indicated in Fig. 4.8. Initially, there are some trace amounts of charged particles that exist in a gas system. This could be due to the ionization caused by cosmic rays. The charged particles can be accelerated by the electrical field. Assuming an electrical field $\vec{E}$, acting on an initially stationary single-charged particle, and accelerating it to travel a distance, X, then the work, W, done

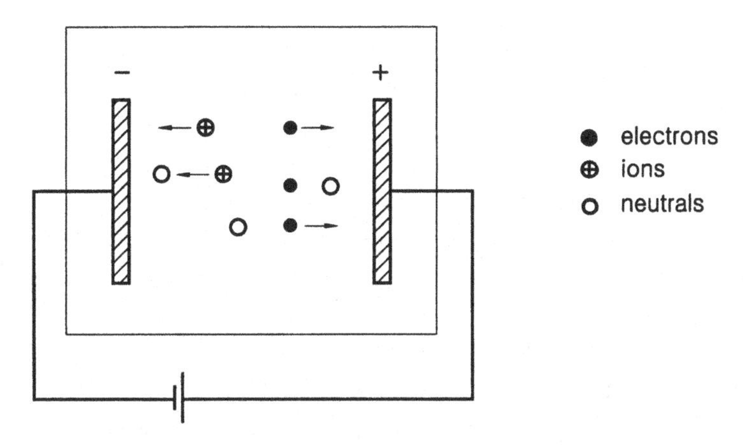

Fig. 4.8. A gas system contained between two electrodes in a closed system.

by the field can be measured as

$$W = e\vec{E}X = \frac{(e\vec{E}t)^2}{2M_p}\,, \tag{4.7}$$

where M_p is the mass of the charged particle. Since the mass of an electron is much smaller than that of an ion, the electrical field mainly provides energy to the electrons. On the other hand, the moving charged particles, being accelerated by the electrical field, can undergo collisions with other particles, either charged ones or neutrals. On collision, the energy gets transferred. Assuming that an electron collides on an initially stationary particle, the amount of energy transferred from the electron to the particle, F_{ie}, is

$$F_{ie} = \frac{4M_eM_i}{(M_e + M_i)^2}\,, \tag{4.8}$$

where F_{ie}, is the maximum fraction of energy that can be transferred from the impinging electron to the target particle after an elastic collision. Because the mass of an electron, M_e, is much smaller than that of the particle, M_i, the amount of transfer energy is negligibly small, $F_{ie} \approx 4M_e/M_i$. Conversely, ions due to their comparable masses readily transfer energy to neutral particles on elastic collision. Consequently, electrons pick up more energy from the electrical field and lose less energy due to collisions than ions do. As a result, the plasma system used in microelectronic processing has an average electron temperature (T_e) as high as 10^4–10^5 K, while the neutrals (T_g) stay at low ambient temperatures, around room temperature to 500 K. Such a plasma state is called *cold plasma*. In a high-pressure system, the frequency of electron–particle collision increases, and the electrons may lose a significant amount of energy. The electron temperature approaches the neutral temperature as the pressure increases, as shown in Fig. 4.9. A plasma system with T_e approaching T_g is called a *hot plasma*. Examples of hot plasma are the sun, stars, or a nuclear fusion reactor. The electron temperature can be estimated from its kinetic energy, $(1/2)mv^2 = (3/2)kT_e$. In a plasma system, the electron density and electron energy are very critical parameters, and they can be measured experimentally.

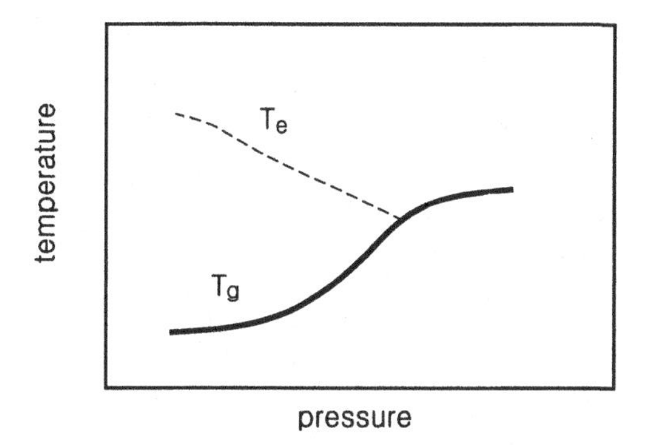

Fig. 4.9. In a plasma system, the gas temperature approaches the electron temperature as system pressure increases.

The highly energetic electrons can also undergo inelastic collisions with other particles. As shown in Fig. 4.10, when the electron temperatures are relatively low, they will excite low-level electrons of the target particle to a higher level. There is a certain probability that the excited electrons can go back to the low levels, in which the energy can be released in the form of emitting photons. This is why the plasma glows. The electrons with higher electron energy levels can cause the target molecules to dissociate into atoms or molecules with unpaired electrons, or radicals. Owing to the unpaired electrons

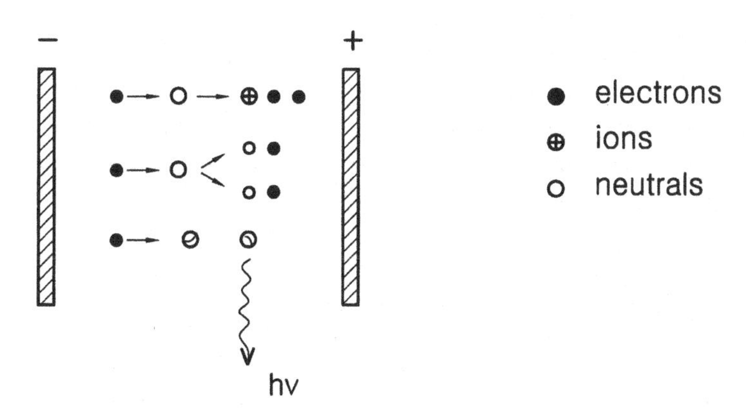

Fig. 4.10. Electron impacts that lead to excitation, dissociation, and ionization.

in the outer shells, radicals are extremely reactive. This very characteristic can enable the plasma to initiate CVD or etching reactions at low ambient temperature (200–350°C), which would otherwise be initiated at much higher temperatures (around 700–1100°C) with thermal activation. The reactive radicals are the key to CVD and etching reactions. Electrons with even higher energies can inelastically collide with particles and knock the electrons out of their outer shells, causing ionization. The ionization generates ion–electron pairs and therefore sustains the plasma state. In particular, it compensates for the electrons and ions that are consumed on colliding on the container walls. In plasma, the total number of electrons roughly equals that of ions and is called the *plasma density*. The plasma density is much smaller than the number of neutrals (molecules, atoms, and free radicals) in cold plasma.

In plasma, electrons move much faster than ions. When a floating object or a container wall is exposed to plasma, the surface will be quickly charged up negatively due to the incoming electron flux. The negatively charged surface will then drive the positive ions to bombard onto it, inducing an ion influx. As more and more electrons flow out of the plasma bulk, the bulk becomes more positive and therefore slows down further electron out-flux. Meanwhile, the ion flux also decreases with the electron flux. A steady state can be reached, at which the electron out-flux equals the ion out-flux. Also, there is a voltage drop built up across the plasma bulk to the floating object or the container wall. This built-up voltage drop is called the *sheath voltage*. The voltage applied to the system nearly equals the voltage drop across the two plasma sheaths near the cathode and anode, as depicted in Fig. 4.11. The sheath voltage dictates the ion bombardment energy in a plasma process chamber.

4.3. Introduction to Plasma Physics

In the previous section, we introduced the concept of plasma and how it is formed. Now, we will introduce the characteristics of a plasma system, the electron temperature, and density. We will also look at

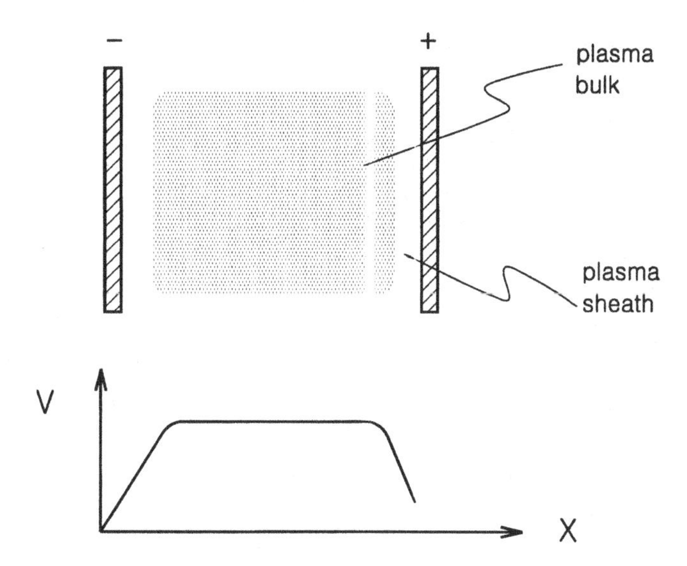

Fig. 4.11. The plasma sheaths account for most of the voltage drops between electrodes.

the concept of collision cross-section and the collision processes that lead to excitation, dissociation, and ionization.

A neutral gas system can be characterized by its molecular density and system temperature, whereas a plasma system can be characterized by its electron temperature and density. Just like the energy of each molecule in a neutral gas system, the electrons in the plasma system may not have a uniform electron temperature. Instead, the energy of each electron can be different. If one looks at the entire plasma system, the energy distribution of electrons can be obtained by solving the Boltzmann equation:

$$\frac{\partial f}{\partial t} + \vec{v} \cdot \nabla_r f + \vec{X} \cdot \nabla_v f = \left(\frac{\partial f}{\partial t}\right)_{\text{coll}}, \tag{4.9}$$

where $f = f(r, v, t)$ is the number of electrons at a given time t, in position r and with energy v. Therefore, by solving Eq. (4.9), the electron population profile with respect to space and energy levels can be obtained. Consequently, the total number of electrons in a defined system can be evaluated with

$$\int_{dv}\int_{dr} f\,dr\,dv = \text{total number of electrons in the system} \,. \quad (4.10)$$

The first term in Eq. (4.9) represents the variation of f with respect to time; the second term represents the electron streams in and out of the infinitesimal space ($dx\,dy\,dz$); the third term represents the variation of f in energy space, $dv_x dv_y dv_z$, as a result of the external force, X. The right-hand side of the equation accounts for the changes of electron population due to various collisions. The Boltzmann equation is a six-dimensional nonsteady state problem. It is very difficult to obtain an exact solution. However, approximated solutions can be achieved by making certain assumptions. Assuming a steady state and uniform spatial distribution, the two most commonly used energy distributions are Maxwellian and Druyvestein, as shown in Fig. 4.12. The Maxwellian energy distribution is obtained in the absence of an electrical field and by making several assumptions. First, the electrons are in thermal equilibrium with others such as neutrals and ions. Second, the inelastic collisions are negligible

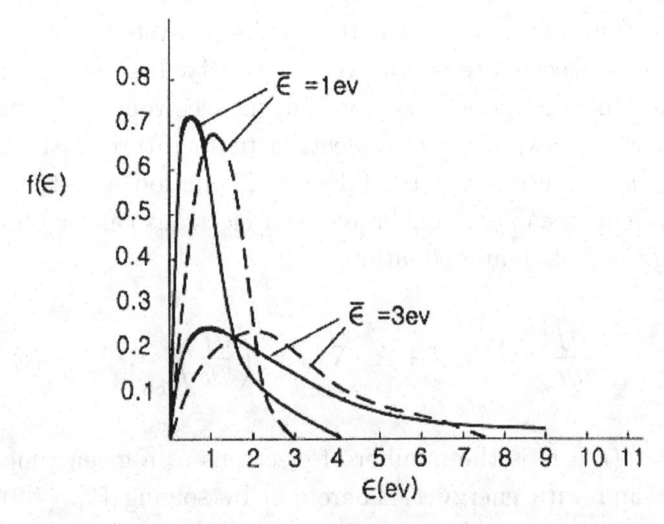

Fig. 4.12. The Maxwellian (—) and Druyvestein (---) electron energy distributions.

when they are compared to the elastic ones. Third, columbic (electrostatic) collisions are negligible. If one further assumes that the elastic collision cross-section is independent of electron energy, the Druyvestein energy distribution can be obtained. The latter has less high-energy electrons than the former, and it accounts for the fact that high-energy electrons tend to travel fast toward the container wall and recombine on colliding on it. It has been proven that the Druyvestein electron energy distribution is closer to the actual electron energy distribution in real systems than the Maxwellian. One even more meaningful parameter for a plasma system is the average electron temperature. With the energy distributions, one can further evaluate the average electron temperature, which is a valuable parameter in a plasma system.

The spatial distribution of electrons plays a significant role in PECVD and etching since it dictates the deposition or etching uniformity. Instead of solving the Boltzmann equation, the concept of the ambipolar diffusion model is often used in literature. The ambipolar diffusion is the hypothesis that in a plasma system, the electrons and ions move collectively with the same diffusivity. The electron motion following the ambipolar diffusion model is described as

$$\nabla^2 N_e + \nu_i N_e = 0\,, \tag{4.11}$$

where the first term represents the electron diffusion, while the second term represents the generation due to ionization. The equation coupled with proper boundary conditions can be solved for the spatial distribution of electron density. Assuming that the electrons disappear on colliding onto the reactor wall in a closed system, the following solution can be obtained:

$$N_e = N_{eo}\cos\left[\left(\frac{Z - Z_0}{H} - \frac{1}{2}\right)\pi\right], \tag{4.12}$$

where N_{eo} is the electron density at the mid-plane of the reactor. This spatial electron density distribution is shown in Fig. 4.13; it is symmetric with respect to the reactor centerline. The central electron density is experimentally correlated to the system power input and

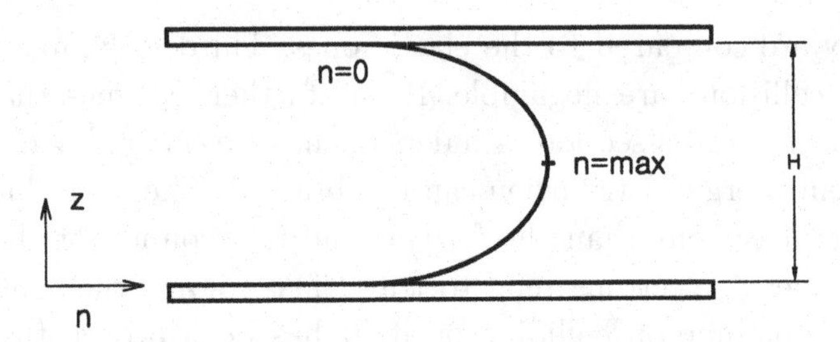

Fig. 4.13. The electron density distribution is symmetrical with respect to the center of the plasma chamber.

inversely proportional to the system pressure:

$$N_{eo} = cT_g \frac{P_w}{p},$$

where T_g is the gas temperature, P_w is the system power input, and p is the system pressure.

4.4. Electron Impact Phenomena

Although plasma systems have been widely used in semiconductor manufacturing applications, such as etchings and depositions, the exact reaction mechanisms are extremely complicated in nature and are not well known in general. However, derived from the species, as detected *in situ* with instrumental analysis, there are some fundamental reaction pathways that can be identified amid the complicated plasma reaction nature. All of these have resulted from either elastic or inelastic collisions of the energetic electrons in the plasma with other particles.

(a) *Excitation and Relaxation*

$$\text{Excitation: A} + \text{e}^- \longrightarrow \text{A}^* + \text{e}^-$$
$$\text{Relaxation: A}^* \longrightarrow \text{A} + \text{photons}\,.$$

During excitation, after receiving the energy from the impinging energetic electron via an inelastic collision, one of the electrons in the target particle gets elevated to a higher energy level. There is then a certain probability for the elevated electron to fall back to its original energy level, by releasing energy in the form of emitting light: photons. This is the very mechanism that keeps the plasma glowing.

(b) *Ionization*

$$e^- + A \longrightarrow A^+ + 2e^- .$$

Essentially, in an ionization process, the impinging energetic electron knocks out one of the outer shell electrons, generating an ion–electron pair. To keep the plasma on, the density of electrons and ions must be kept above a certain level; in other words, the generation and loss of electrons must be balanced. Electrons can be generated through ionization and a secondary electron generation mechanism, in which an electron is generated when an ion collides onto a container wall. On the other hand, electrons can be lost from the body of the plasma through recombination or loss to a wall.

(c) *Recombination.* On colliding with a positive ion, a negative electron can give rise to a recombination process:

$$e^- + B^+ + B \longrightarrow 2B .$$

In this process, the plasma system loses an electron–ion pair.

(d) *Electron attachment.* On an inelastic collision, an energetic electron can combine with an atom or molecule to form a negative ion. This process is called electron attachment:

$$e^- + A \longrightarrow A^- .$$

This process often occurs with atoms or molecules with high electron affinities such as oxygen and halogen atoms.

(e) *Dissociation.* In a dissociation or fragmentation process, a molecule is dissociated into two equal parts; each has an unpaired electron:

$$e^- + F_2 \longrightarrow e^- + F + F .$$

F stands for an atom with an unpaired electron; it is also called a free radical, which is very chemically active even at low ambient temperatures. The radicals are the very species that enable high-temperature chemical reactions to occur at low ambient temperatures — a common characteristic of plasma etching and deposition.

The probability of each collision leading to a specific event strongly depends on the electron energy levels and the threshold of each of the events. Figure 4.14 shows the excitation cross-section for the hydrogen molecules, Fig. 4.15 illustrates the ionization cross-section of xenon molecules, and finally, Fig. 4.16 demonstrates the dissociation cross-section of ammonia molecules. The rate of each can be determined experimentally. Unlike rigid body collision, when an electron collides on a target molecule, the molecule is seen as a cloud of negatively charged electrons surrounding a positively charged nucleus. Intuitively, the electron impact cross-section is a function of electron energy. Physically, the electron impact cross-section is a quantity that describes the probability of a target particle encountering an impact leading to the occurrence of the process *i*. Taking PECVD of silicon nitride as an example, the dissociation

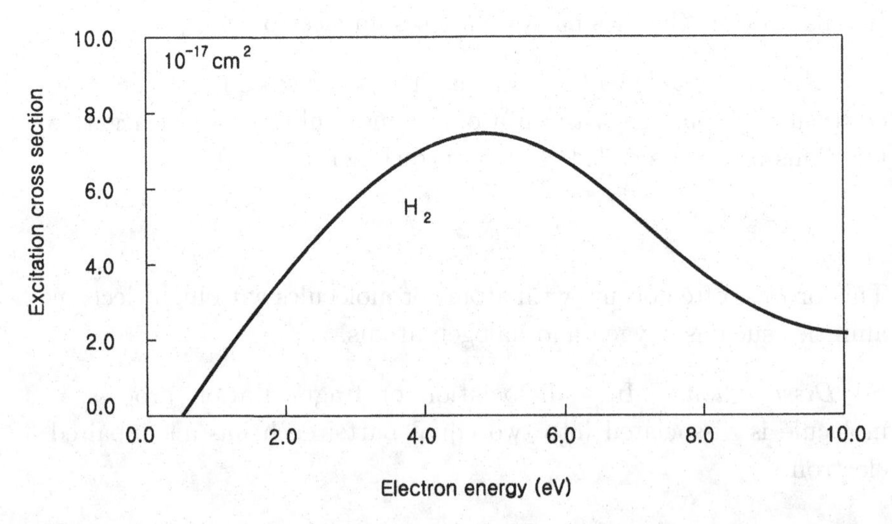

Fig. 4.14. The excitation cross-sections of hydrogen molecules as a function of the electron energy.

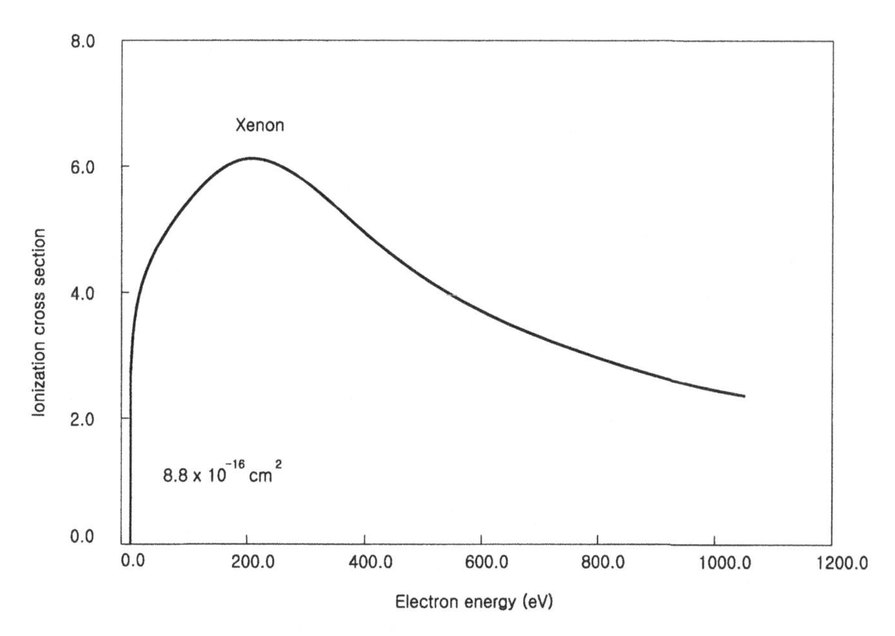

Fig. 4.15. The ionization cross-sections of xenon molecules as a function of the electron energy.

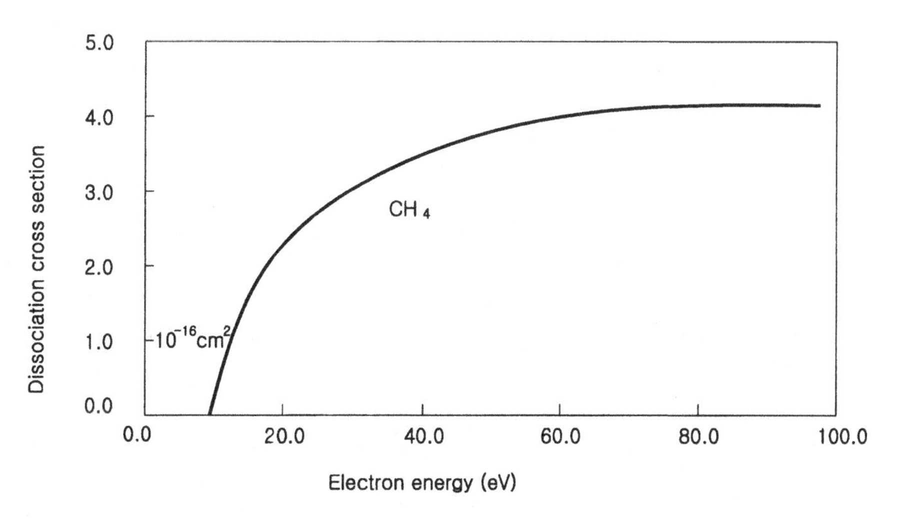

Fig. 4.16. The dissociation cross-sections of methane molecules as a function of the electron energy.

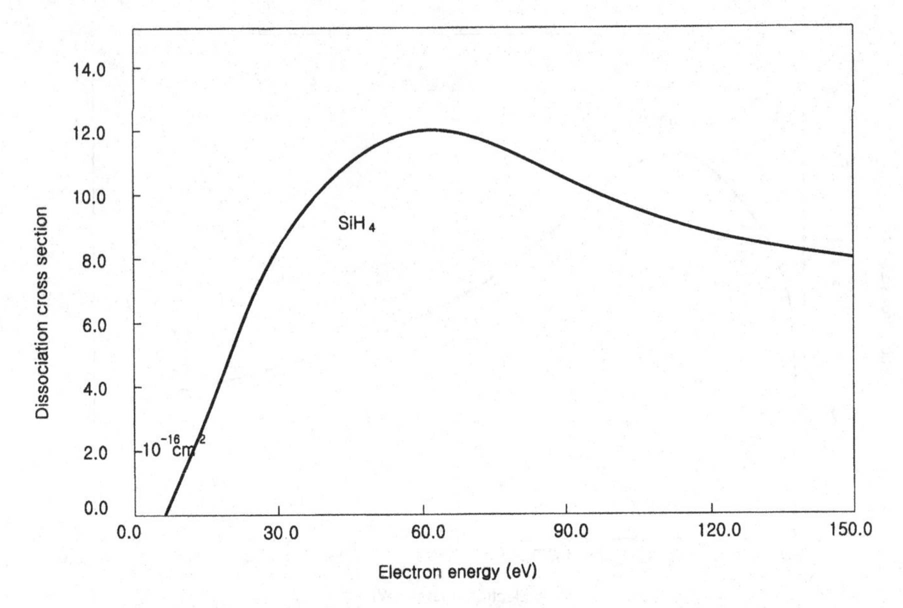

Fig. 4.17. Electron impact dissociation cross-sections of silane.

cross-section of SiH_4, NH_3, and N_2 can be obtained from related literature, as shown in Figs. 4.17, 4.18, and 4.19.

In literature, the rate constant of each process is often reported in terms of the collision cross-section:

$$K_i(\bar{\varepsilon}) = \int_0^{\infty} \left(\frac{2\varepsilon}{M_e}\right) \sigma_i(\varepsilon) f(\varepsilon, \bar{\varepsilon}) d\varepsilon \,, \tag{4.13}$$

where K_i is the rate constant for the process i, which has an impact cross-section σ_i. By integrating Eq. (4.13), the dissociation rate constants in terms of the mean electron energy of the system can be obtained, as shown in Fig. 4.20.

Looking at the collisions between electrons and target molecules, the rate of the process i can be expressed as

$$i = K_i N_e C \,, \tag{4.14}$$

where N_e is the electron density and C is the target molecule density. The rate of each process can also be determined experimentally.

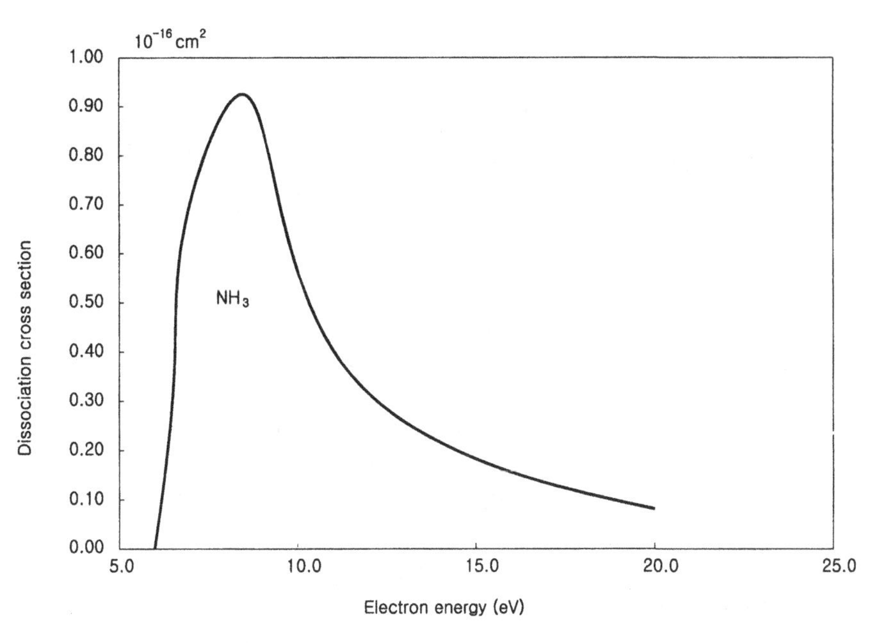

Fig. 4.18. Electron impact dissociation cross-sections of ammonia.

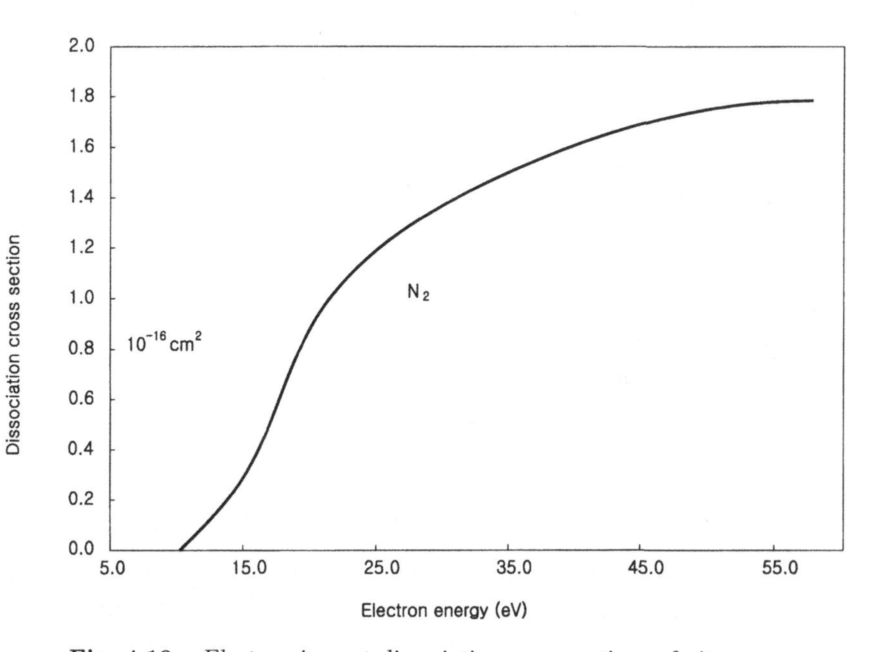

Fig. 4.19. Electron impact dissociation cross-sections of nitrogen.

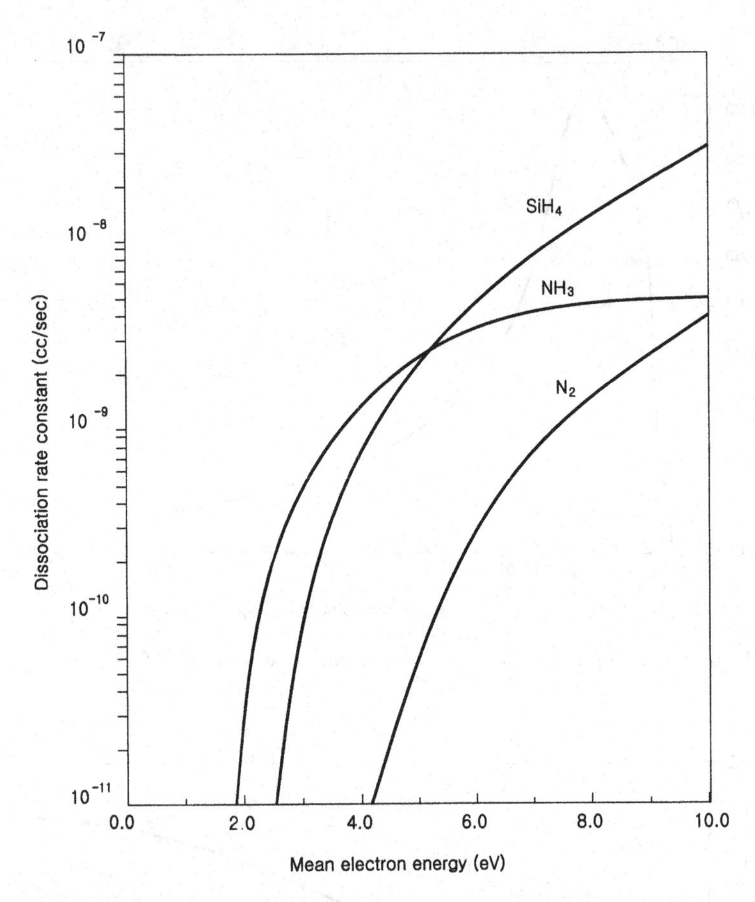

Fig. 4.20. Mean electron energy dependence of electron impact dissociation rate constants for silane, ammonia, and nitrogen.

Assuming that i designates a dissociation process, it can be seen that the dissociation rates increase as the mean electron temperatures increase. With the other parameters being fixed, the mean electron temperatures increase with the electrical fields.

4.5. Fundamental Plasma Reactor Configurations

In semiconductor processing, there are three major electrode configurations for plasma deposition or etching reactors. These configurations are capacitively coupled, inductively coupled, and magnetically enhanced reactors. The capacitively coupled plasma systems can be equipped with DC or AC. A DC plasma is formed by applying an electrical field across the two electrodes in an enclosed chamber, as shown in Fig. 4.8. Initially, a small amount of charged particles, generated through various means, such as irradiation of cosmic rays in the gas system, would respond to the electrical field. When increasing the electrical field strength to a certain level, the gas system breaks down and starts to glow. Electrons are accelerated toward the anode, and positive ions are accelerated toward the cathode, resulting in a current flow. A DC plasma operates at a relatively high pressure ($> 3 \times 10^{-3}$ Torr) to ensure short electron mean free paths, which allow the electrons to generate enough impact events before diminishing on colliding with the wall. In fact, at low pressures, it is difficult to maintain the plasma if the electron mean free path is larger than the electrode spacing, unless other means are applied such as a magnetic field confinement to slow down the electron loss rate. This feature will be discussed later. Except for sputtering, the DC plasma is rarely used for deposition and etching processes.

As shown in Fig. 4.21, an AC plasma system can be formed by applying an alternating electrical field across an enclosed gas system. When the electrical field is increased to a certain value, the plasma is ignited, and the charged particles are accelerated toward the electrode of opposite polarity. As the field changes its polarity, the charged particles change their moving directions. At low frequencies,

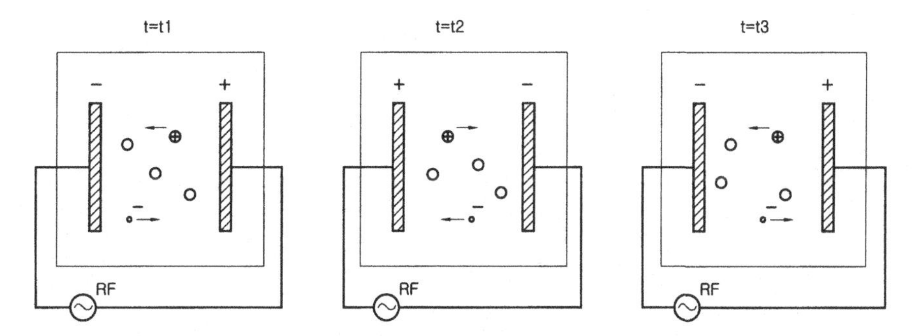

Fig. 4.21. Charged particles move toward electrodes, which alternate in polarity with time. The neutral particles' motion remains random.

both electrons and ions can respond to the electrical field changes. However, as one increases the frequency to a certain value, ions will no longer be able to respond to the field changes. In other words, ions are accelerated in one electrical field polarity and are slowed down in the other. But electrons continue to respond to the field changes due to their much smaller mass. Therefore electrons can efficiently pick up energy from the alternating electrical fields. When the electron moving direction is synchronized with the electrical field change, the electron can pick up energy from the electrical field with best efficiency. In other words, for a fixed frequency, an optimal gas pressure exists at which the electrical field can most efficiently impart energy to electrons, as shown in Fig. 4.22. The charged particles, on the average, have a shorter lifetime in a DC plasma when compared to an AC plasma. Therefore an AC plasma can be operated at a lower pressure than a DC plasma.

One obvious advantage of an AC plasma over a DC plasma is that the AC plasma can handle cases with an insulating electrode surface or insulating reacting material built up on the electrode surface. With a DC plasma, the cathode with an insulating surface is subject to continuous positive ion bombardment, and the plasma can be turned off. With an AC plasma, the positive ions bombarded in one half-cycle can be neutralized by bombarding electrons in the subsequent half-cycle. This explains why most plasma systems in semiconductor manufacturing processes are of

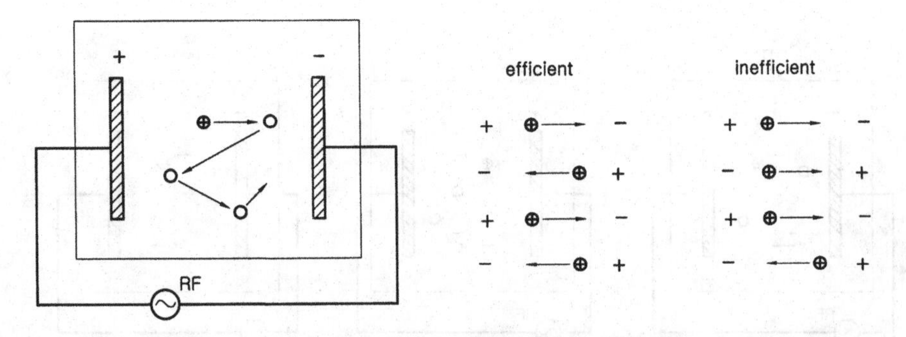

Fig. 4.22. The electrical field can most efficiently impart energy to the charged particles if the moving direction of the charged particles is synchronized with the electrical field.

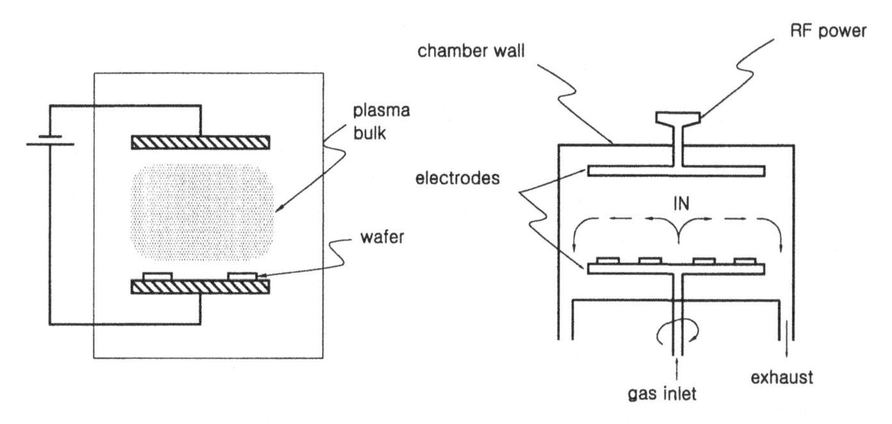

Fig. 4.23. DC (left) and AC (right) plasma reactor configurations.

the AC type. Figure 4.23 illustrates DC and AC plasma reactor configurations.

AC and DC plasmas are essentially capacitively coupled plasmas. They have electrode surfaces in the process chamber, which often become the source for contaminations. Furthermore, for such reactor configurations, it is difficult to maintain the plasma state at low pressure.

In an inductively coupled plasma (ICP), the plasma is generated by an inductive coil around a quartz tube or by the planar coils on the top of the quartz chamber; in other words, physically, there are no electrodes, as shown in Fig. 4.24. As the AC current flows through the coil, it generates an oscillating magnetic field, which in

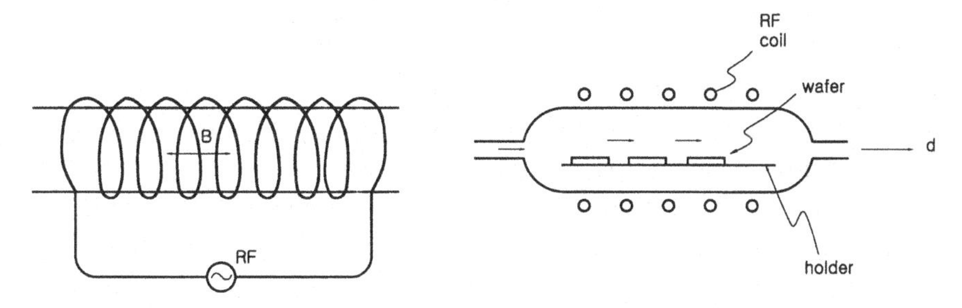

Fig. 4.24. The inductively coupled plasma (ICP) reactors, with wafers placed flat on holders.

turn induces a circumferential electrical field that imparts energy to charged particles in the plasma. Electrons generated in this manner also circle around in the plasma, instead of drifting toward the reactor wall directly; therefore longer electron lifetimes and more collisions can be expected, compared with those in the capacitively coupled plasma system. As a result, one can operate the plasma reactor with enough plasma density at low pressures.

4.6. Magnetic Field Confinement

As discussed earlier, electrons can pick up energy from the electrical field and pass it on to other particles through collisions. Collisions that lead to ionization result in the generation of ion–electron pairs. On the other hand, the energetic electrons are continuously lost from the plasma bulk once they hit the reactor walls or undergo inelastic collisions. Increasing the electrons' lifetime will result in more particle collisions and higher plasma density. This can be realized by taking advantage of the fact that imposing a magnetic field on a moving charged particle can bend its moving direction. A charged particle that moves along an electrical field at a velocity, $\vec{v}$, under the influence of a magnetic field, $\vec{B}$, experiences a force that can be obtained as

$$\vec{F_B} = q(\vec{V} \times \vec{B})\,. \tag{4.15}$$

The direction of $\vec{F_B}$ is reversed for negatively charged particles such as electrons. Owing to random scattering, the electrons' velocity may or may not be perpendicular to the magnetic field. Some electrons that move perpendicular to the magnetic field circle around the field lines. Other electrons that do not move perpendicular to the magnetic field have a helical path around the field line, as illustrated in Fig. 4.25. As a result of the magnetic field confinement, the plasma density can be significantly increased.

A new type of plasma reactor that is currently gaining popularity in semiconductor manufacturing is the electro cycletron resonance

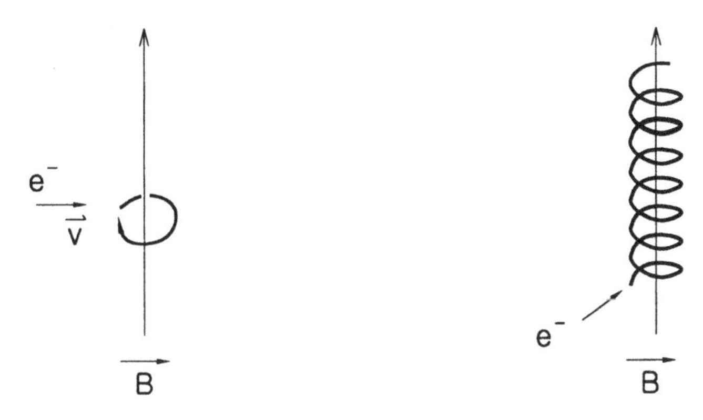

Fig. 4.25. A moving electron enters a magnetic field.

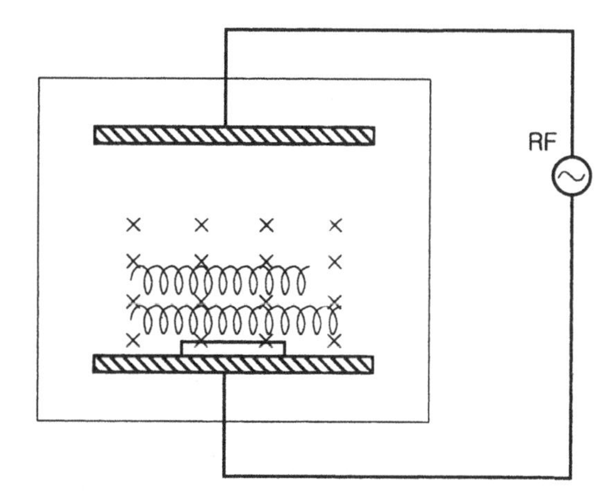

Fig. 4.26. Electrons are confined around the wafer surface with a magnetic field.

(ECR) reactor. It is electrodeless, and it has a high plasma density and can be maintained at very low pressures, below the mTorr range. Observing a moving electron in a magnetic field, as shown in Fig. 4.26, one can see that the electrons can be confined to the local area by circling around the magnetic field lines. If a circling electron is not scattered by other collisions, the electron moves around the field line with a centrifugal force

$$F = \frac{mV^2}{r}. \tag{4.16}$$

Equating Eqs. (4.15) and (4.16), the radius of the circling, r, can be obtained:

$$r = \frac{mV}{qB}. \tag{4.17}$$

It is apparent that since the mass of an ion is about 1000 times larger than the mass of an electron, its circling radius is much larger than that of an electron. In other words, for the plasma bulk of interest, the magnetic field has a marginal effect on the ions.

With Eqs. (4.16) and (4.17), the Larmor frequency can be derived:

$$f = \frac{qB}{m}. \tag{4.18}$$

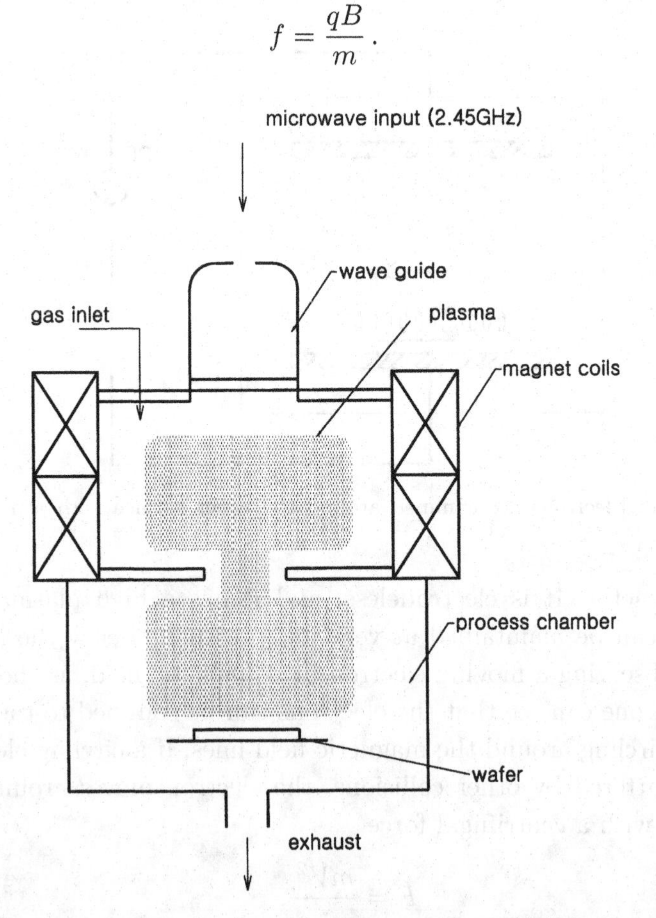

Fig. 4.27. Schematic for an electro cycletron resonance (ECR) plasma reactor.

The Larmor frequency is the frequency at which a charged particle circles around the influencing magnetic field lines. The Larmor frequency is proportional to the imposed magnetic field strength, and it is inversely proportional to the mass of the particle. If the frequency of the input microwave source equals the Larmor frequency, the magnetic field can efficiently add energy to the electrons as they circle around the magnetic field lines. This is the underlying principle that an ECR, CVD, or etching system is based on. The circling of electrons significantly increases the electron lifetimes in the plasma bulk, and more collisions can result; therefore high-density plasma can be obtained even in a low-pressure regime. Figure 4.27 demonstrates an ECR system, the microwave (2.45 GHz) inputs from the top of the reactor, through a dielectric window and the wave guide and into the plasma bulk. The circular motion of the electrons in the plasma is controlled by the magnetic coils to match the microwave frequency and allow for efficient energy transfer from the microwave power source to the plasma.

Chapter 5

CHEMICAL VAPOR DEPOSITION

A semiconductor chip is composed of a number of different conducting layers. Each of these layers is intended to serve a specific purpose. Each layer and patterns of the same layer need to be perfectly isolated. The isolation is accomplished using dielectric materials to fill up the empty space in the intralayers and the interlayer structures, as shown in Fig. 5.1. The dielectric materials physically support and electrically isolate the conducting structures. Chemical vapor deposition (CVD) is a widely used technique to form dielectric and conducting films in device manufacturing.

CVD occurs when a gas mixture is passed over a heated substrate, and chemical reactions are initiated in the vicinity of the substrate surface to form a nonvolatile solid film. In this chapter, we will introduce the classification of the CVD processes, deposited films, and CVD reactors. This is followed by the CVD reactor design concept illustration for various applications. CVD reactor modeling is also illustrated. This helps readers to understand the CVD system in production processes. The characterization of a film is discussed in terms of its compatibility with other films in the device structure. Finally, the chemical reactions for various films and their applications are discussed.

5.1. Classification of CVD Reactors and Films

CVD can be divided into several different groups, based on the energy source, pressure, and wall temperature, as shown in Fig. 5.2. Thermal CVD is a CVD process in which the energy source is purely derived

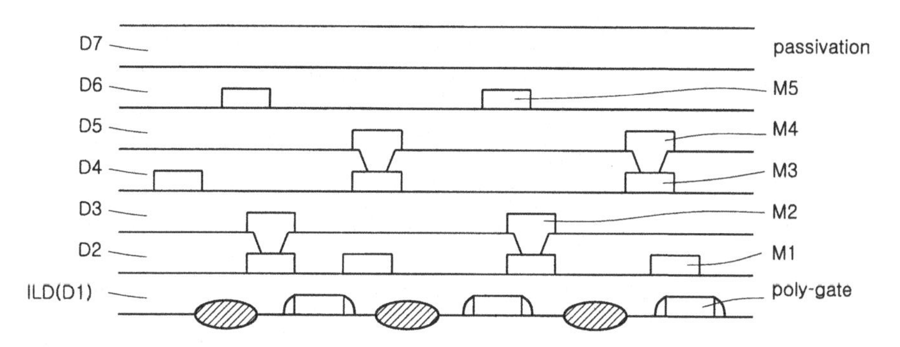

Fig. 5.1. A cross-sectional view of a five-metal device with five metals and seven dielectric layers.

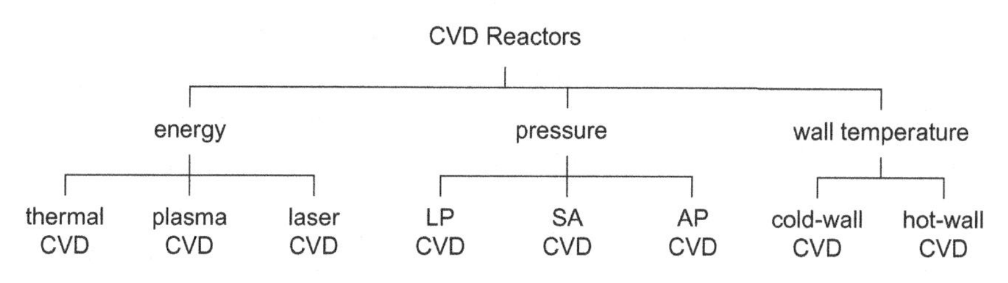

Fig. 5.2. Classification of CVD reactors.

from thermal energy such as energy from lamps or heated substrates. The CVD reactions in a plasma-enhanced CVD (PECVD) are initiated by highly energetic electrons in the plasma system. On the other hand, laser-assisted CVD uses laser energy to initiate chemical reactions. Thermal CVD and PECVD are the two most widely used CVD approaches in semiconductor manufacturing. Thermal CVD runs at high temperatures, ranging from 600°C to 1100°C. It cannot be used after metal layers are deposited since aluminum melts at this temperature range. Thermal CVD, due to its high operation temperatures, produces dense films with low hydrogen content. On the contrary, PECVD operates at low temperatures, ranging from room temperature to 400°C. PECVD is widely used for postmetal planarization and passivation processes.

Another method of classifying CVD reactors is based on their operation pressures, such as atmospheric pressure (APCVD), low pressure (LPCVD), and subatmospheric pressure (SACVD), as shown in Fig. 5.2. APCVD operates at atmospheric pressure; it has a high gas density and therefore a high film deposition rate. The system can be designed in a continuous mode by using a conveyor instead of the closed-chamber batch mode. The diffusivity is inversely proportional to the system pressure, as follows:

$$D_{ij} \propto \frac{aT^b}{p}, \tag{5.1}$$

where D_{ij} is the diffusivity of species i in j, T is the absolute temperature, and p is the system pressure. By having a lower system pressure in the milli-Torr regime, LPCVD results in a film that has better thickness uniformity as compared to APCVD. Because of its highly diffusive precursors, the wafer in a deposition system can be arranged in an equally spaced stand-up fashion, while providing good wafer uniformity, as shown in Fig. 5.3. The reason is that the precursor gas mixture, in a low-pressure system, is able to diffuse into the interwafer narrow space. It can also be observed from Fig. 5.4 that as the pressure increases, the wafer film thickness tends to be thicker in the outer ring and thinner at the center. The nonuniformity is aggravated with system pressure. The disadvantage of LPCVD reactors is that the deposition rate is lower due to its lower gas density.

SACVD operates at pressure ranges around 600 Torr. A common application in the semiconductor manufacturing process is SACVD

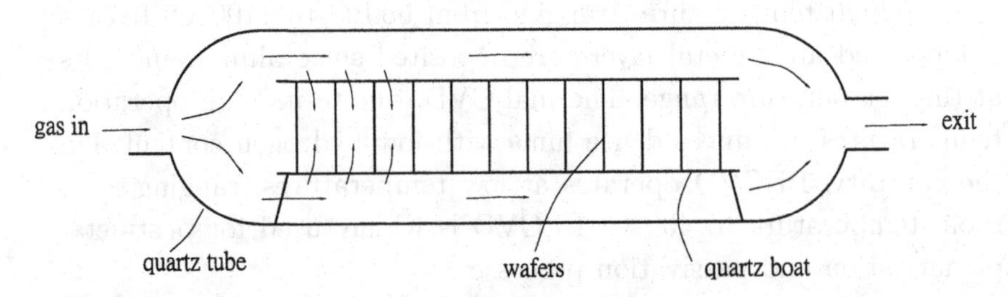

Fig. 5.3. Wafers in an equally spaced stand-up arrangement in an LPCVD system.

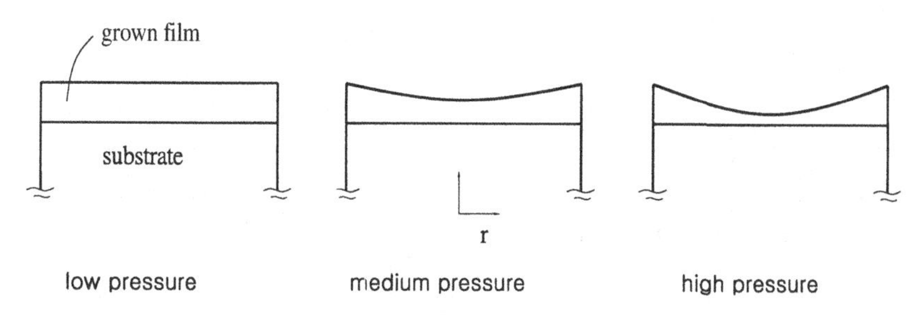

Fig. 5.4. The radial film thickness variation as a function of depositing pressures in a LPCVD system.

of O_3/TEOS for silicon dioxide formation. This process provides a high deposition rate and excellent uniformity as well as good step coverage in filling high aspect ratio trenches.

The CVD reactors can also be divided into cold-wall and hot-wall reactors, according to the wafer and reactor wall temperatures, as illustrated in Fig. 5.2. If the wafer temperature is higher than the wall temperature, the reactor is called a cold-wall reactor. If the wall temperature is higher than or the same as the wafer temperature, the reactor is called a hot-wall reactor. In general, a cold-wall reactor is used for very high temperature deposition processes, such as epitaxial growth, which is often operated at temperatures above 1100°C. For high-temperature deposition processes, the cold-wall reactor is employed to avoid excessive deposition on the reactor wall. As indicated in Fig. 5.5, heating methods, such as inductive heating, radiant heating, and resistive heating, are commonly used for CVD reactors. Radiant heating by means of high-powered lamps, such as halogen lamps, and resistive heating are known as nonselective heating. In other words, any objects that are under the exposure range are heated simultaneously. However, the heat capacity of different objects could give rise to different ultimate object temperatures. Inductive heating takes advantage of the Faraday effect. When a conductive material is placed in an oscillating magnetic field, it generates an electrical field and, consequently, an electrical current. The Joule heating effect, I^2R, results in temperature increases in the conductive material. The inductive heating is a selective heating method that is only effective on conducting materials.

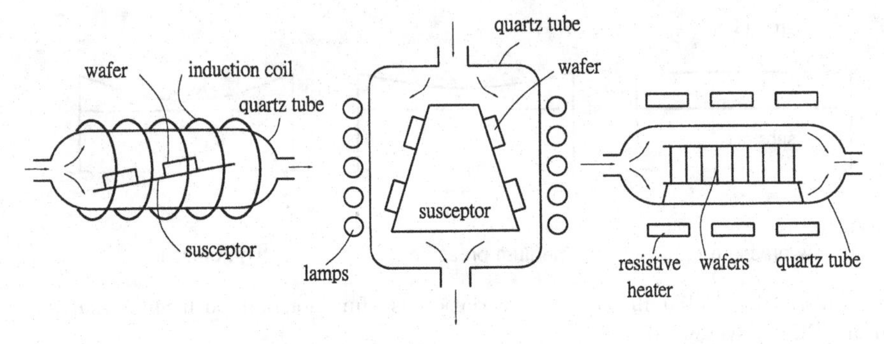

Fig. 5.5. Configurations of inductive heating, radiant heating, and resistive heating for CVD reactors.

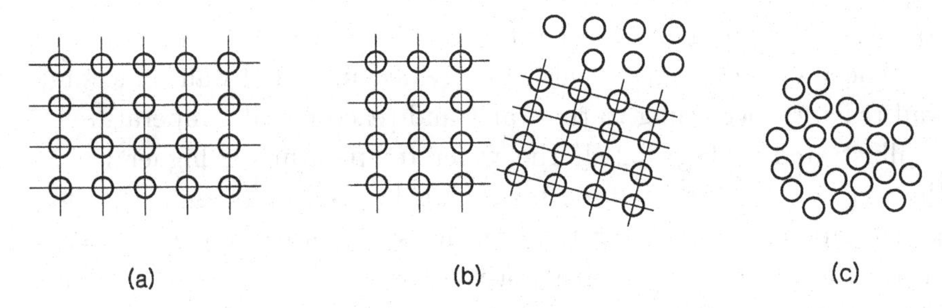

Fig. 5.6. Atomic arrangement for (a) single-crystalline, (b) poly crystalline, and (c) amorphous films.

Different CVD approaches or operating conditions may result in different deposited film characteristics. Film morphology can be classified into three categories in terms of the order of atom arrangement: the single-crystalline, polycrystalline, and amorphous, as shown in Fig. 5.6. In a single-crystalline film, the order of atom arrangement is fixed throughout the bulk. In a polycrystalline film, the order is short-range, or local; in other words, a number of different atom arrangement orders exist in the solid. An amorphous film has no fixed atom arrangement order. As indicated in Fig. 5.7, in CVD, the substrate temperature provides energy (ability) to the adsorbed species to migrate around on the substrate surface. On the other hand, the gas concentration dominates the frequency of gas species that strike on the substrate surface and hinder the surface species' migration. A high substrate temperature tends to provide the adsorbed species

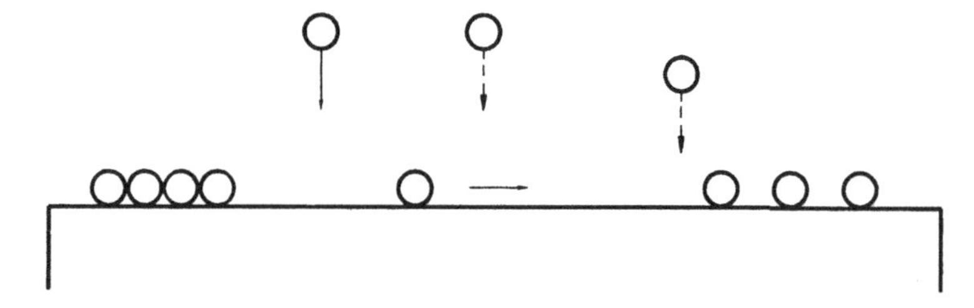

Fig. 5.7. The surface migration and the particles landing occur simultaneously on the surface.

with more energy to migrate on the substrate surface to look for a preferred site to settle. On the other hand, a low precursor gas concentration provides the adsorbed species with more time to migrate around before being stopped by the next impinging precursor gas species. A deposition condition with high temperature and low gas pressure has a higher tendency to form single-crystalline films such as epitaxial growth at 1100°C. On the contrary, a low substrate temperature and a high gas concentration tend to result in an amorphous film since the adsorbed species are less mobile and have less time to migrate around before settling down. Almost all PECVD films, deposited at around 250°C–350°C, are amorphous films. The intermediate reactor conditions tend to produce polycrystalline films, for example, polycrystalline silicon is grown at 550°C–650°C.

5.2. CVD Reactor Design Concepts

The design goal for a CVD reactor is essentially to optimize its operating conditions and configurations to maximize the throughput and to achieve good film uniformity and properties. To achieve this goal, a few concepts are critical for the reactor. For example, if a high-throughput reactor is desired, the wafers must be in a stand-up arrangement and narrowly spaced, as shown in Fig. 5.8. In this scenario, to achieve good uniformity across wafer and wafer to wafer, a low-pressure condition has to be used to enhance the species' diffusivities. Furthermore, if the wafers are in a stand-up arrangement, a cold-wall reactor configuration cannot be used since

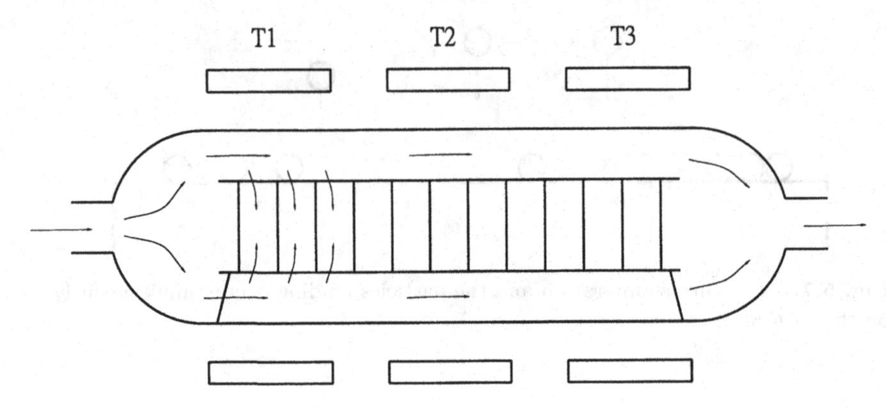

Fig. 5.8. A hot-wall LPCVD system with a three-zoned resistive heating.

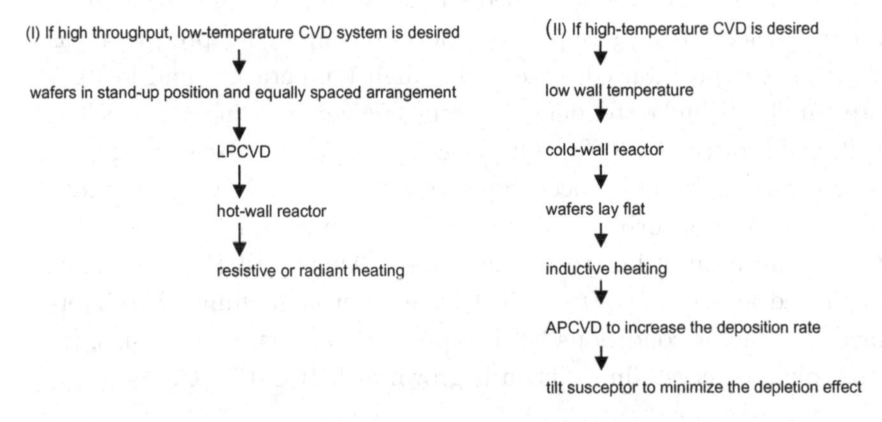

Fig. 5.9. The general concepts in CVD reactor design.

it will generate an unacceptably large temperature gradient across a wafer and cause the wafers to crack. With the demand of hot-wall reactors, either resistive heating or radiant heating can be used; however, inductive heating cannot be used. Figure 5.9 illustrates the conceptual flow of a CVD reactor design. If a high-temperature process is desired, such as epitaxial growth at 1100°C, a hot-wall reactor configuration should be avoided since its high temperature can result in excessive deposition on the reactor wall. The undesired wall deposition wastes chemicals during the deposition, and it generates particles during open-chamber load–unload steps. Because the thermal expansion coefficients are different between the film and the reactor wall, the wall deposits crack and peel off — particles are formed when temperature fluctuates. Therefore a cold-wall reactor is required. The cold-wall reactor calls for a selective heating method (the inductive heating method) since the reactor wall is not to be heated. To maintain the reactor wall at a controlled low temperature, a cooling jacket can be used to surround the reactor wall. Owing to the use of a cold-wall reactor, the wafers have to be laid flat on a conducting substrate holder to avoid wafer cracking due to excessive temperature gradient. If wafers are laid flat, one can use APCVD to increase the deposition rate. Furthermore, to achieve good wafer-to-wafer film thickness uniformity, the substrate holder surface can be tilted to enhance the mass transfer rate uniformity across the substrates in the reactor. Figure 5.10 shows a cold-wall APCVD reactor with a cooling jacket on the reactor wall.

In semiconductor manufacturing, if a dielectric film is to be deposited on top of an aluminum layer, the temperature is limited to below 500°C. This is due to the fact that the metal quality degrades significantly at such high temperatures. This thermal

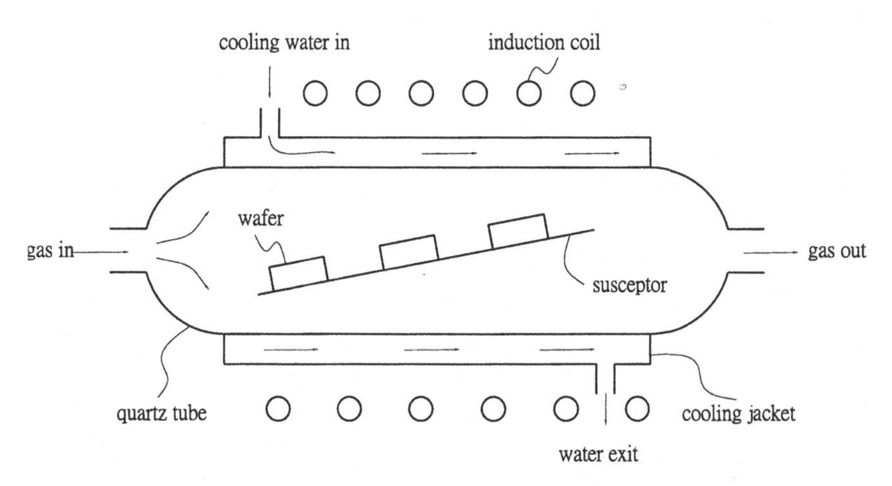

Fig. 5.10. An APCVD cold-wall reactor with induction coil heating and water cooling jacket surrounding the chamber.

budget mandates the use of the PECVD process, which is normally conducted at low temperatures ranging from room temperature to 450°C. The reactive species are the free radicals in the plasma, which are extremely reactive even at low temperatures. To avoid excessive deposition on the reactor wall, *in situ* cleaning and purge cycles are normally performed to etch off the undesired film deposition on the chamber wall. This should be done periodically. Owing to the fact that the reaction energy source is not from the thermal energy, using either a hot-wall or cold-wall configuration for a PECVD is not as critical as it is in a thermal CVD reactor. However, a PECVD reactor is often kept at around 300°C to 400°C to obtain good film qualities since higher deposition temperatures result in lower hydrogen contents in the film. The other type of plasma-enhanced etching or CVD reactor is the downstream reactor. The radicals are extracted from the plasma bulk between the electrodes and are moved into another chamber, where the etching or CVD reactions occur. By doing so, the ion bombardment or ion-induced damages on the substrate can be eliminated, and only the radical reactions that are isotropic in nature will be left.

5.3. CVD Reactor Modeling

CVD has versatile applications in semiconductor processing. Most films used in the semiconductor industry can be deposited by using CVD approaches. CVD film formation often involves both homogeneous and heterogeneous reactions. Modeling of a CVD system helps to explain the film growth mechanism and phenomena as well as correlating the system conditions to film growth rate, uniformity, and composition. It also helps to predict the film growth characteristics. There are two approaches for CVD reactor modeling. One is the macroscopic approach, while the other is the microscopic approach. The macroscopic model correlates the average film growth rate to the CVD reactor operating conditions. On the other hand, the microscopic model predicts the local film growth rate, and therefore it is able to predict not only the film growth rate, but also the thickness uniformity.

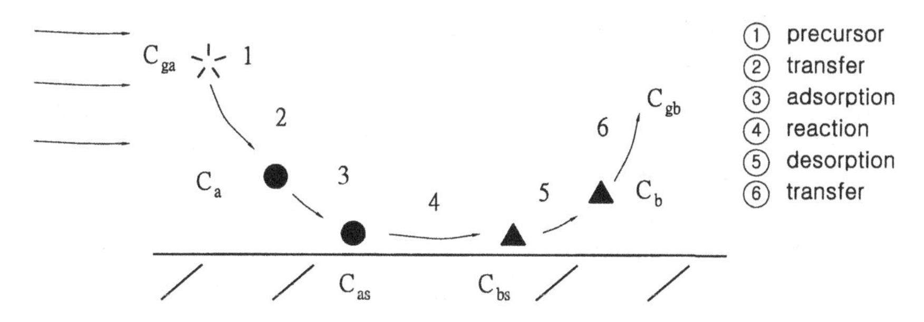

Fig. 5.11. The mass transport steps involved in a CVD process.

5.3.1. *Macroscopic modeling*

Despite the complicated nature of CVD reaction mechanisms and kinetics for practical application purposes, the transport and reaction pathways can be divided into six steps. Figure 5.11 shows a schematic of the mass transport steps that are involved in a CVD process, as described in the following:

(a) *Homogeneous gas phase reactions to form precursors.* Reactant molecules in a gas bulk can undergo numerous possible reactions. Some lead to desired precursors, while others lead to undesired by-products. Observing the elementary reaction of the reactant that leads to depositing precursors, the gas phase reaction rate is

$$R_1 = k_g C_g \,, \tag{5.2}$$

where k_g is the gas phase reaction rate constant and C_g is the concentration of the reactant.

(b) *The precursor transport onto the substrate surface.* The desired precursors are continuously generated as a result of the homogeneous reactions and are consumed at the substrate surface or reactor walls. A concentration gradient builds up from the gas bulk to the substrate surface and reactor walls and results in a mass transfer flux, F_1, toward the film growing surfaces:

$$F_1 = h(C_{ag} - C_a) \,, \tag{5.3}$$

where h is the mass transfer coefficient and C_{ag} and C_a are the concentrations of the depositing precursors in the gas bulk and in the vicinity of the substrate surface.

(c) *The precursors adsorb on the substrate surface.* The precursors that land on the substrate surface can adsorb on the surface. There are two types of adsorption — dissociative adsorption and nondissociative adsorption, as illustrated in Fig. 5.12. For the former, a molecule splits into two fragments when adsorbed onto a solid surface; for the latter, the molecule adsorbs directly without splitting. For a reversible nondissociative adsorption,

$$A + S \xleftrightarrow{k_a} A \cdot S . \tag{5.4}$$

The rate of adsorption is

$$R_a = k_a C_a C_v - k_d C_{as} , \tag{5.5}$$

where the second term represents the desorption rate of the adsorbed species. Variable k_a is the adsorption rate constant, and k_d is the desorption rate constant; C_v is the concentration of the vacant sites on the substrate surface, and C_{as} is the concentration of the depositing species that are adsorbed on the substrate surface.

The adsorbed precursors may desorb from the substrate surface. Desorption rate increases with substrate temperatures.

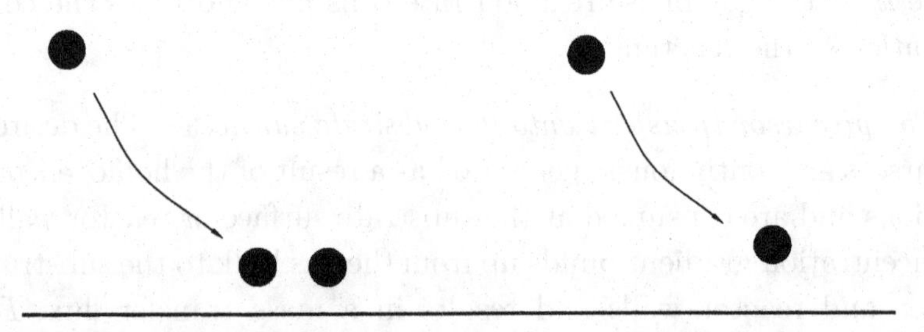

Fig. 5.12. Schematic representation of dissociatiative (left) and nondissociative (right) adsorptions.

(d) *The adsorbed species undergo dissociation reactions or surface reactions with other adsorbed species to form a nonvolatile film.* For the case of silicon deposition from SiH_4,
In gas phase

$$SiH_4 \rightarrow SiH_2 + H_2 \tag{5.6}$$

On substrate surface

$$SiH_{2(ad.)} \rightarrow Si_{(s)} + H_{2(ad.)} \,. \tag{5.7}$$

The adsorbed hydrogen then desorbs from the substrate surface.

The case of silicon nitride CVD involves dissociation reactions and surface reactions:
In gas phase

$$SiH_4 \rightarrow SiH_2 + H_2 \tag{5.8}$$

$$NH_3 \rightarrow NH + H_2 \tag{5.9}$$

On the substrate surface

$$3SiH_{2(ad.)} + 4NH_{(ad.)} \rightarrow Si_3N_4 \cdot H_x + \frac{(10-x)}{2} H_{2(ad.)} \,. \tag{5.10}$$

At high substrate temperatures, the dissociation reactions may proceed as

$$\begin{aligned} SiH_2 &\rightarrow Si_{(ad.)} + H_{2(ad.)} \,, \\ NH &\rightarrow N_{(ad.)} + \frac{1}{2} H_{2(ad.)} \,, \end{aligned} \tag{5.11}$$

then

$$3Si_{(ad.)} + 4N_{(ad.)} \rightarrow Si_3N_4(S) \,.$$

For simplicity, it is assumed that the surface reaction is nondissociative, irreversible, and first order with respect to the adsorbed species:

$$\text{A.S} \xrightarrow{k_s} \text{F} + \text{B.S} \tag{5.12}$$

$$R_3 = k_s C_{as} \,, \tag{5.13}$$

where F is the nonvolatile film formed by the reaction, k_s is the surface reaction rate constant, and C_{as} is the surface concentration of the adsorbed reacting species.

(e) *By-products desorb from substrate surface.* The dissociation reactions or the surface reaction by-products can either desorb from the surface or be incorporated into the grown film. Desorption is reversible:

$$\text{B.S} \overset{k'_d}{\longleftrightarrow} \text{B} + \text{S} \,. \tag{5.14}$$

Desorption rate can be expressed as

$$R_d = -\, k'_d C_{bs} + k'_a C_b C_v \,, \tag{5.15}$$

where k'_d is the desorption rate constant of the adsorbed reaction by-product, C_b is the concentration of the product in the vicinity of the substrate surface, and C_{bs} is the surface concentration of the adsorbed by-product.

(f) *The by-products out-diffuse into gas bulk:*

$$-\, F'_1 = h'_g (C_{gb} - C_b) \,, \tag{5.16}$$

where h'_g is the mass transfer coefficient of the by-product, and C_{gb} is the gas phase concentration of the by-product.

The above consecutive steps comprise the overall deposition process. Certain steps may proceed faster than the others. The slow steps are called rate-determining steps. They dictate and characterize the deposition process. At steady state, the rates of all steps are equal. The step with a smaller rate constant would build up a larger concentration gradient. On the other hand, the probability of being able to detect an intermediate species is small if it is associated with a large rate constant. To obtain the deposition rate, the related rate constants or concentrations must be determined experimentally.

For example, if one looks at relatively slow steps (b) and (d) at steady state,

$$\begin{aligned}\text{mass transfer in gas bulk} &= \text{surface reaction rate}\\ &= \text{deposition rate}\,,\end{aligned} \tag{5.17}$$

namely,

$$F_1 = h(C_{ag} - C_a) = R_3 = k_s C_{as}\,. \tag{5.18}$$

Since step (c) is relatively fast, $C_{as} \sim C_a$, we can replace the C_{as} with

$$C_{as} = \frac{hC_{ag}}{h + k_s}\,. \tag{5.19}$$

As a result, the surface reaction rate can be expressed in terms of measurable parameters:

$$R_s = k_s C_{as} = \frac{hk_s C_{ag}}{h + k_s}\,. \tag{5.20}$$

Now, if one desires to speed up the deposition rate, one can either increase h or k_s. For the former, one can increase the gas flow rate; for the latter, one can increase the substrate temperature. For CVD reactions at very high temperatures, such as 1100°C–1200°C for silicon expitaxial growth, the surface reaction rate is so fast that one can further assume $k_s \gg h$; therefore

$$R_s = hC_{ag}\,. \tag{5.21}$$

Then, the film growth rate, V (cm/min), is simply expressed as

$$V = \frac{R_s}{N}\,, \tag{5.22}$$

where N is the film density.

To calculate the film deposition rate of a certain CVD process, the values of k and h need to be evaluated. The value of k can be estimated from studying the reaction kinetics, while h can be estimated from the stagnant film model and the boundary layer model.

The stagnant film theory states that the mass transfer resistance from a gas bulk to a solid surface resides in a thin stagnant film, right next to the solid surface. The mass transfer coefficient is expressed as

$$h = \frac{D}{\delta}, \tag{5.23}$$

where D is the diffusivity, and δ is the hypothetical stagnant film thickness, which can be evaluated experimentally or estimated with the boundary layer theory as

$$\bar{\delta} = \frac{2}{3}\frac{L}{\sqrt{\mathrm{Re}}}, \tag{5.24}$$

where $\overline{\delta}$ is the average boundary layer thickness, L is the characteristic length of the CVD reactor system, Re is the Reynolds number of the fluid flow, and

$$h = \frac{3}{2}\frac{D}{L}\sqrt{\mathrm{Re}}, \tag{5.25}$$

$$\mathrm{Re} = \frac{LV\rho}{\mu}, \tag{5.26}$$

where V stands for the linear gas flow velocity; ρ stands for the gas density.

5.3.2. *Microscopic modeling*

The macroscopic model predicts the bulk average growth rate in a reactor system. However, the effects of fluid flow patterns and temperature profiles, which are often observed in a cold-wall reactor, as illustrated in Fig. 5.13, are not accounted for. A rigorous microscopic model, consisting of coupled continuity, momentum, energy, and mass

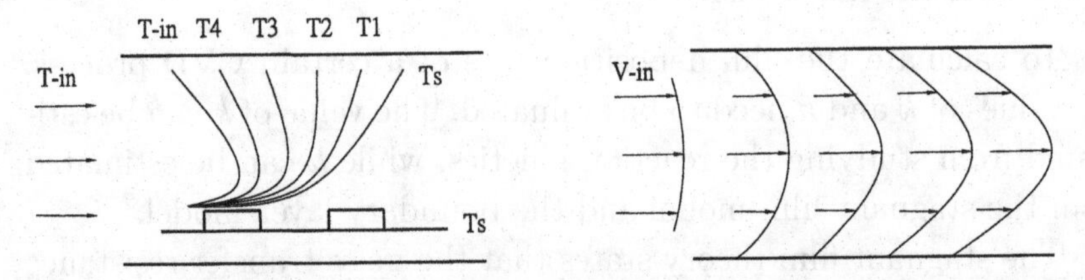

Fig. 5.13. The temperature (left) and velocity profile (right) develop as the fluid flows into a cold-wall CVD reactor.

balance equations, accounts for all the flow patterns and temperature profile effects. In reality, it can be tedious to solve the comprehensive microscopic model. Instead, some simplifications can be made. For example, for a hot-wall reactor, the system can be assumed to be reasonably isothermal. That eliminates the energy balance equation. If one further assumes that the flow pattern is a plug-flow type or fully developed, the momentum balance equation can be omitted, which leaves the mass balance equation as the reactor model.

Applying the above concepts to a hot-wall CVD reactor, such as a multiradial flow CVD reactor, as demonstrated in Fig. 5.14, the CVD reactor model for silicon epitaxy can be written in a dimensionless form as

$$\frac{(r_w - r_o)V_r}{D_{im}}\frac{\partial x_i}{\partial r} = \frac{\partial^2 x_i}{\partial r^2} + \left[\frac{r_w - r_o}{r(r_w - r_o) + r_o}\right]\frac{\partial x_i}{\partial r} + \left(\frac{r_w - r_o}{H}\right)^2 \frac{\partial^2 x_i}{\partial z^2}, \tag{5.27}$$

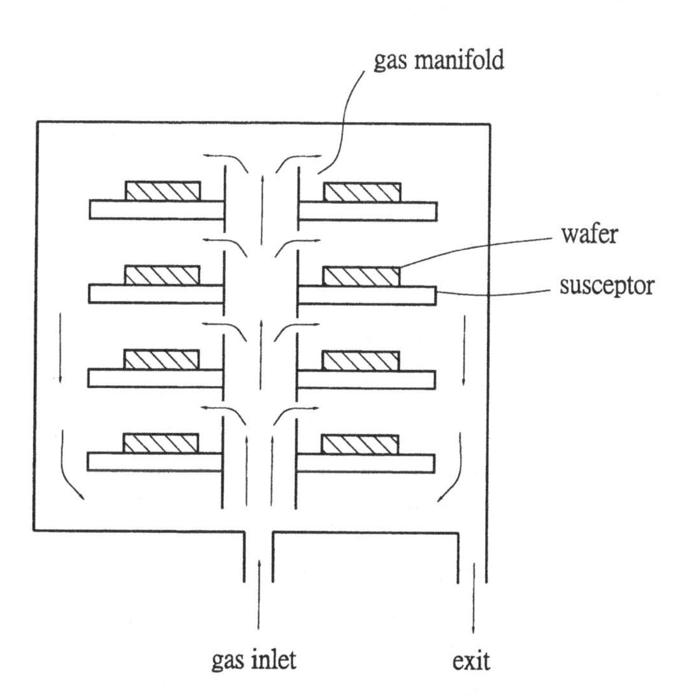

Fig. 5.14. Schematic of a multiradial flow CVD reactor system.

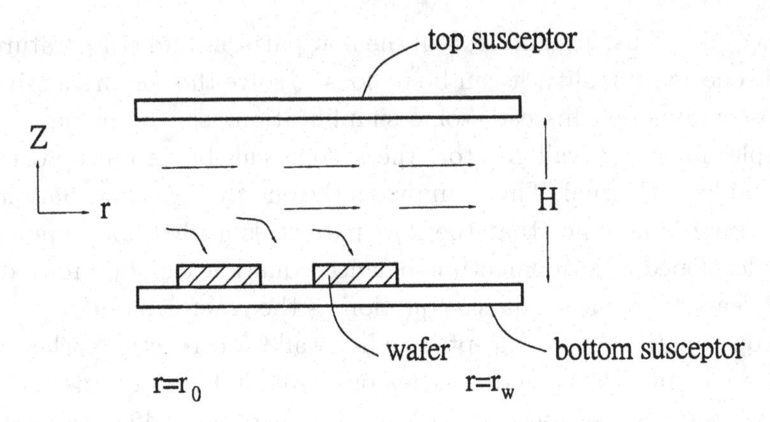

Fig. 5.15. Enlarged view of the cross section of the deposition chamber between two susceptors.

where x_i is the concentration fraction of the depositing species, D_{im} is the diffusivity of $SiCl_4$ in hydrogen, H is the spacing of the two parallel pates, and r_w, r_o are the radii of the disc outer edge and inner edge, respectively. The model describes the mass transfer between two parallel discs, in which wafers are placed, as shown in Fig. 5.15.

The boundary conditions are

$$\begin{aligned} r = 0: &\quad -\frac{D_{im}}{r_w - r_0}\frac{\partial x_i}{\partial r} = V_r(x_{i0} - x_i)\,, \\ r = 1: &\quad \frac{\partial x_i}{\partial r} = 0\,, \\ z = 0: &\quad \frac{\partial x_i}{\partial z} = 0\,, \\ z = 1: &\quad -\frac{CD_{im}}{H}\frac{\partial x_i}{\partial z} = \frac{Kx_i p^{3/4}}{1 + Bp}\,. \end{aligned} \tag{5.28}$$

The model can be solved numerically to obtain the concentration profile. Given the concentration profile, the film deposition rate at the wafer surface can be evaluated by

$$-\frac{CD_{im}}{H}\frac{\partial x_i}{\partial z} \quad \text{at } z = 1 \; (\mu\text{m/min})\,. \tag{5.29}$$

The model solution is demonstrated in Fig. 5.16, which indicates that the higher initial radial velocity results in a higher silicon film

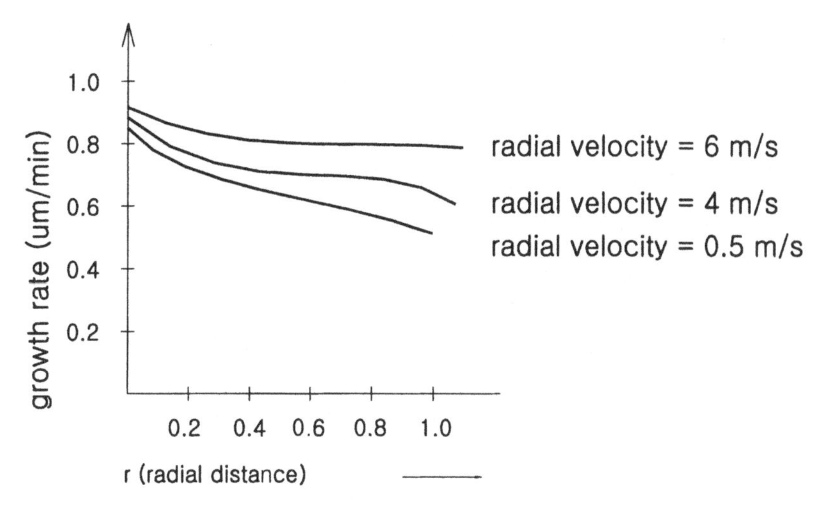

Fig. 5.16. The film growth rate distribution across the susceptor along the radial direction as a function of the inlet gas velocity.

deposition rate and better thickness uniformity. This is easy to understand as the higher initial velocity provides a greater total supply of chemicals for film deposition. Therefore, for the same amount of deposited film, it diminishes the depletion effect. On the other hand, Fig. 5.17 shows that the larger the spacing between the parallel discs,

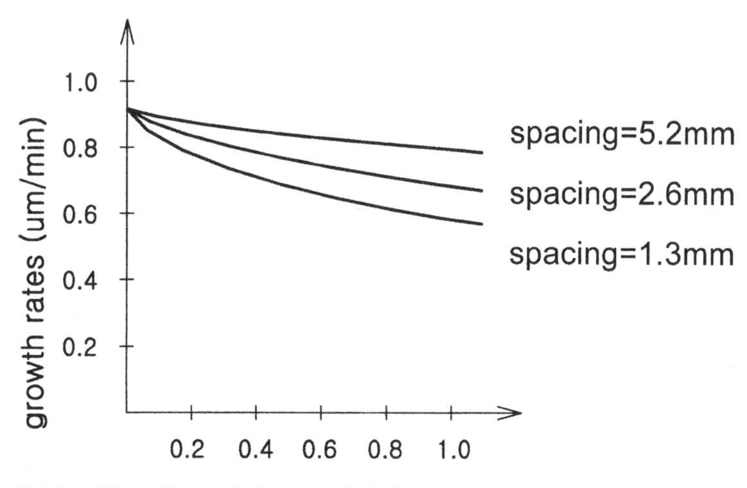

Fig. 5.17. The effect of the parallel disc spacing on the film growth rate distribution.

the better the film uniformity; again, this is due to more chemical supply and therefore a reduced depletion effect.

5.4. Characterization of Thin Films

A film is often characterized against its immediate upper or lower layers. For example, the stress of aluminum on a doped silicon oxide layer can be different from that on an undoped oxide. Film characteristics are strongly correlated to the film's composition. The composition in turn is a function of film deposition conditions such as temperature, pressure, and gas ratio; namely, the variations of the film characteristics often reflect the production machine stability. Some film characteristics that are often used for monitoring purposes will be discussed next.

Film thickness uniformity is the most basic requirement for a film to be used in device processing. Within a wafer and wafer to wafer, a uniformity better than $\pm 5\%$ is required. For a conducting film, this requirement is clear as the thickness uniformity directly influences the local resistance of the conducting layer. A dielectric film etching, such as contact or via etching, can leave unopened holes in the thick film area, even though the holes are completely opened in the areas with normal film thickness, as indicated in Fig. 5.18.

Pinhole density on a deposited film is of critical concern, especially when the film deposition is followed by a wet etching process. This is because the etching chemicals can penetrate through

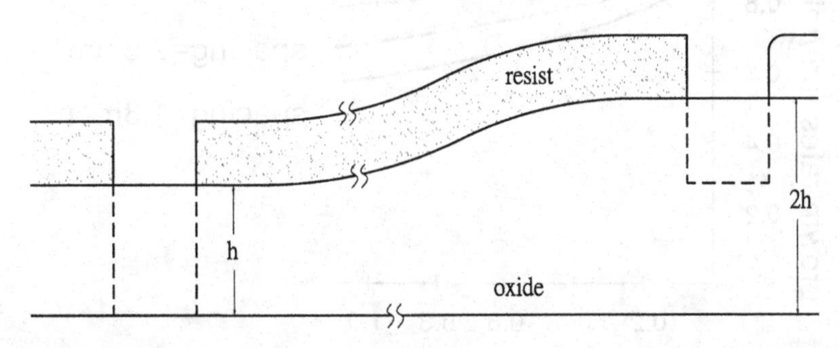

Fig. 5.18. Two contact holes are to be etched on the oxide surface with different thicknesses.

the pinholes and attack the underlying layers. An example of this occurs when a wet–dry via etching step is applied to open a via layer on top of an oxide-on-metal structure. The buffered HF solution can penetrate through the pinholes and attack the underlying metal, which results in pits on the metal surface. The pits affect the metal line resistance and can cause reliability concerns.

Step coverage is a critical measurement when a film is deposited over a substrate with topography. The step coverage, S, is defined as

$$S = \frac{h}{H}, \tag{5.30}$$

where H is the film thickness at the field surface, and h is the thickness at the bottom corner. Figure 5.19 shows the step coverage of an Al PVD over a contact hole versus a W-CVD. S is about 10% for the former and about 80% for the latter. In general, CVD has an inherently better step coverage than PVD. The capability of achieving good step coverage hinges on three factors. First, the system pressure needs to be low enough to render long enough mean free paths so that the reactants can reach the bottom and corners of the features. Second, the system needs to have a low sticking coefficient so that the impinging reactants can be scattered off and possibly redirected toward the bottom. Third, the surface must provide enough surface migration energy so that the surface precursors can migrate toward the bottom and corners. By doing so, the deposited molecules have

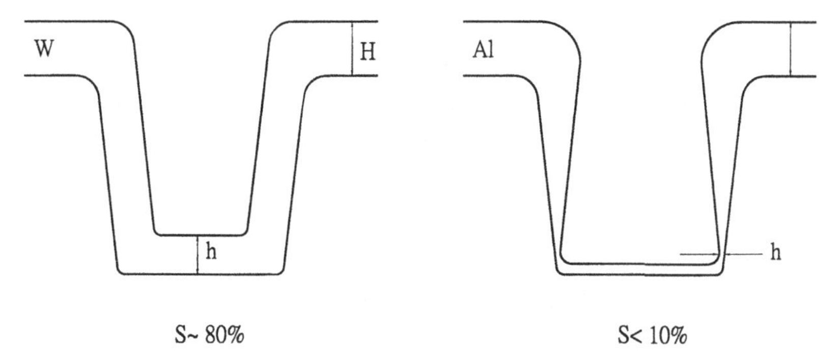

Fig. 5.19. The step coverage of W-CVD (left) and Al PVD (right) over 0.8-μm contact holes are approximately 80% and 10%, respectively.

a lower tendency to clog up the opening and allow more impinging molecules to reach the bottom of the substrate features, as shown in Fig. 5.20(a). On the contrary, two overhangs can be formed at the opening, causing the negligible bottom deposits, as shown in Fig. 5.20(b). With a moderate mean free path, sticking coefficient, and surface migration, the resulting film can be similar to the one shown in Fig. 5.20(c).

The compositional analysis is critical because it plays a decisive role in other film properties such as density, refractive index, stress, and so on. Raman backscattering spectroscopy (RBS) is a technique that is often used in determining film compositions. RBS is an off-line monitoring approach. One of the most convenient ways of monitoring the film characteristics through a refractive index measurement using an ellipsometer. The film composition, refractive index, and density are correlated by means of a Lorentz–Lorenz correlation:

$$\frac{n^2 - 1}{n^2 + 2}\frac{M}{\rho} = \frac{4\pi}{3}A\alpha_e\,, \tag{5.31}$$

where n is the film refractive index, M is the average molecular weight, ρ is the film density, A is Avogadro's number, and α_e is the average polarizability. In other words, in a diatomic film, by knowing the film refractive index and density, the film composition can be derived. The Lorentz–Lorenz correlation of a silicon nitride film is shown in Fig. 5.21.

Monitoring the resistivity of a conducting film sheds light on its composition. For example, tungsten silicide film is often used in device manufacturing. The silicide is formed from the reaction of WF_6 and SiH_4. With fixed flow rates for both reactants, high reaction temperatures tend to drive more silane dissociation and result in a silicon-rich film. The silicon-rich films provide higher resistivities than those of tungsten-rich films. A consistent resistivity trend in a production line implies good stability of the deposition process in terms of film composition.

The wet etch rate of a film is a good indication of the film density and hydrogen contents. Silicon oxide or nitride films that are grown with LPCVD provide a highly steady film composition with

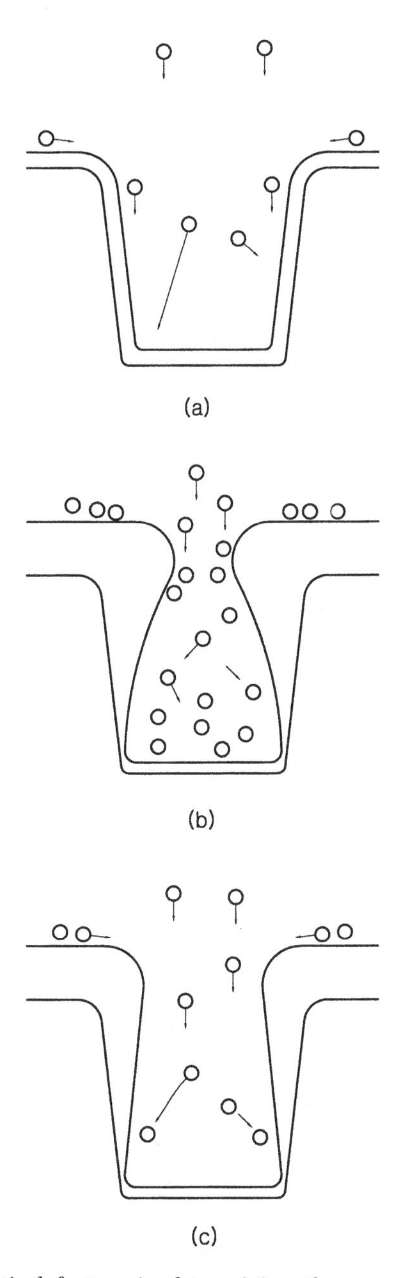

Fig. 5.20. The critical factors in determining the step coverage over a hole or trench pattern. (a) Long mean free path (L), high surface migration energy (E_s), and low sticking coefficient (S_k); (b) short L, low E_s, and high S_k; and (c) moderate L, E_s, and S_k.

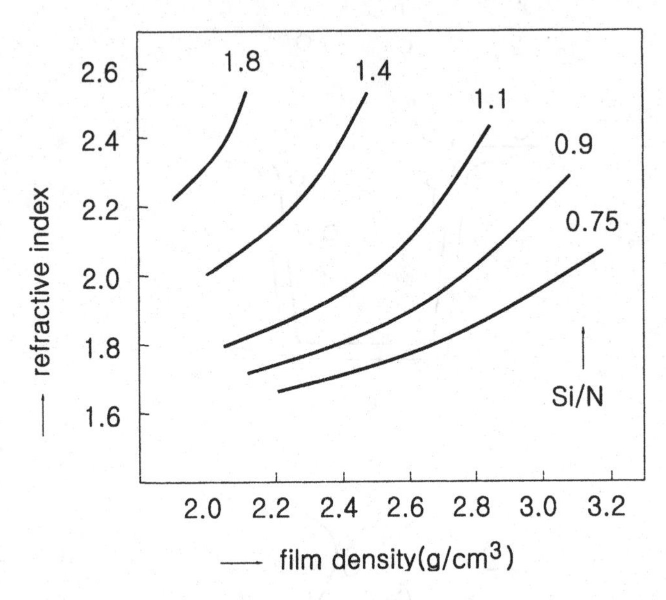

Fig. 5.21. The Lorentz–Lorenz correlation for PECVD silicon nitride films.

the hydrogen contents, which decrease when the substrate temperatures are increased, as indicated in Fig. 5.22. The trend of decreasing hydrogen content with increasing temperatures complies with the film density, which increases with increasing temperature. On the other hand, thermal oxide films tend to be denser than their PECVD counterparts; as demonstrated in Fig. 5.23, the denser films result in lower wet etching rates.

The stress of a thin film may stem from internal, external, or thermal effects. The internal stresses may result from crystal defects, while the external stresses may result from lattice mismatching between the substrate and the film itself. The thermal stress stems from variations in temperature. The difference in thermal expansion coefficients of the film and the substrate gives rise to thermal stress. When the film's thermal expansion coefficient is larger than that of the substrate, as the temperature decreases, the film contracts more than the substrate, and therefore the film experiences a pulling-apart force from the substrate (a tensile stress). On the contrary, the film would experience a pushing-in force (a compressive stress). Figure 5.24 shows the effects of temperature on film stress for different thermal expansion coefficients. It is clear that the stress

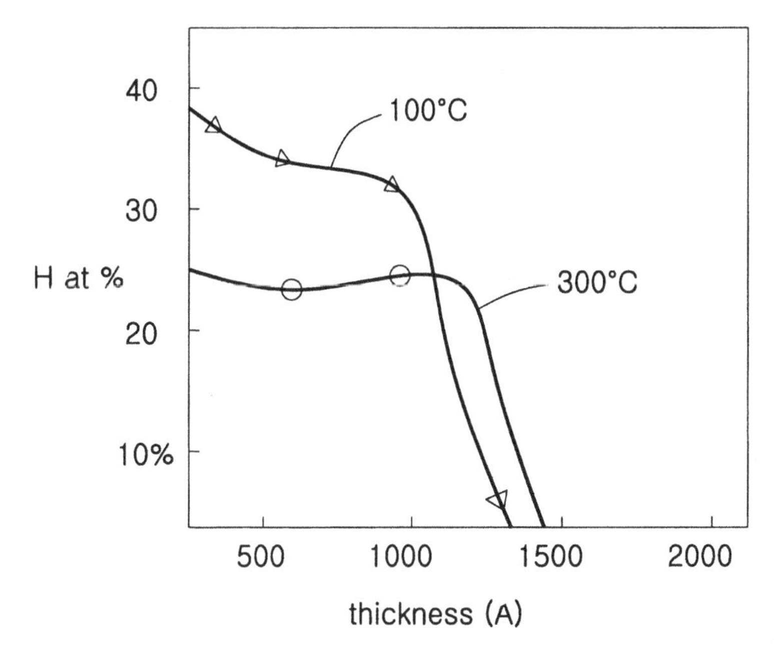

Fig. 5.22. Hydrogen contents in the bulk of a silicon nitride film with different deposition temperatures.

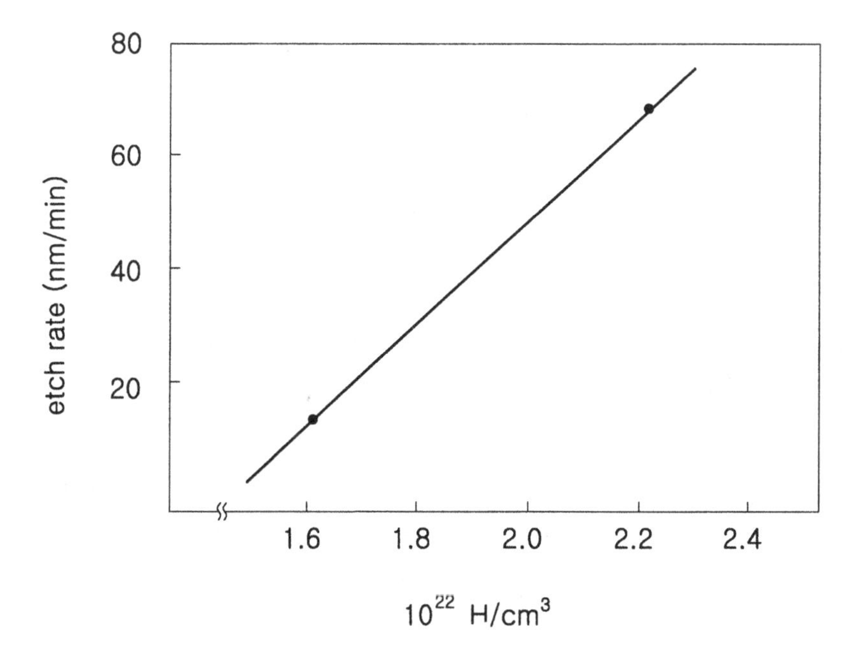

Fig. 5.23. The wet etch rate of a PECVD nitride film increases with its hydrogen content.

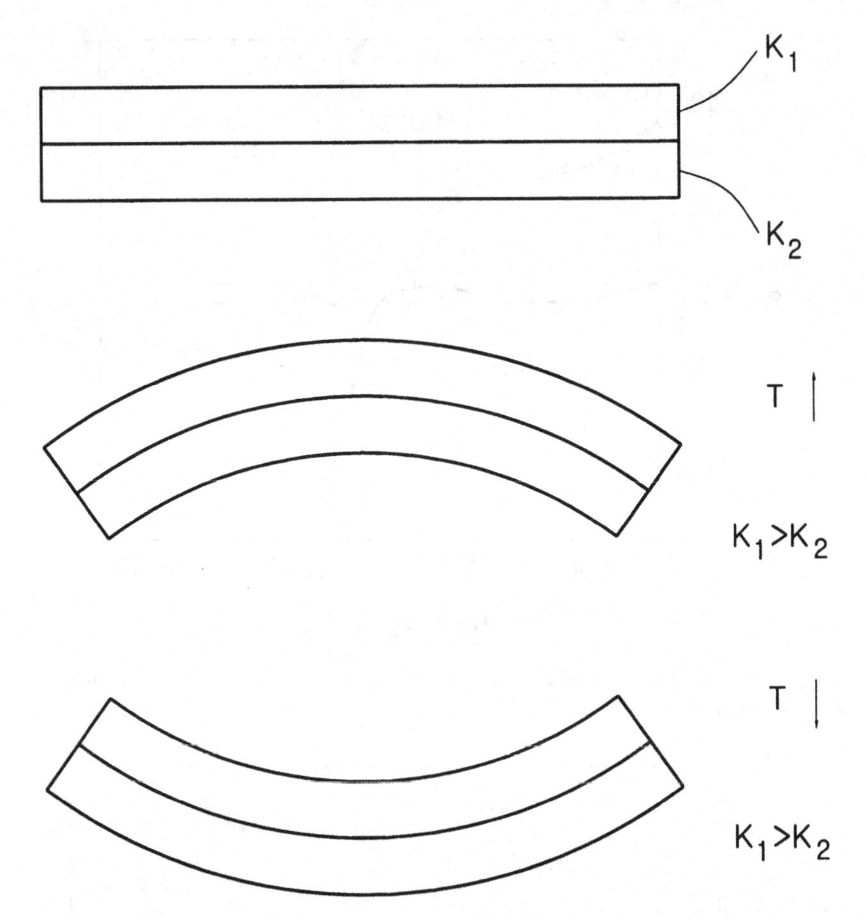

Fig. 5.24. A two-slab material with thermal expansion coefficients of k_1 and k_2. The material bends as temperature changes.

can change from tensile to compressive as the temperature varies. Excessive stress can affect the film adhesion on the substrate surface. The film may peel, buckle up, or crack when it is under tensile stress or may form hillocks when it is under compressive stress. Figure 5.25 demonstrates a peeling tensile-stressed tungsten silicide film. The film peeling leads to both contamination in the process line and device failures. Sometimes it may not show up in the device yield, but it will show up in the product qualification stage and will cause qualification failures.

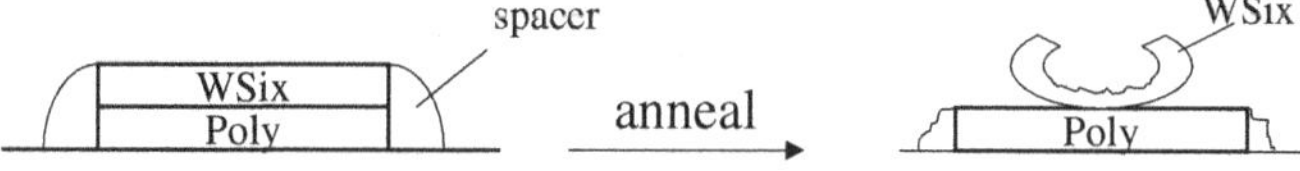

Fig. 5.25. The tungsten silicide film buckles up after a high temperature annealing: (top) SEM; (bottom) schematic flow.

5.5. Applications of CVD Films

In this section, we will introduce the reactions and applications of various films that are used in device manufacturing.

Silicon oxide films grown with CVD approaches are widely used as polysilicon spacers and as underlayers for interlayer dielectrics. The spacer shown in Fig. 5.26 is a silicon oxide layer along the polysilicon gate structure with a width ranging between 250 and 1500 Å; the widths vary with different technologies. The purpose of using a spacer is to form a lightly doped drain (LDD) structure to reduce the hot-carrier effects. Figure 5.27 shows the spacer structure around a polysilicon gate and the implant sequence of forming an LDD device. The requirement for the spacer film is primarily twofold: one is the thickness uniformity, which dominates the spacer width uniformity, while the other is the film purity, especially the hydrogen content, which can affect the device function as the film is immediately on the

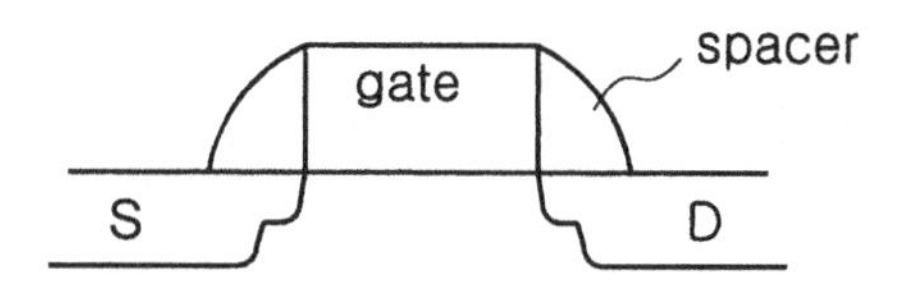

Fig. 5.26. The spacer along the sidewall of the polysilicon gate with LDD structure.

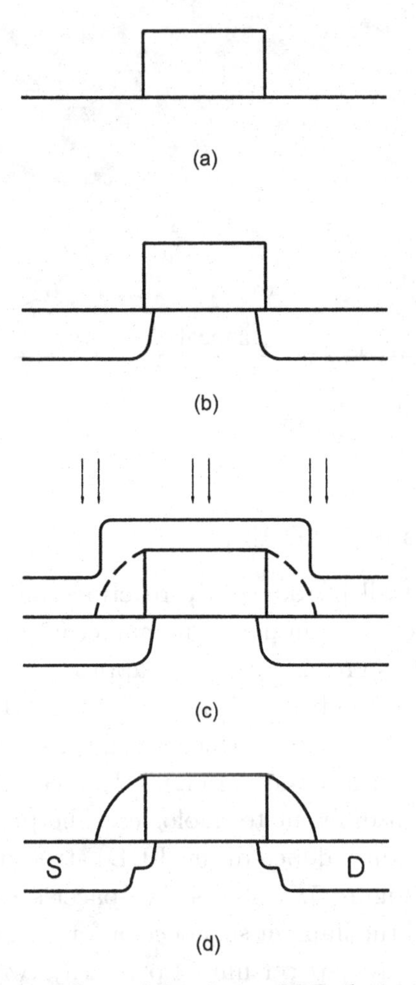

Fig. 5.27. The process flow of forming a LDD device structure: (a) after gate etching; (b) LDD implant and drive-in diffusion; (c) TEOS deposition; and (d) after anisotropic spacer oxide etching, followed by S/D implant and anneal.

silicon surface. For spacer applications, tetraethyloxylsilane (TEOS)-based oxide films are preferred due to their superior quality:

$$Si(OC_2H_5)_4 \xrightarrow{650^\circ C-750^\circ C} SiO_2 + \text{others}\,. \tag{5.32}$$

Decomposing TEOS at high temperatures gives rise to a TEOS oxide film with excellent step coverage, uniformity, and properties such as low impurity levels and hydrogen contents.

As a part of the interlayer dielectric (ILD), the CVD oxide is used between the polysilicon gate and the BPSG planarization layer. The goal is to block dopant (boron or phosphine atoms) diffusion from the BPSG layer into the device structure, as shown in Fig. 5.28. It can also be used as a blocking layer for ion implantation. The principle behind this is that the oxide defines the areas that are not to be implanted. The oxide simply blocks the impinging ions by trapping the ions within the film bulk, as demonstrated in Fig. 5.29. Another important application of silicon oxide is its use as the blocking layer for dopant diffusion. This takes advantage of the fact that most dopants have low diffusivities in the oxide layer. A patterned oxide defines the areas to be and not to be diffusion doped. Using a proper thickness for the oxide layer, by the time the diffusion process is complete, the dopants are still trapped within the oxide bulk.

There are a few approaches to grow the CVD silicon oxide, as follows:

(a) $$SiH_4 + O_2 \xrightarrow{< 500°C} SiO_2 + 2H_2 : \quad (5.33)$$

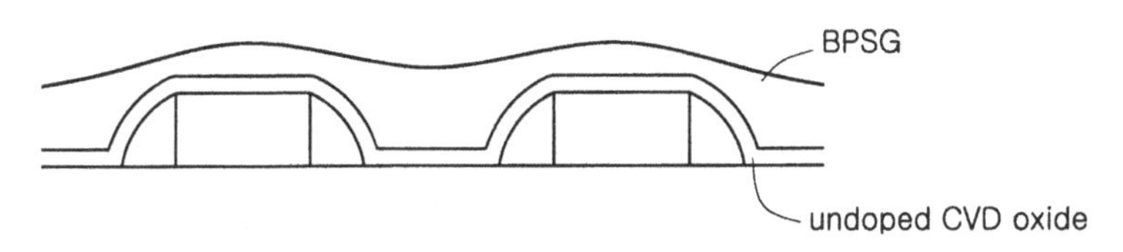

Fig. 5.28. A layer of undoped CVD oxide is often used under the BPSG layer to prevent dopants from diffusing downward to the substrate.

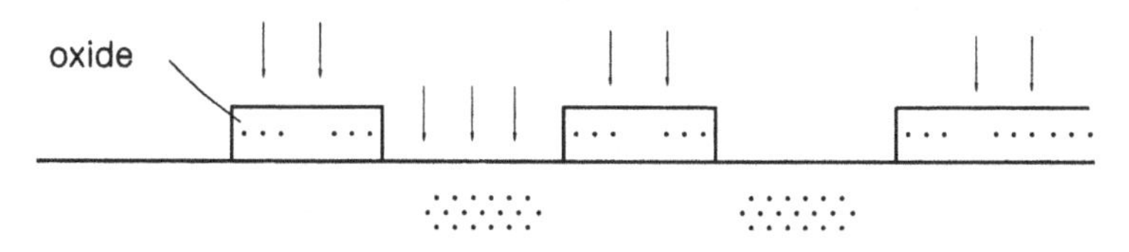

Fig. 5.29. Dopants are implanted onto a silicon substrate covered with a patterned oxide, which functions as a blocking layer.

This reaction takes place at low ambient temperatures and therefore it provides poor step coverage.

$$\text{(b)} \qquad SiCl_2H_2 + N_2O \xrightarrow{900°C} SiO_2 + 2N_2 + 2HCl : \qquad (5.34)$$

This approach provides much better step coverage and uniformity. However, the grown films may contain some chlorine atoms, which could react with polysilicon or cause the film to crack.

SACVD with a TEOS/O_3 reaction is often employed to planarize severe topography or deposit films in high aspect ratio trenches or contacts. The reaction proceeds as

$$Si(OC_2H_5)_4 + 8O_3 \xrightarrow{400°C} SiO_2 + 10H_2O + 8CO_2 . \qquad (5.35)$$

The reaction proceeds at pressures in the vicinity of hundreds of Torrs. The temperature is significantly lowered to 400°C, as compared to the thermal decomposition of TEOS at above 700°C. Like most other film deposition processes, the wet etch rate monotonically decreases with deposition temperatures, indicating that the film density increases. The flow rate ratios, O_3/TEOS, also affect the film deposition and wet etch rates. The important feature of this reaction is its excellent film step coverage for filling narrow gaps, as shown in Fig. 5.30. The step coverage for filling a deep trench with

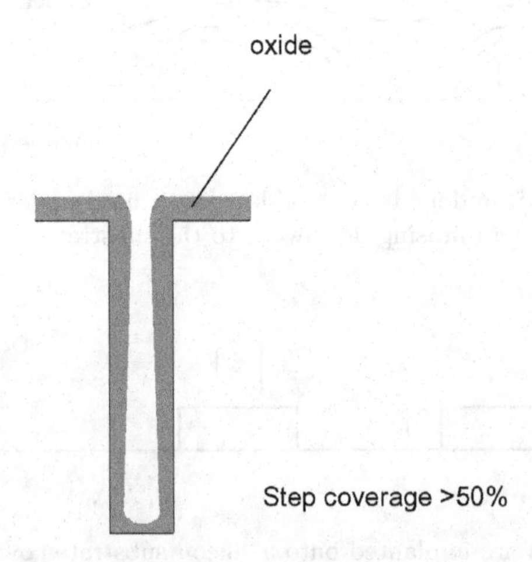

Fig. 5.30. Step coverage of SACVD ozone-TEOS in a high aspect ratio trench.

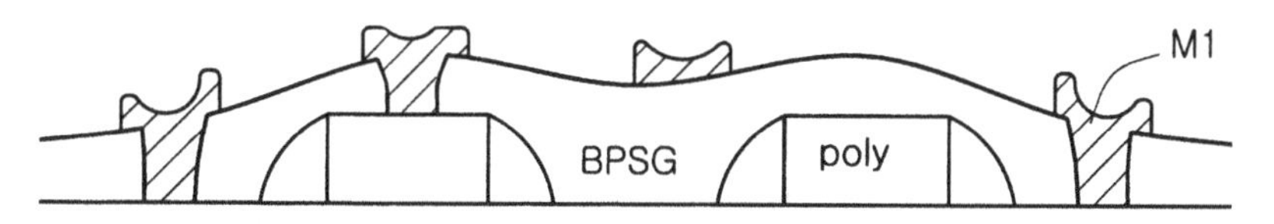

Fig. 5.31. BPSG is commonly used as the ILD between the gate and the first metal.

an aspect ratio of >10 varies from 35% at 415°C, to 60% at 465°C, to 80% at 490°C, respectively.

BPSG is used in a wide variety of applications such as the insulating layer between the first metal layer and the gate structure, as shown in Fig. 5.31. To ensure good step coverage and good resistance for the first-layer metal, it is critical to have a planarized surface immediately underneath the first metal layer. The pure silicon oxide layer step coverage is not good enough to fill up the narrow gap between the two neighboring gate structures. Furthermore, the flow-to-planarize approach cannot be applied on the pure oxide since it requires extremely high temperatures. A viable approach is to dope the oxide with either boron or phosphorous or both into the oxide to lower its flow temperatures to the 700°C–850°C range. To form a BPSG, the oxide formation reaction, Eq. (5.33), proceeds simultaneously with

$$4PH_3 + 5O_2 \longrightarrow 2P_2O_5 + 6H_2 \tag{5.36}$$

and

$$2B_2H_3 + 6O_2 \longrightarrow 4B_2O_3 + 3H_2 \, . \tag{5.37}$$

At about 400°C, the resulting product is essentially a mixture of oxides SiO_2, P_2O_5, and B_2O_3.

The as-deposited BPSG then flows in a furnace at around 750°C–850°C to obtain a planarized surface. The flow angle or the extent of BPSG planarization is defined as the angle shown in Fig. 5.32. The flow angle is a function of dopant concentration, delay time, flow ambient, flow temperature, and time. In general, the flow angle increases with the delay time, which is the time interval between the film leaving the deposition chamber and going into the furnace

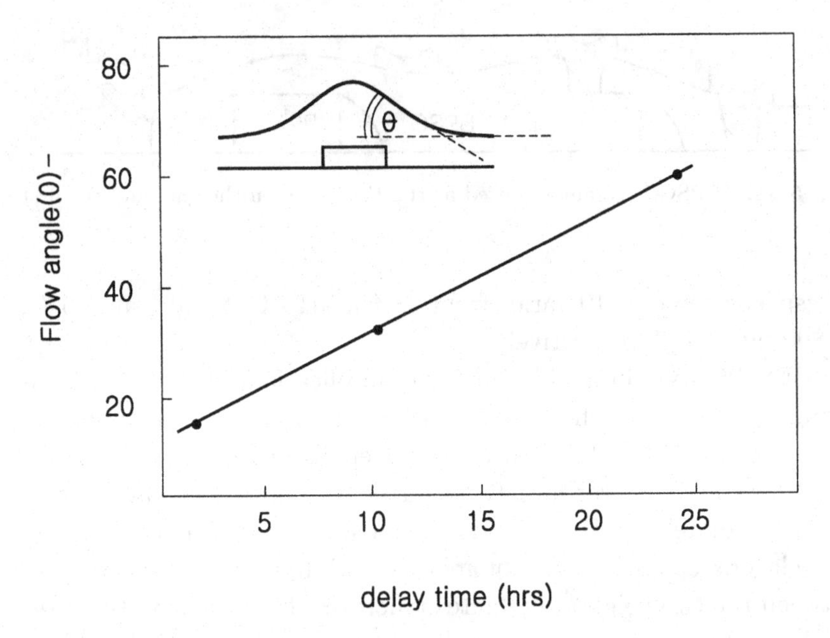

Fig. 5.32. The flow angle of BPSG increases with delay time.

for flow, as indicated in Fig. 5.32. The delay causes the BPSG surface to be converted into hydroxide due to moisture absorption, and it results in reduced flow capability. To achieve small flow angles, the delay time must be minimized. In particular, the production line should be managed properly such that as soon as the films leave the deposition chamber, they should be placed into the furnace for flow. The flow angles also decrease with the dopant concentration, as illustrated in Table 5.1. However, excessive boron concentration causes the film to be hygroscopic and results in crystal formation on the film surface. As discussed earlier, an undoped oxide is often used underneath the BPSG to prevent the dopants from diffusing downward into the silicon substrate and affecting the device performance.

Silicon nitride films are good diffusion barriers for mobile ions, moisture, and oxidation. Furthermore, they hardly oxidize in an oxidizing ambient. The major application of a CVD silicon nitride film is an oxidation mask. Nitride films can be formed when ammonia

Table 5.1. Flow angles of BPSG decrease with dopant concentrations.

Wafer	B (wt %)	P (wt %)	Flow angles (deg)
A	1.99	3.5	36
B	2.17	4.82	15
C	2.7	5.4	13

reacts with silane or dichlorosilane:

$$3SiH_4 + 4NH_3 \xrightarrow{700^\circ C-900^\circ C} Si_3N_4 + 12H_2 \qquad (5.38)$$

$$3SiCl_2H_2 + 4NH_3 \xrightarrow{700^\circ C-800^\circ C} Si_3N_4 + 6HCl + 6H_2\,. \qquad (5.39)$$

The deposited films are stoichiometric in nature, with some percentage of hydrogen. Thermal CVD nitride inherently has a relatively high stress level. As a result, when used as an antioxidizing layer, it is often used on top of a buffer oxide layer, to prevent wafers from being cracked. Table 5.2 compares the physical properties of thermal CVD nitride with plasma CVD nitride. Compared with CVD nitride films, PECVD films are nonstoichiometric in nature, less dense, and lower in stress. Most important, PECVD is conducted at much lower temperatures.

Polycrystalline silicon films are often used as the transistor gate material, as the capacitor plates in DRAM, or as the resistor load in SRAM. For all of these applications, the films are grown with

Table 5.2. Comparison of LPCVD and EPCVD silicon nitride film properties.

Condition	LPCVD	PECVD
Temperature (°C)	700–800	250–350
Composition	$S_3N_4(H)$	$Si_xN_yH_z$
Atomic % H	4–8%	20–25%
Refractive index	2.01	1.8–2.5
Density (g/cc)	2.8–3.1	2.4–2.8
Dielectric constant	6–7	6–9
Stress (10^9 dyn/cm^2)	10 (tensil, T)	2(c)–5(T)

LPCVD using silane pyrolysis (chemical decomposition with heat in the absence of oxygen):

$$SiH_4 \xrightarrow{600^\circ C-650^\circ C} Si + H_2 \,. \tag{5.40}$$

The film structure is strongly influenced by the deposition temperatures, dopant concentrations, and conditions of the postdeposition heat treatments. The silane decomposition can be carried out in a wide range of temperatures (from 560°C and up). At temperatures lower than 580°C, the grown films are amorphous. At around 590°C, the grown films are partially crystalline, with amorphous islands. Above 630°C, the films become polycrystalline, with columnar structures with well-defined grain boundaries, as demonstrated in Fig. 5.33. The average grain size increases with deposition temperature. In the meantime, the *in situ* doping significantly affects the grain structures. At the same temperature, the grain sizes of the *in situ* doped polysilicon films are larger than those of the postdeposition doped films. There are different approaches to dope the polysilicon layer such as diffusion, implantation, or even *in situ* doping.

The resistance ranges of polysilicon films can vary from GΩ/□ to mΩ/□, depending on the doping levels. The high-resistance film is used in applications such as polyloads in SRAM, while the low-resistance film is used as gate interconnects for devices.

The above discussion primarily focuses on CVD dielectric films. Some CVD conducting films, such as tungsten silicide and tungsten films, are widely used in semiconductor manufacturing. As the device geometry continues to shrink both laterally and vertically, the resistance of the doped polysilicon gate is increasingly high and

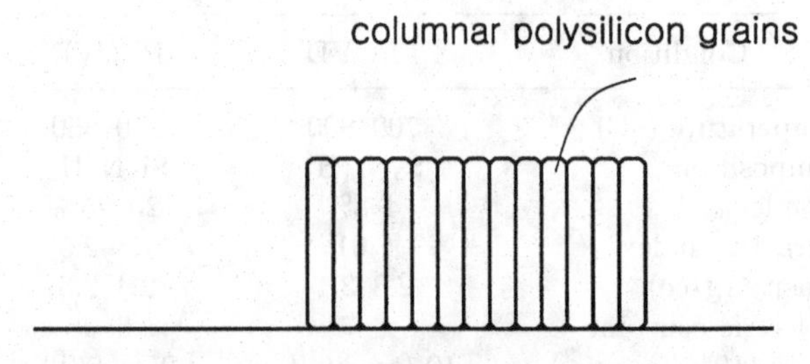

Fig. 5.33. The columnar polysilicon grain structure.

causes the signal speed in the gate layer to slow down. Tungsten silicide films are used to shunt the polysilicon gate (a polycide structure), as shown in Fig. 5.34. By doing so, the gate layer resistance can be reduced by more than an order of magnitude. There is a whole spectrum of silicides to choose from such as the silicides of Ti, Ni, Ta, or Mo. However, the only commercially available CVD process is the WSi_x film. The reaction is carried out in a LPCVD cold-wall reactor, as shown in Fig. 5.35. The reaction involves silane and tungsten hexafluoride:

$$WF_6 + SiH_4 \xrightarrow{350^\circ C-400^\circ C} WSi_x + HF + \cdots . \qquad (5.41)$$

The gas flow rate ratio, WF_6/SiH_4, is very small (around 12 sccm/2000 sccm). The reaction is dominated by WF_6. The resulting film is

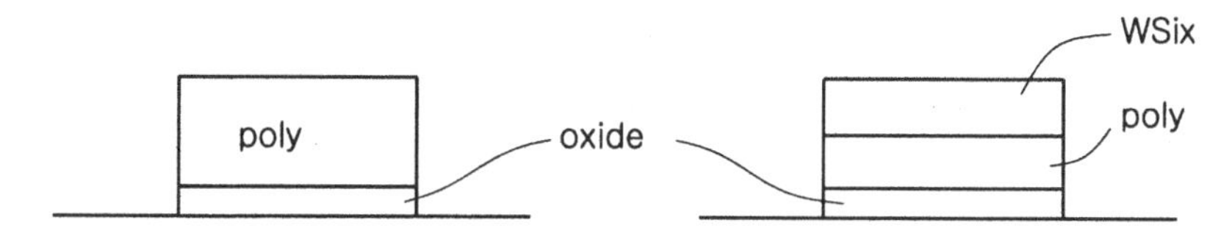

Fig. 5.34. Conventional polysilicon gate structure in comparison to tungsten polycide structure.

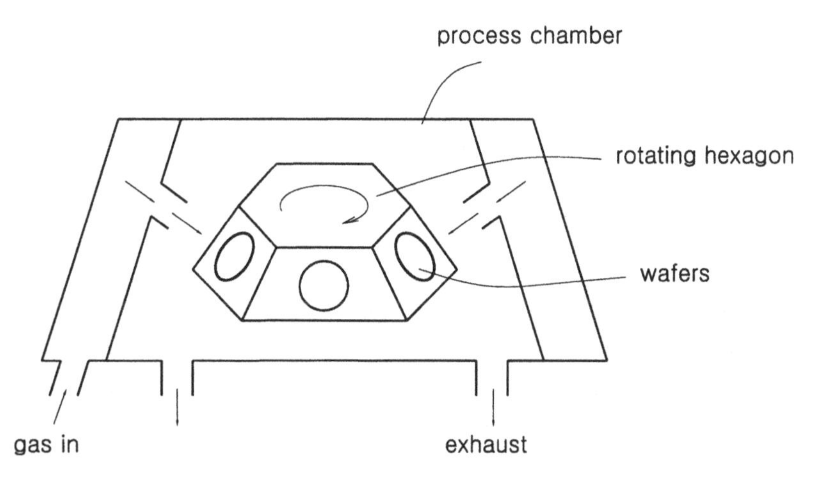

Fig. 5.35. LPCVD cold-wall reactor for tungsten silicide film deposition.

not stoichiometric; the atomic ratios vary with temperature and WF_6 flow rates. Increasing the temperature enhances the silane decomposition and gives rise to silicon-rich films with high resistances. On the other hand, increasing the WF_6 flow rates results in tungsten-rich films, with lower resistances. There is a trade-off between the film resistance and the film stress. A tungsten-rich film has a low resistance but high film stress, and therefore it is more prone to film peeling. The film stoichiometry must be optimized for several factors such as final film resistance, adhesion, and postdeposition thermal budget. The as-deposited film is amorphous, and thereby on annealing, it crystallizes, and the resistance drops by an order of magnitude, as illustrated in Fig. 5.36. In addition, the film stress increases significantly, as shown in Fig. 5.37.

During device manufacturing, the silicide film experiences several thermal cycles, some with oxygen and others without. The oxidation behavior is quite interesting. The WSi_x film cannot tolerate wet oxidation even at 700°C; it peels right away. The oxide formation on the silicide surface is faster compared to oxide formation on a silicon surface, which could be due to its amorphous nature. RBS reveals that on oxidizing the amorphous WSi_x film, the silicon atoms in the

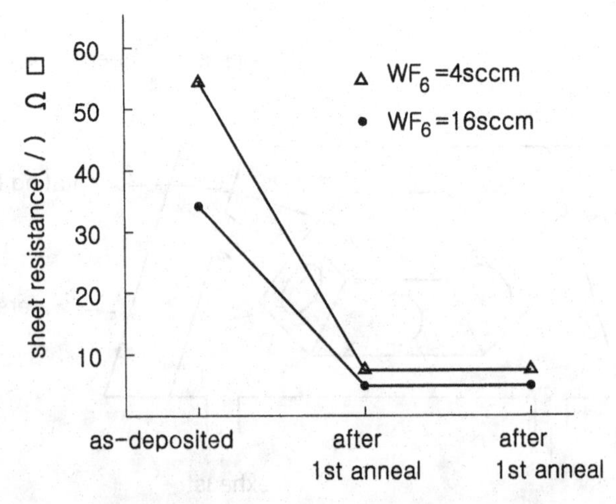

Fig. 5.36. Sheet resistances of tungsten polycide films decrease after annealing.

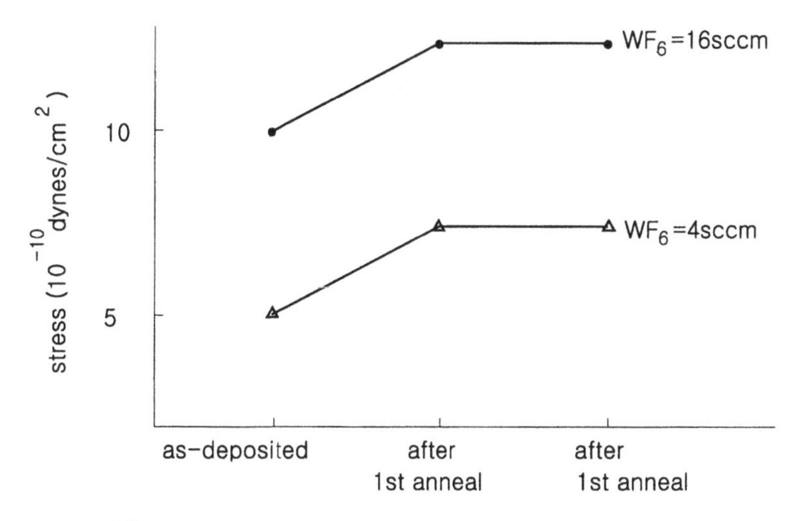

Fig. 5.37. WSi_x film stress increases after annealing.

silicide bulk actually diffuse outward toward both silicide surfaces to form silicon oxide and toward the silicide–polysilicon interface and pile up there. After a long oxidation time, or when the silicide film reaches a stable stoichiometry, x equals 2.2. The silicon atoms start to be supplied from the underlying polysilicon, as can be clearly seen from Fig. 5.38. It has also been demonstrated that the annealed silicide surface cannot be exposed to a high temperature ambient, or the film will peel. Tungsten silicide film peeling seems to be one of the painful experiences in most process lines. To avoid peeling, the following precautions can be taken:

(a) The polysilicon surface must be completely clean and free of residual oxide, moisture, and organic contaminations before silicide deposition.
(b) No wet oxidation is allowed.
(c) An annealed or oxidized silicide surface cannot be exposed to a high temperature ambient unless it is capped with a capping layer or the surface is supplemented with extra silicon atoms to make up for the silicon that was consumed during the previous oxidation.

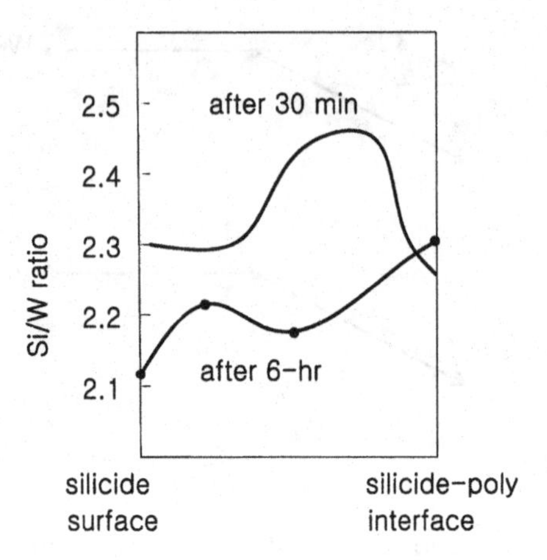

Fig. 5.38. The silicon-to-tungsten atomic ratios change across silicide bulk after anneals.

Tungsten is one of the most popular conducting CVD films used in semiconductor manufacturing. It has an even lower resistance than do the tungsten silicide films. Furthermore, owing to its superior step coverage, it is often used as the tungsten plug to fill holes: either contact or via holes. The film can be deposited first by using a nucleation step, followed by bulk deposition:

$$\begin{aligned} WF_6 + 3H_2 &\xrightarrow{400^\circ C-430^\circ C} W(s) + 6HF\,, \\ 2WF_6 + SiH_4 &\xrightarrow{400^\circ C-430^\circ C} 2W(s) + 4HF + 2SiF_4 + \cdots\,. \end{aligned} \tag{5.42}$$

The hydrogen reduction reaction appears to be independent of the hexafluoride flow rate for temperature ranges between 594 and 621 K. The step coverage varies with the gas flow ratio, as indicated in Table 5.3.

Table 5.3. Effects of gas flow ratios on the tungsten film step coverage (430°C, 200 mTorr).

WF_6	SiH_4	H_2	Step coverage/dep. rate (%/Å/min)
100	0	2000	80/250
100	60	1940	55/588
100	100	1900	25/1170

Table 5.4. Effects of gas flow ratios on the Si/W ratios of the resulting films.

SiH_4/WF_6	Si/W ratios
0.55	0.03
1.3	0.03
1.5	0.08
1.6	0.11
3.8	1
15	1.28
166	2.2

The second step, the silane reduction reaction, is also used for the tungsten silicide deposition, as discussed earlier, except that it is used with a different gas flow ratio. Table 5.4 shows the gas flow ratios versus the silicon-to-tungsten atomic ratios of the resulting films. Excessive silane in the reactants can result in a silicon-containing tungsten film, which is not desirable. In manufacturing, by choosing proper chemistry, 90% film step coverage can be achieved with little silicon in the films.

Tungsten films provide good adhesion on metal or silicon surfaces such as TiN, TiW, and polysilicon, but poor adhesion on oxide surfaces. Contact plug applications require a stack structure consisting of Ti/TiN/W or Ti/TiW/W. As mentioned earlier, precautions must be taken to avoid tungsten film peeling. However, unlike tungsten silicide on the gate, the W-plug does not need to go through high temperature cycles (over 700°C) for either contact or via filling; therefore the thermal stress is less of a concern. Figure 5.39 shows a plug application for the tungsten films.

Tungsten hexafluoride also reacts with silicon:

$$2WF_6 + 3Si \xrightarrow{200^\circ C-300^\circ C} W(s) + SiF_4\,. \tag{5.43}$$

The reaction can be used to form selective tungsten in the contact holes (connecting the silicon substrate to the first level metal)

Fig. 5.39. Tungsten plug for contact filling.

since the oxide surface does not have the silicon source for the reaction. A very thin and uniform oxide layer on top of the silicon substrate is required to form a uniform plug deposition. It can be devastating if the thin oxide layer is not uniform. Tungsten hexafluoride can react with the exposed silicon in the oxide pinhole and consume the silicon. As a result, the silicon atoms continue to be dug out, and therefore they form a worm hole, which can extend as long as several microns into the substrate. This definitely is a junction leakage path. For the silicon reduction reaction, the grown tungsten film thickness seems to be self-limited. The grown tungsten film acts as a diffusion barrier for the silicon diffusion. The selective deposition can also be made possible with hydrogen and tungsten hexafluoride, but the selectivity strongly depends on the temperature; above around 350°C, the process is nonselective. In general, for selective tungsten deposition processes, the process window is very small in terms of surface cleanliness and reaction condition control (temperature, pressure, and gas composition controls); the selective tungsten deposition has not been realized in volume production.

WF_6 is used for both W and WSi_x film growth. It is very sensitive to residual moisture in the reaction chamber or piping lines:

$$\begin{aligned} WF_6 + 3H_2O &\longrightarrow WO_3 + 6HF \\ WO_3 + 2WF_6 &\longrightarrow 3WOF_4 \,. \end{aligned} \tag{5.44}$$

It is a bluish yellow powder that can inhibit the tungsten deposition process. To avoid this, the chamber or pipelines should be kept as dry as possible.

Chapter 6

PLASMA-ENHANCED CHEMICAL VAPOR DEPOSITION AND ETCHING

In device manufacturing, high-temperature procedures can be detrimental to the underlying device performance and the previously deposited materials if they are not properly integrated into the overall device manufacturing flow. For example, after the metal layer is deposited, a process temperature of higher than 550°C is no longer appropriate. In such circumstances, plasma processing comes to the rescue. The thermal effect resulting from plasma reactions is often negligible from device and material viewpoints. In this chapter, bearing in mind the concepts introduced in Chapter 4, we will introduce the plasma reaction pathways and the complexity of a plasma system in terms of its variables and responses. The following two sections focus on plasma deposition as well as film characteristics and applications. Once familiarized with the plasma deposition, we will then introduce the etching counterparts. Finally, plasma process modeling is illustrated, with the plasma deposition of silicon nitride as an example.

6.1. The Plasma Reaction Pathways and System Variables

Like most CVD processes discussed earlier, plasma etching or deposition processes involve a series of substeps, as illustrated in Fig. 6.1. What is unique is that they are initiated by electron impacts. First, the reactant gases must be dissociated in plasma through electron impacts to form reactive radicals. Second, the reactive radicals are transported across the gas bulk to the substrate vicinity. Third, the

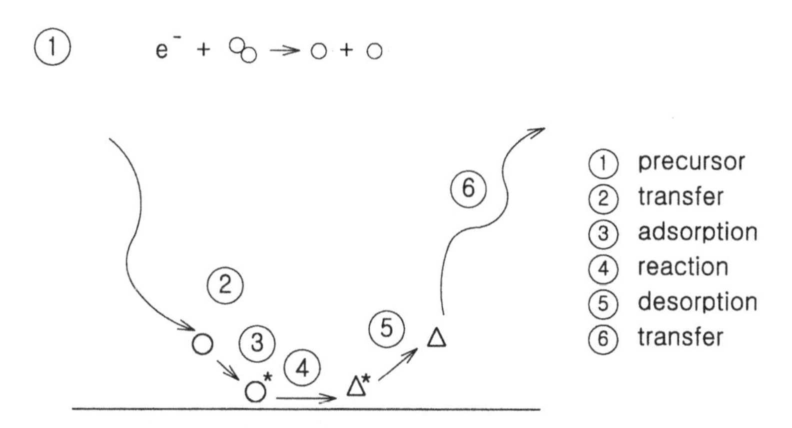

Fig. 6.1. The steps involved in a plasma deposition or etching process.

radicals are adsorbed onto the substrate surface. Fourth, the surface adsorbed atoms or adsorbed molecules undergo surface reactions. At this point, the plasma deposition and plasma etching are differentiated by the volatilities of their reaction products. In plasma depositions, the reactive radicals form nonvolatile products, such as a thin film on the substrate surface, while in plasma etching, the radicals react with the substrate surface to form volatile products. Fifth, the adsorbed products desorb. Finally, the desorbed products are transported to the gas bulk and carried out of the process chamber by the pumping mechanism. In the meantime, homogeneous reactions resulting from the radicals' reaction in the gas phase occur. Generally, these side reactions are undesirable. The process conditions should be properly controlled to minimize the homogeneous gas phase reactions since they may cause particle contamination.

The variables of a plasma system include substrate temperature, power input, pressure, gas flow rate, gas ratio, and so on. Each of these can influence the system outputs of a deposition or etching process, as shown in Fig. 6.2. Although until now, there has been no theory that can predict a plasma reactor's behavior (the effects of parameters can vary from one system to the other), there are general trends that can help understand the reactor behavior. Decreasing the pressure increases the particle mean free path. Long particle

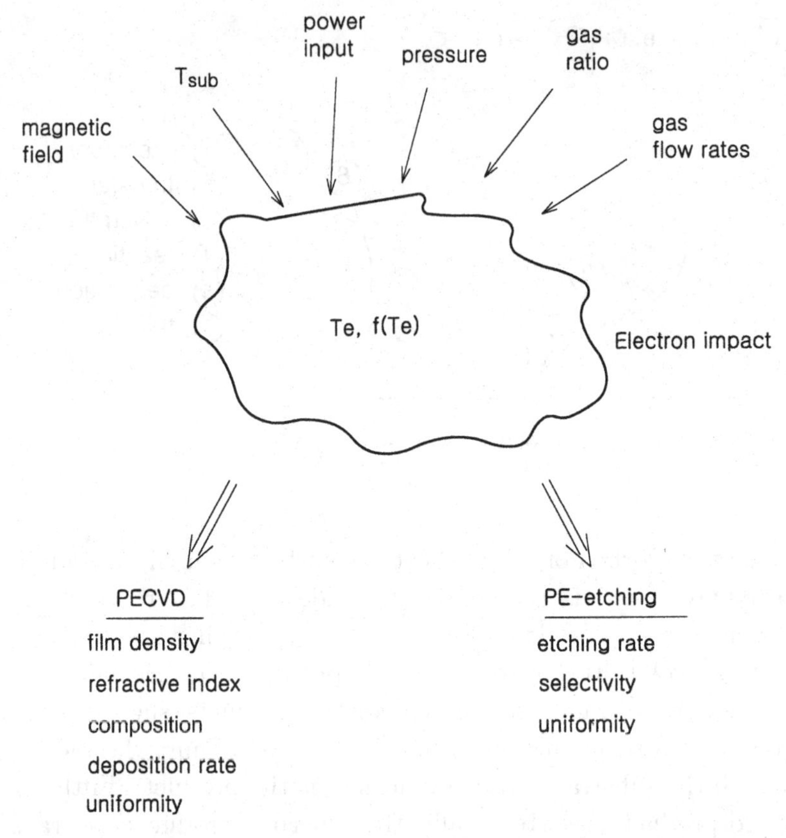

Fig. 6.2. The system variables in a plasma system and their influences on PECVD and PE-etching process characteristics.

mean free path allows particles to diffuse over a long distance and possibly travel across the process chamber without colliding with other particles. As a result, the etching or deposition uniformity can be improved. However, lower pressures mean lower gas density, and therefore lower electron density. As a result, lower etching or deposition rates are expected. At low pressures, owing to the long mean free path, the ions have a better opportunity to reach the substrate surface and thereby increase the ion bombardment effects. For etching, this means stronger anisotropic etching characteristics. In terms of etching rates, if the etching is physical sputtering-dominated, a lower pressure can result in higher overall etch rates. On the other hand, if the etching is chemical-dominated, the ion sputtering has marginal effects. The radical generation rate, R, is proportional to the gas and electron density and is expressed as

$$R = kn_eC\,, \tag{6.1}$$

where k is the dissociation rate constant and is a function of the mean electron energy and electron energy distribution, n_e is the electron density, and C is the gas density. Changes in pressure affect all of the three items on the right-hand side, some positively and some negatively. The end result depends on the factor that is most dominant.

The power input to the plasma system will eventually heat up the gas system and the substrate through collisions and bombardment. In addition, the back of the wafer can be heated or cooled, depending on the process optimization. When compared to these two heating sources in cold plasma processes, the thermal effects from the endothermic or exothermic reactions are often negligible. For deposition, the increase in the substrate temperature causes the film density to increase; this is partly due to desorption of hydrogen from the film bulk. Increasing the substrate temperature may or may not increase the etching or deposition rate; it depends on whether reactant adsorption or product desorption is the rate-determining step. A more significant effect on etching are the selectivity changes with substrate temperature. The etching rates for the to-be-etched film and the layer beneath it are

$$R_1 = k_1\mathrm{e}^{-A/RT} \tag{6.2}$$

and

$$R_2 = k_2\mathrm{e}^{-B/RT}, \tag{6.3}$$

respectively, where k_1 and k_2 are the rate constants, A and B are the activation energies, and T is the absolute temperature. Therefore the selectivity of the film with respect to the underlying layer is

$$S = \frac{k_1}{k_2} \exp\left(\frac{B - A}{RT}\right). \tag{6.4}$$

It is obvious that the temperature can be manipulated to change the etching selectivity.

Increasing the power input boosts the electron temperature and density, which in turn will increase the reactive radical concentration and lead to potentially high deposition and etching rates. As the thermal temperature changes impact the reaction selectivity, the electron temperature changes affect the relative impact dissociation rates of different species. Increases in power also enhance the ion bombardment energy. As discussed in Chapter 5, the magnetic field is often employed around the plasma reactor to increase the electron lifetime; therefore a higher electron density can be obtained. This gives rise to higher precursor gas dissociation rates, and higher radical concentrations will result.

Unlike thermal CVD reactors, plasma reactors do not have universal trends of parameter effects. The major reason is that in the case of plasma reactors, electric fields and charged particles exist. For plasma reactors, such factors as electrode design, reactor geometry, and the wafer size-to-electrode area ratio all influence the effective plasma conditions that initiate the reactions on the wafer surface. For example, in a conventional CVD, the pressure effect on a CVD reaction is similar for different reactors, which is not the case for plasma CVD reactors. This is because the way the pressure affects the plasma reaction is much more complicated than in the conventional CVD. In a thermal CVD reactor, pressure changes the species diffusivity and gas density, but in a plasma CVD reactor, the pressure could affect the electron density, ion density, electron energy, and ion bombardment energy, in addition to the diffusivities and gas density. Furthermore, the relative size of the plasma bulk with respect to the reactor size can change the effective power input to the system. For example, Fig. 6.3 shows a situation where, owing to the charged particles' out-diffusion, the plasma can outgrow the reactor electrode

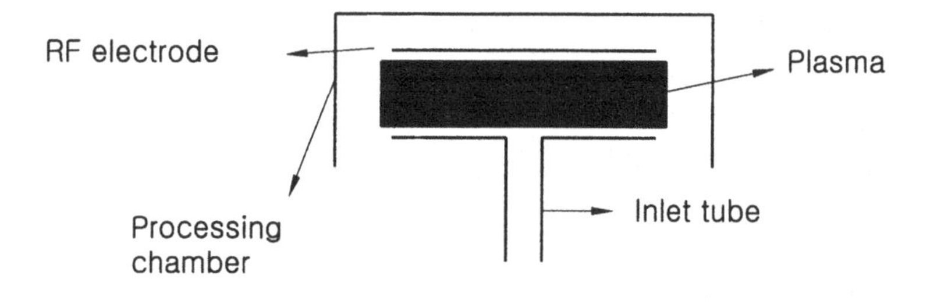

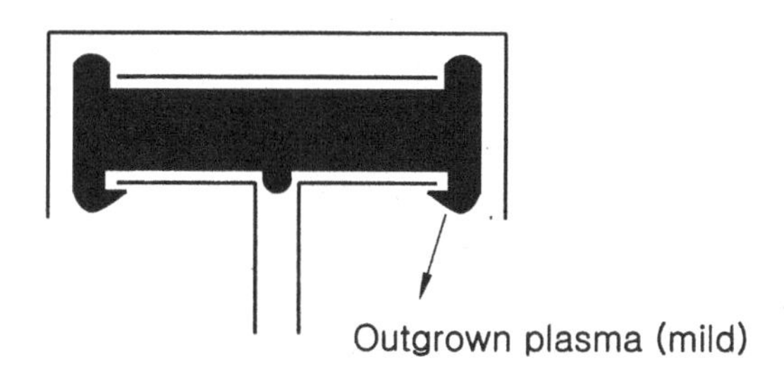

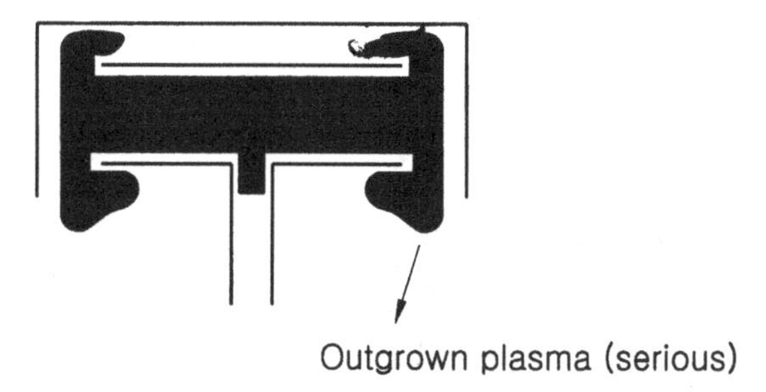

Fig. 6.3. Plasma outgrowth at various power input levels.

area. The reactor A can utilize the power input better than reactor B, and reactor B can utilize the power input better than reactor C. As a result, the deposition rate for reactor A increases with the power input, while for reactors B and C, it may possibly level off.

6.2. Introduction to PECVD and Film Characterization

Most thermal CVD dielectric films can be grown with the PECVD approach whenever the thermal budget is of concern. Nonetheless, the PECVD films have different characteristics when compared with their thermal CVD counterparts. PECVD film stoichiometry changes with system variables, and the trends can be system-dependent. In the case of silicon nitride deposition, the film stoichiometry varies with system inputs, and so do such film properties as refractive index, stress, and density, as illustrated in Fig. 6.4. Another unique feature of PECVD is that the films tend to incorporate significant amounts of hydrogen.

The characteristics of PECVD films differ significantly from those of thermal CVD. Some important PECVD film properties are required:

(a) *Good thickness uniformity and step coverage.* Similar to the requirement for thermal CVD, thickness uniformity and step coverage are important, especially when PECVD films are used for passivation.

In such applications, the poor step coverage along with a thicker film will make it more prone to form a void or keyhole than the thinner film area, as illustrated in Fig. 6.5. The voids or keyholes can cause reliability issues. Furthermore, the better the film thickness uniformity, the better the control over the void formation.

(b) *No particle contamination.* If a PECVD process is not properly controlled, it can result in massive particle formation due to excessive homogeneous reactions, in which radicals swiftly combine with

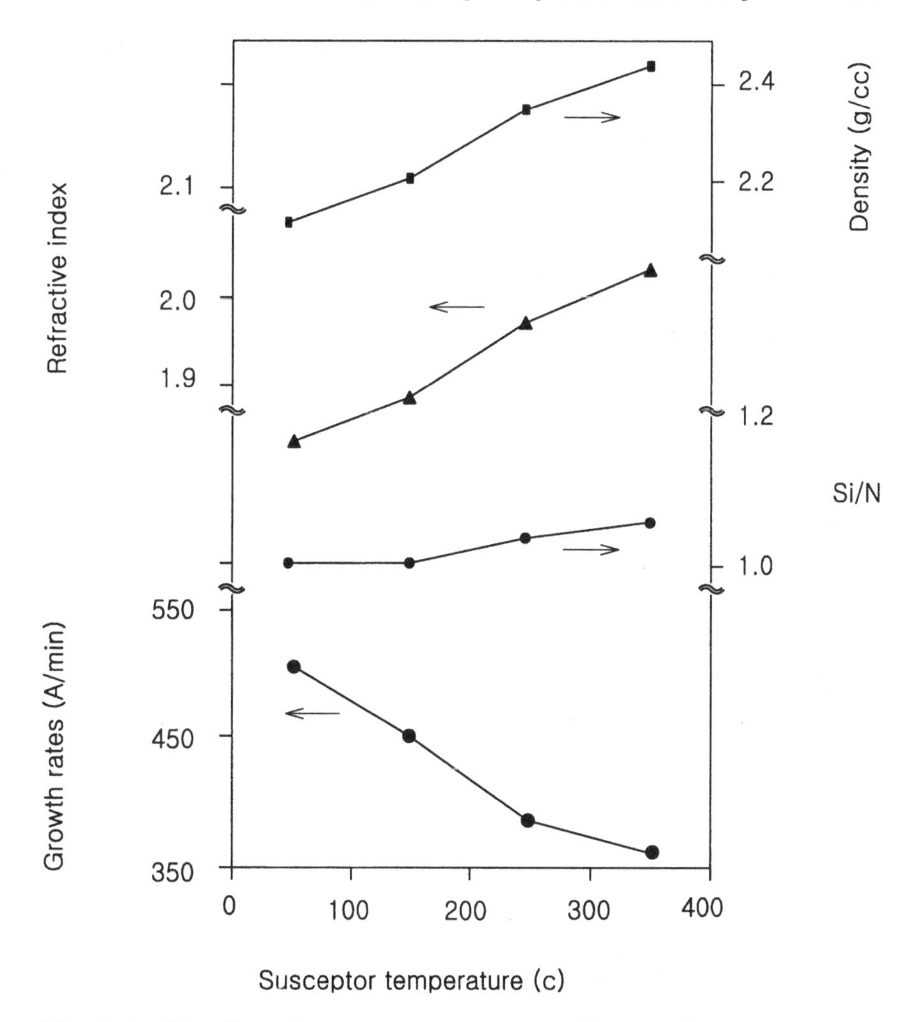

Fig. 6.4. The effect of susceptor temperature on film growth rate composition and properties.

each other. For example, an excessively high pressure in PECVD gives rise to massive particles in the PECVD SiN plasma chamber. Metal patterns deposited on the particles can lead to humps, which are turned into residues after etching. In all cases, particles are not desired in semiconductor processing.

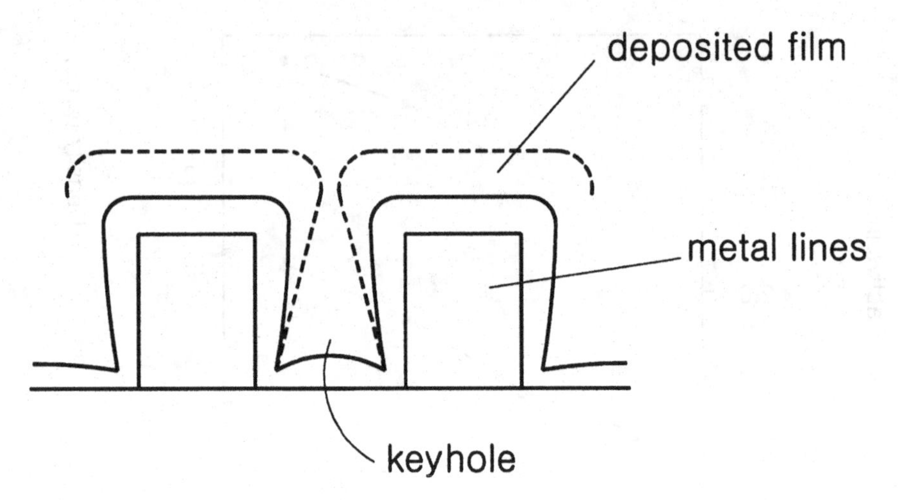

Fig. 6.5. A film deposition process with poor step coverage can lead to a key hole.

(c) *Low thermal cycle.* The thermal cycle refers to a high temperature step at a certain temperature for a certain period of time. The temperature and duration time, the thermal cycle, it takes for a thin film deposition process could alter the underlying device characteristics by pushing the dopants further. Therefore the thermal cycle has to be kept as small as possible. In general, a thermal CVD generally runs above 700°C and has a higher tendency to alter the device characteristics than does PECVD, which generally runs below 450°C. Oftentimes, in device design, the thermal budgets of thermal CVD steps are taken into account, while those of PECVD are not.

(d) *Low film stress.* Film stress is a critical factor in affecting device performance because high film stress can cause silicon defects or warpage and affect the device performance. If applied on top of metal layers, excessive stress may degrade the metal line integrity and result in hillocks, extrusions, or pits, as shown in Fig. 6.6. The film stress is a function of film composition, which changes with the

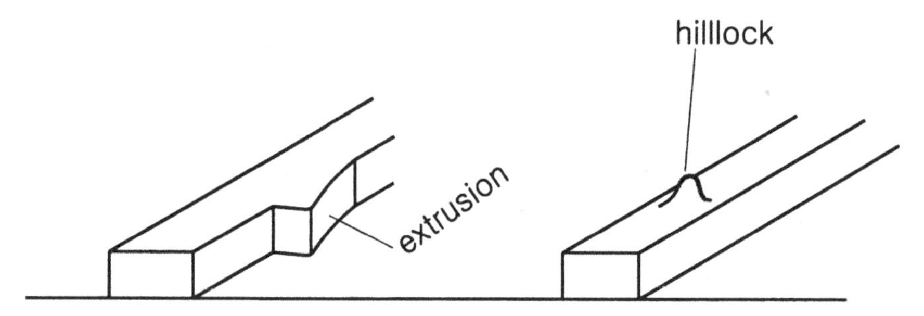

Fig. 6.6. A hilllock or extrusion can occur when excessive film stress is applied on the metal lines.

deposition parameters such as power, pressure, flow rate, and so on etc., in a PECVD.

(e) *Low hydrogen contents.* Unlike thermal CVD films, PECVD tends to produce films with higher hydrogen contents. This is because the hydrogen-containing precursors in the plasma end up being incorporated into the film. Hydrogen is a very light element, and it is very difficult to detect with instrumental analysis. Even worse, hydrogen can freely diffuse around in dielectric films toward the gate oxide and affect the device characteristics. Therefore the hydrogen content must be kept as low as possible. One way to do this is to raise the substrate temperatures to allow the incorporated hydrogen to diffuse out.

(f) *Low charge contents.* With film deposition in a plasma reactor, the wafer surface is constantly experiencing harsh plasma conditions such as ion bombardment. The ion bombardment causes structure damage and charge trappings on the film surface. The charges contained in the films can adversely affect the device. For example, a field device can be electrically inverted by the charges in the PECVD oxide films, as illustrated in Fig. 6.7. The concern about the trapped charges prevents the use of PECVD films in direct contact with the device surface.

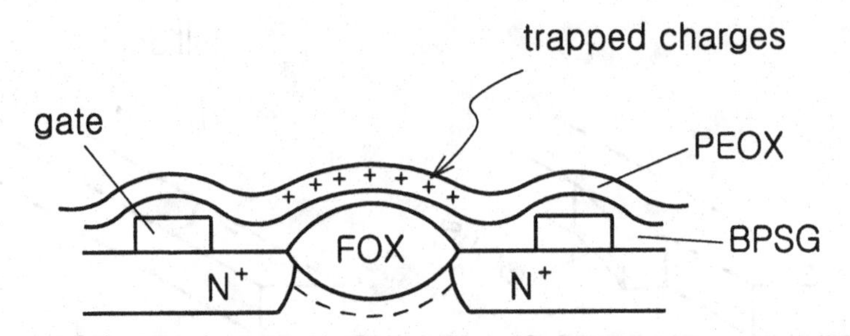

Fig. 6.7. The excessive trapped positive charges in the PEOX may turn on the parasitic field device.

6.3. Applications of PECVD Films

The fact that PECVDs can be carried out at low ambient temperatures further increases their widespread applications in back-end processes of semiconductor manufacturing. Unlike thermal CVD films, most thin films, which are grown with the PECVD approach, are nonstoichiometric. Also, the film characteristics rely on such deposition parameters as gas flow rate ratios, power input, pressures, and temperatures. Nevertheless, as long as the deposition conditions are fixed, the film characteristics are quite reproducible in a given system.

6.3.1. *PECVD oxide*

Silicon oxide film grown with PECVD, $PESiO_x$, is nonstoichiometric; it is commonly used as the top and bottom of the sandwich layers in a spin-on-glass (SOG) planarization scheme (PEOX/SOG/PEOX). It is also used as the bottom layer of a dual-passivation scheme (PEOX/PESiN), as illustrated in Fig. 6.8. The most commonly used chemical reaction to form the PECVD oxide films is as follows:

$$SiH_4 + O_2 \xrightarrow{e^-} Si_xO_y \cdot H_z\,, \tag{6.5}$$

where e^- represents the chemical reaction in the PECVD, which is initiated by the highly energetic electrons in the plasma through electron impacts. The relative atomic composition ratios x, y, and z vary according to reaction conditions. Some residual amount of hydrogen always exists in the deposited films because there are

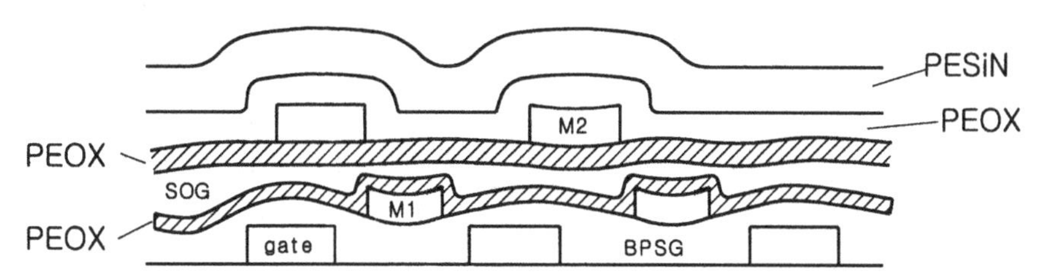

Fig. 6.8. PECVD silicon oxide is used in SOG sandwich and dual passivation schemes.

massive radicals in the form of SiH_x that are incorporated into the deposited film. In general, the PECVD oxide film does not have very good step coverage. Hence, when the PECVD oxide is used as the SOG underlayer, it often leads to poor planarization. As a result, the SOG sandwich layer is often optimized with a thin bottom PEOX and a thick top PEOX. For passivation applications, a thinner PEOX (5000 Å) is often used, and it is followed by a thicker PESiN layer (7000 Å). The PEOX is meant to be a stress buffer under the nitride. It is certain that the thickness should be adjusted when the metal spacing is shrunk. Otherwise, voids can easily form in between the metal lines and cause reliability and contamination concerns.

Extensive studies have been done on characterizing the PEOX films. Figure 6.9 shows that the deposition rate of PEOX films increases with system power input. This is because the high power inputs produce high electron densities and electron temperatures, and thereby, more radicals are generated for the deposition process. Figure 6.10 shows that the Si/O ratios in the deposited films follow those in the gas during deposition; this is also true in the case of film refractive indices. Hydrogen contents in the deposited films are of great concern as they might interact with SOG and cause reliability issues.

6.3.2. *PECVD nitride*

PECVD silicon nitride film, PESiN, was the first successful material deposited with the PECVD approach on a large scale. It has

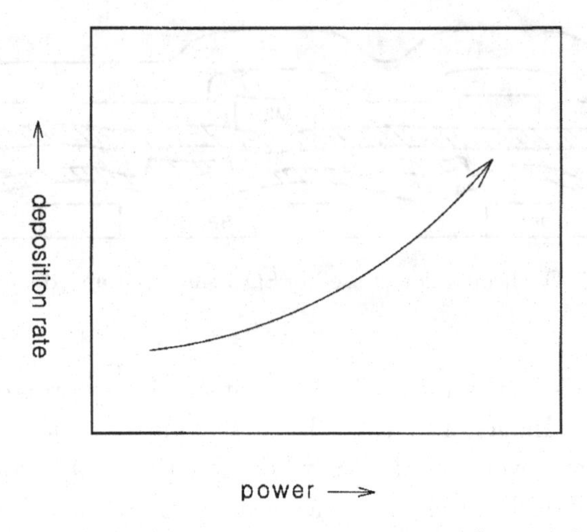

Fig. 6.9. PEOX deposition rate increases with power input.

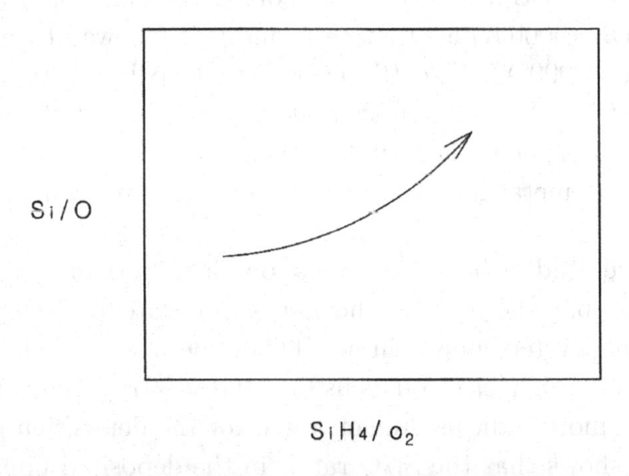

Fig. 6.10. The film Si/O atomic ratio follows the gas flow ratio.

extensive applications in passivation, which protects the device structure from mechanical scratching. It also prevents mobile ions, such as sodium and potassium ions, and stops the moisture from diffusing into the device. As a passivation layer, it is often used with an underlying stress buffer such as a PEOX layer, as discussed earlier. Typical reactions for forming the PESiN layers are as follows:

$$SiH_4 + NH_3 + N_2 \xrightarrow{e^-} Si_xN_y \cdot H_z \tag{6.6}$$

$$SiH_4 + NH_3 \xrightarrow{e^-} Si_xN_y \cdot H_w \tag{6.7}$$

$$SiH_4 + N_2 \xrightarrow{e^-} Si_xN_y \cdot H_v\,. \tag{6.8}$$

Each reaction has its own set of characteristics. When silane reacts only with ammonia, it produces films with higher hydrogen content; however, when it reacts only with nitrogen, it produces lower Si:N ratios. This is because the bond strength of N–H is weaker than that of N–N. As a compromise, ammonia and nitrogen are used at the same time. Again, the resulting film characteristics are complicated functions of the deposition system conditions. Figure 6.11 shows the deposition rates of silicon nitride films as functions of wafer holder temperatures, power inputs, and pressures used by different authors. As explained earlier, the trends are system-dependent. Figure 6.12 illustrates how the film stoichiometry, Si:N ratios, vary with the operation conditions.

In addition to the passivation application, the PECVD nitride film can also be used as a passivation layer for the EPROM device. The UV light must be able to shine through the EPROM device to erase the signals (electrons) that are already stored in a gate. Regular PECVD nitride with a Si:N atomic ratio near 1 and a refractive index near 2.0 results in extremely low UV light transmittance. By changing the reaction conditions, the Si:N atomic ratios in the PECVD nitride films can be altered, as shown in Fig. 6.13. The reaction proceeds as

$$SiH_4 + NH_3 + N_2 \xrightarrow{e^-} SiN_xH_y\,. \tag{6.9}$$

The key to obtaining a UV-transparent film is to lower the SiH_4/NH_3 gas ratio to achieve a nitride-rich film. The chamber pressure also plays a role in film stoichiometry. The UV transmittance increases with system pressure, which indicates a change toward nitride-rich films. Normally, a PECVD nitride film with a refractive index of 1.9 can reach about 65% UV transmittance.

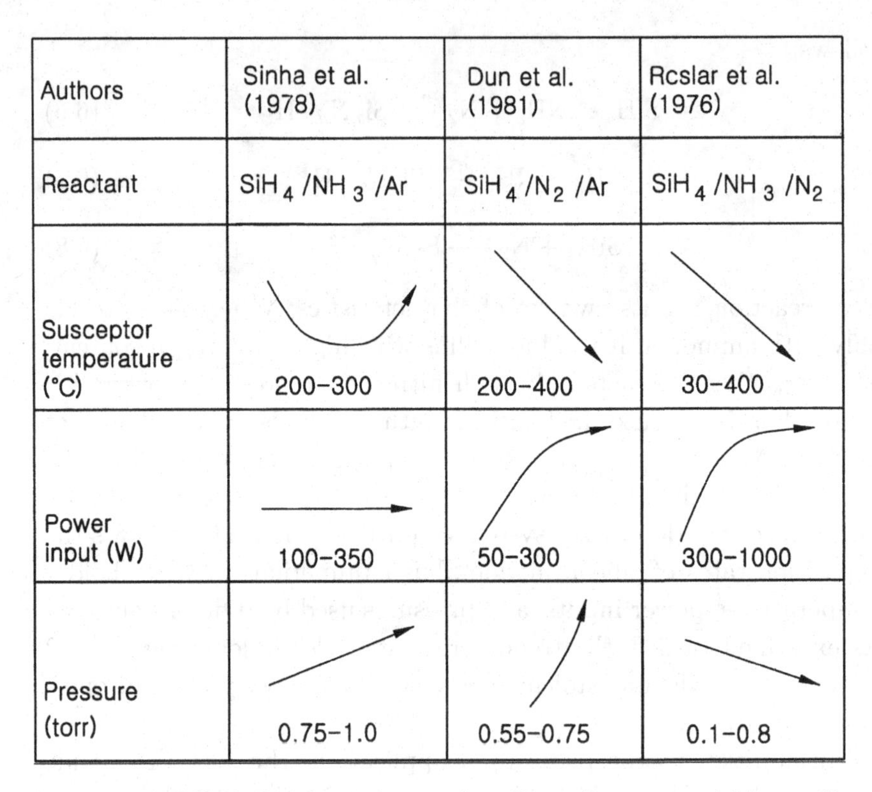

Authors	Sinha et al. (1978)	Dun et al. (1981)	Rcslar et al. (1976)
Reactant	$SiH_4/NH_3/Ar$	$SiH_4/N_2/Ar$	$SiH_4/NH_3/N_2$
Susceptor temperature (°C)	200–300	200–400	30–400
Power input (W)	100–350	50–300	300–1000
Pressure (torr)	0.75–1.0	0.55–0.75	0.1–0.8

Fig. 6.11. PECVD SiN deposition — the effects of system variables on film deposition rates.

6.3.3. *PETEOS*

Plasma-enhanced TEOS (PETEOS) has wide applications as inter-metal planarization layers. The TEOS oxide film step coverage is better than that of PEOX owing to high surface mobility of the reactant. As a result, it is often used to sandwich the SOG layer. The better underlying PETEOS step coverage ensures a better SOG planarization, as illustrated in Fig. 6.14. The reaction in a plasma proceeds as

$$(C_2H_5O)_4Si + O_2 \xrightarrow{e^-} SiO_x\,. \tag{6.10}$$

The deposited PETEOS tends to be hygroscopic. An anneal step is often required to stabilize the films. A high oxygen:TEOS ratio in

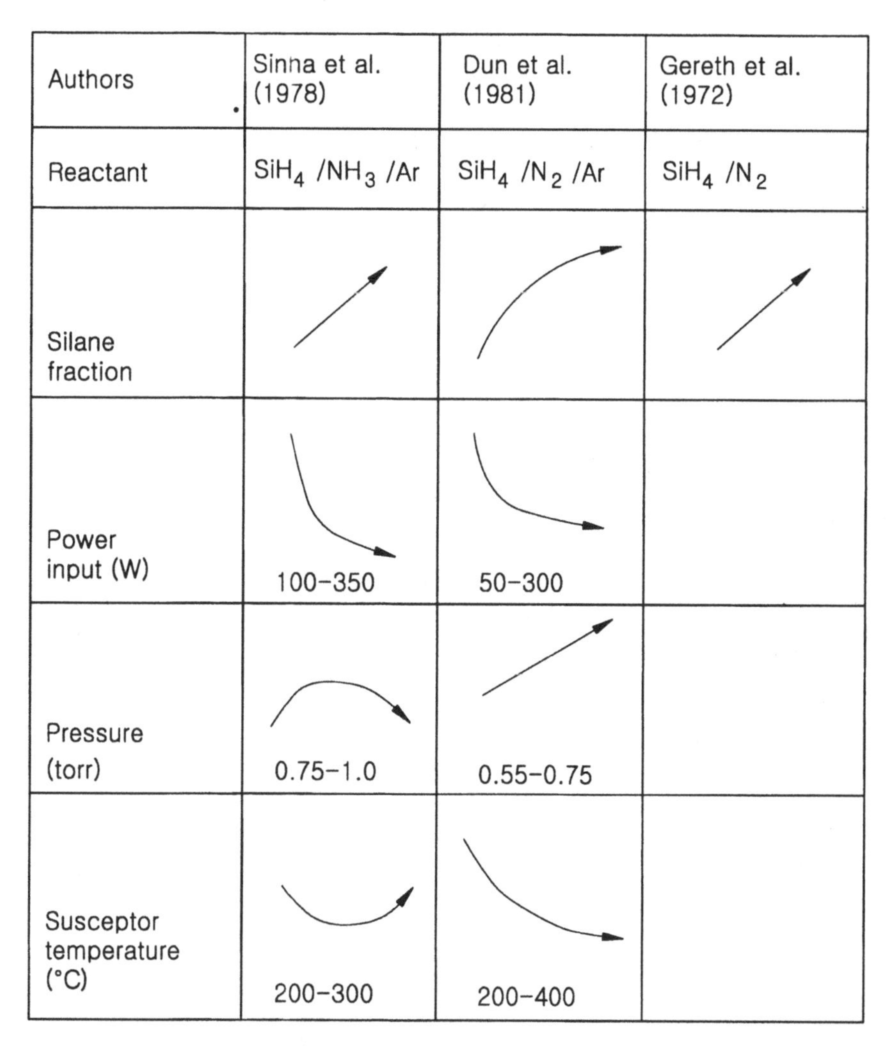

Fig. 6.12. PECVD SiN deposition — the effects of system variables on the film Si:N ratios.

the reaction produces a stable film. On the other hand, a decrease in oxygen:TEOS ratio tends to lead to low deposition rates.

Another important application of PECVD films is the low k value of intermetal dielectric materials for improving circuit speed beyond 100-nm technology. This is accomplished by doping the

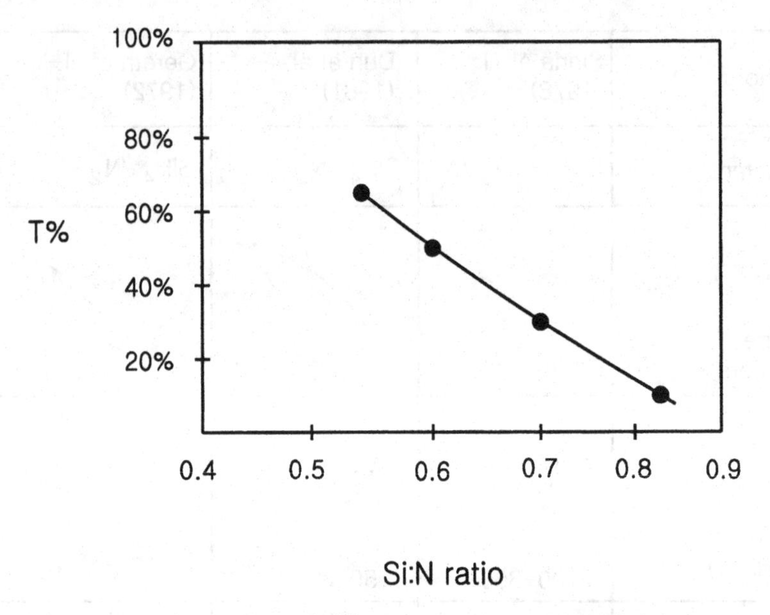

Fig. 6.13. The transmittance (T%) of a PECVD UV-SiN film decreases with film Si:N ratio.

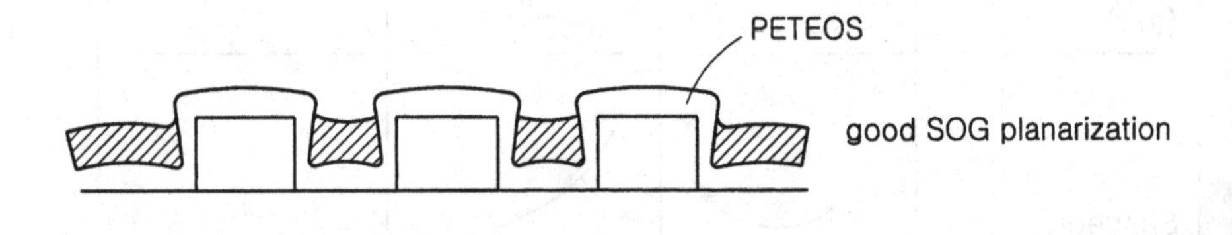

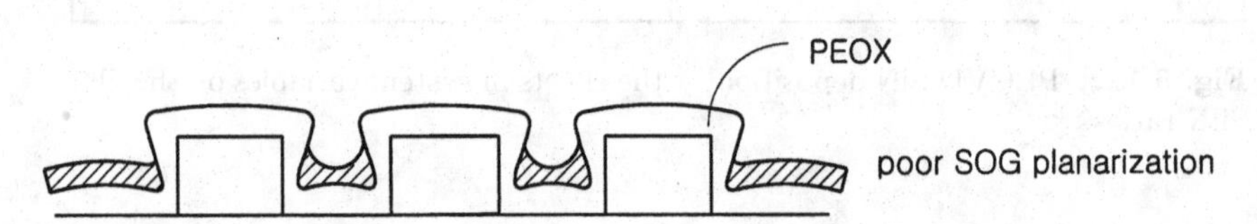

Fig. 6.14. PETEOS has better step coverage than PEOX and hence better planarization with SOG.

PECVD TEOS or SiH_4 oxide films with fluorine atoms from fluorine-containing chemicals such as SiF_4 or C_2F_6:

$$SiH_4 + O_2 + SiF_4 \xrightarrow{e^-} SiO_x \cdot F_y \, . \tag{6.11}$$

Normally, oxide films have dielectric constants ranging from 3.9 to 4.1; however, if they are doped with fluorine, the dielectric constant can be reduced to about 3.5. If doped with carbon, it can be further reduced to about 2.6. High fluorine concentration in the film tends to degrade the film stability due to fluorine outgasing. A proper fluorine concentration is around 3% (atomic%) using the high density plasma (HDP) CVD approach, with the silane as the silicon source. HDP is a type of PECVD with additional magnetic field enhancements such that the ion flux toward the film surface is higher than the depositing radical flux. As a result of intensive ion bombardment on the film surface, the resulting film is much more dense compared to regular PECVD films.

6.4. Introduction to Plasma Etching

Plasma etching is technically the reverse process of plasma-enhanced deposition. Plasma etching, also called dry etching, does not involve any wet chemicals, as required in a wet etching process. In plasma etching, reacting gases dissociate through electron impacts into radicals, which then transport to the underlying substrate surface and react with it to form volatile products. These products are carried away by the vacuum pump systems. Ions can influence the etching process through bombardment in terms of etching rates, selectivities, and sidewall profiles. The key to viable plasma etching processes is to properly select chemical reactions so that the reaction products are volatile (turned into vapors). In semiconductor manufacturing, halides are predominantly used in plasma etching of different films or substrates. This is primarily due to the volatilities (or relatively high vapor pressures at low temperatures) of the etching reaction products. An exception to this trend is the copper halides; instead of being volatile, they form residues on the substrate surface. This explains why copper dry etching still does not prevail in production.

In device manufacturing, photolithography and etching are often employed sequentially. This is the case whenever there is a need to selectively remove surface material, for example, to form conducting lines on a deposited metal layer or contact holes on a dielectric layer. In both cases, photolithography is first used to define the pattern; the etching process is then used to selectively remove the exposed surface and form the desired patterns, as indicated in Fig. 6.15. Of course, to achieve good pattern transfer fidelity from the resist patterns to the underlying substrate, the resist must remain intact as the etching proceeds.

General concerns on plasma etching are critical dimension uniformity, sidewall profiles, selectivity, and plasma damages. For

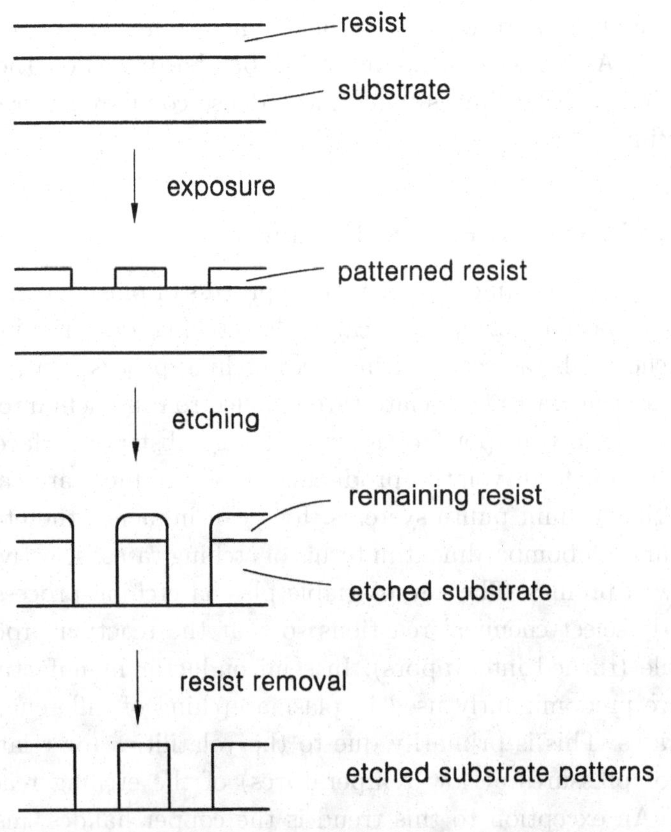

Fig. 6.15. A typical coupled photolithography-etching process to form desired substrate patterns.

conducting line formations, nonuniform CD distribution across a wafer can increase the interconnect conductance variations and therefore lead to poor circuit timing matching issues. For contact hole formation, the CD nonuniformity may lead to a metal–gate shortage, as indicated in Fig. 6.16. Nonuniformity is often expressed in terms of range or standard deviation. The sidewall profiles depend on the ways that the materials are etched away. By reviewing Fig. 6.17, two extremes of etching results are seen: one is isotropic, while the other is anisotropic. In a plasma etching, there are two etching mechanisms that contribute to the material removal, in particular, chemical reactions initiated by the free radicals and the sputtering etching by the ion bombardment. The chemical reactions initiated by radicals are

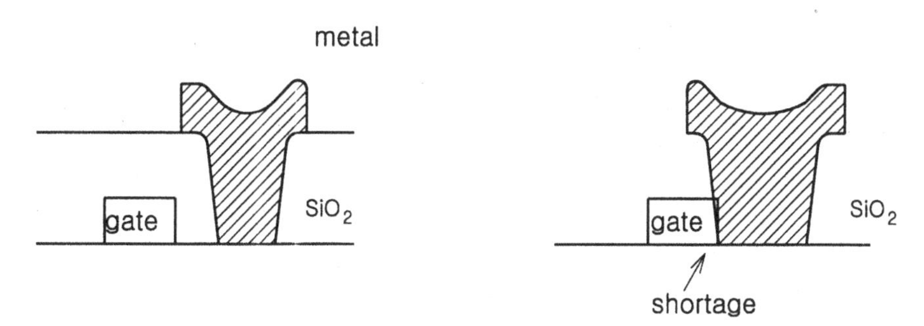

Fig. 6.16. Owing to contact size nonuniformity, an oversized contact hole can lead to metal–gate shortage.

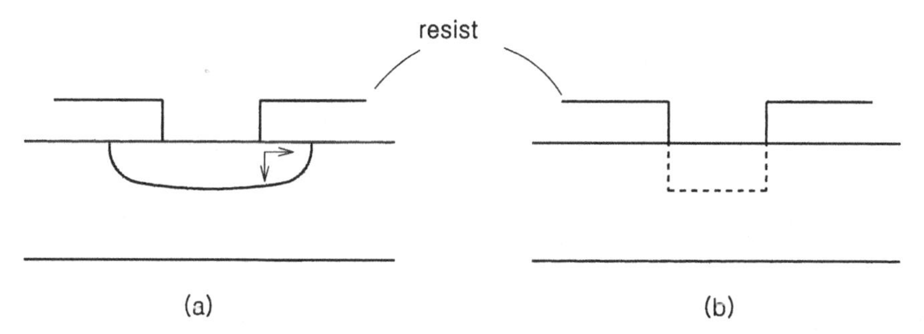

Fig. 6.17. (a) Isotropic etching and (b) anisotropic etching.

isotropic in nature. The etching due to ion bombardment (or sputtering etching) is essentially anisotropic in nature. As the etching mechanism changes from anisotropic to isotropic, the sidewall profile tends to vary from being vertical to being sloped. The sidewall polymers, the etching reaction by-products that form on the sidewall, are the other factor that affect the profile. The polymers passivate the sidewall as the etching proceeds and help reach a vertical sidewall.

The etching selectivity is defined as the etch rate ratio of two materials. In the case of etching, as shown in Fig. 6.18, two selectivities are particularly critical. One is the film etching rate with respect to that of the underlying substrate (SS), and the other one is with respect to that of the etching mask (SM). Both are expected to be as high as possible to achieve the desired etching goals, which are stopping on the underlying layer surface and achieving a vertical profile. A poor SS results in a significant loss of the underlying layer thickness. For the case of metal etching, this aggravates the topographical issue, and it makes it difficult for the ensuing planarization, as indicated in Fig. 6.19. With a poor SM, the etching mask erodes as the etching process proceeds, and it will result in nonvertical sidewall profiles, as illustrated in Fig. 6.20. In general, radical reactions result in high selectivity with isotropic profiles; ion sputtering etchings give low selectivity with anisotropic profiles.

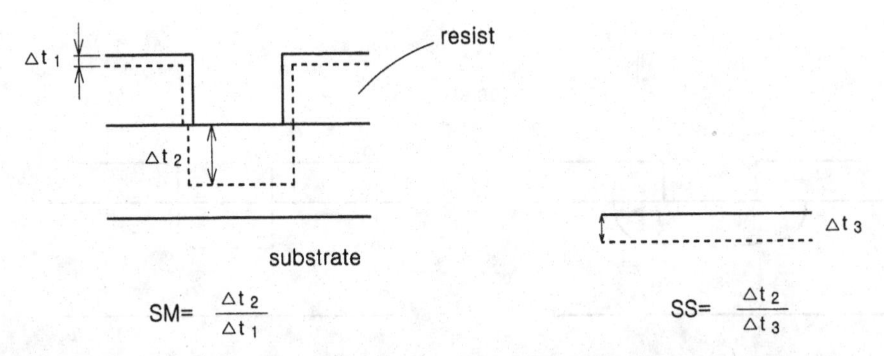

Fig. 6.18. After etching, the thickness loss of film (Δt_2), resist (Δt_1), and substrate (Δt_3) can be used to calculate the etching selectivity.

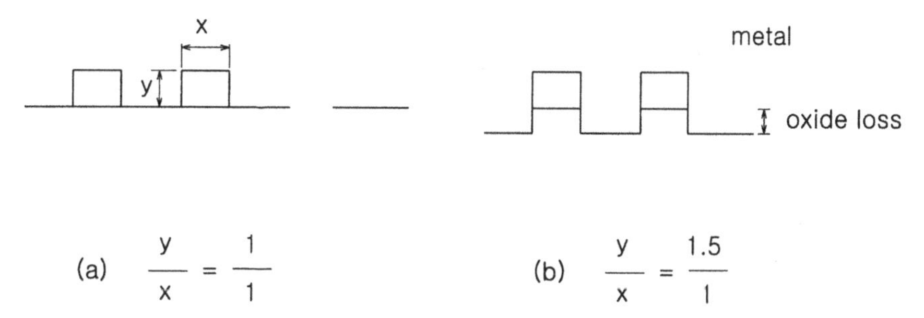

Fig. 6.19. Case (b) has poor SS, resulting in severe oxide loss, increasing the aspect ratio (y/x). That makes it difficult for planarization.

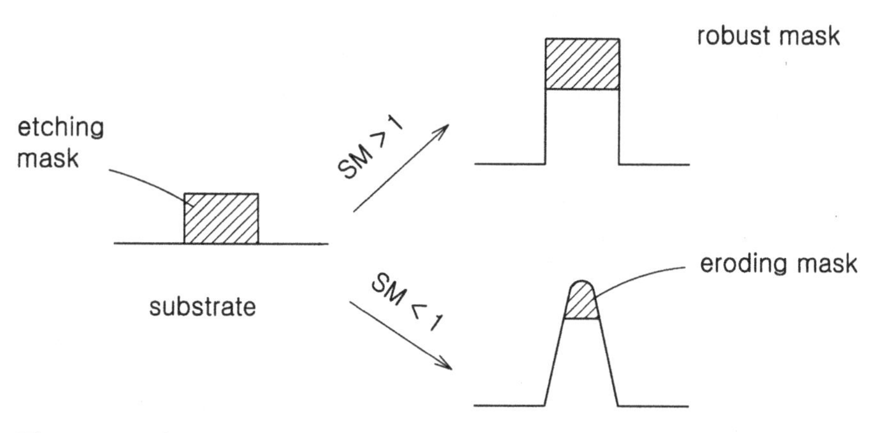

Fig. 6.20. Comparison of etching profiles with robust $(SM \gg 1)$ and eroding $(SM < 1)$ etching masks.

Plasma damages are often observed in either plasma-enhanced etching or deposition as wafers are constantly under charged particle bombardments. For the gate etching, the charges can accumulate in the gate oxide area and thereby damage the gate oxide. An example is demonstrated in Fig. 6.21, in which a large polysilicon pad functions as an antenna and collects charges during plasma etching. Another example is when SOG is exposed to plasma; the bombarding ions can cause the Si–O bond to break and leave excessive charges on the surface. The charges can subsequently affect the field device

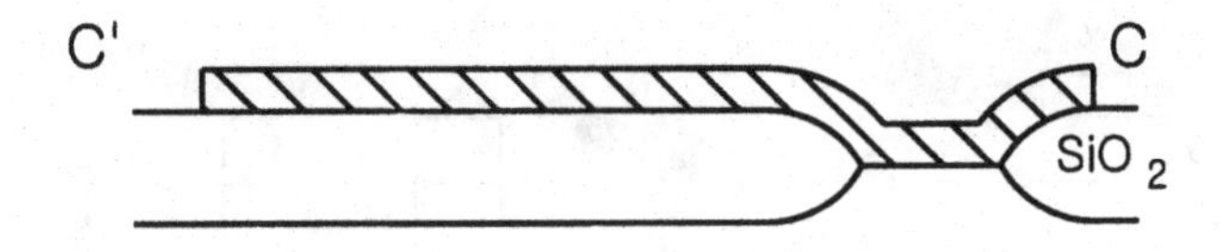

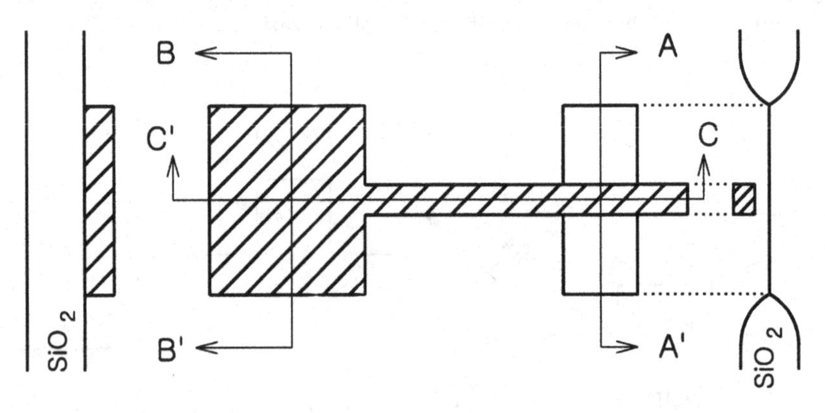

Fig. 6.21. A patterned polylayer functions as interconnects and gates. During poly plasma etching, the large poly pad can collect charges (left), which move to the gate (right) and damage the gate oxide, causing transistor leakage and reliability failures.

threshold voltages. Antenna structures are often designed in the circuit to monitor the charge accumulation.

6.5. Applications of Plasma Etching

To conduct a plasma etching, the substrate surfaces can be either patterned or not patterned. The patterned substrates are masked either with resist or dielectric materials (hard masks). The plasma chemistries used for material etching are primarily halide based. The precursor gases are often a mixture of halogen-containing species with different types of additives, such as oxygen, helium, or nitrogen, to fulfill various purposes in terms of improving the uniformity, selectivity, etching profiles, or etching rates.

6.5.1. *Polysilicon gate etching*

The transistor is the heart of a chip, and a good polysilicon gate-defining process is a prerequisite for excellent performance of the chip. The requirements for gate etching are very stringent. First, the CD control must be very tight. For the state-of-the-art production CD, the requirement is about 3σ of 3 nm for all error sources. Excessive CD variations cause a mismatch in transistor speed and therefore result in circuit function failures. Second, the sidewall profile of the gate must be as vertical as possible. CD control for a vertical sidewall is easier to control than for a sloped sidewall. Furthermore, the nonuniform sidewall angles owing to the sloped sidewall can shadow ion implantation and cause a nonuniform transistor performance across wafer or wafer to wafer, as demonstrated in Fig. 6.22. Third, the underlying gate oxide must remain intact after the gate etching, which is even more critical for the gate oxide thicknesses

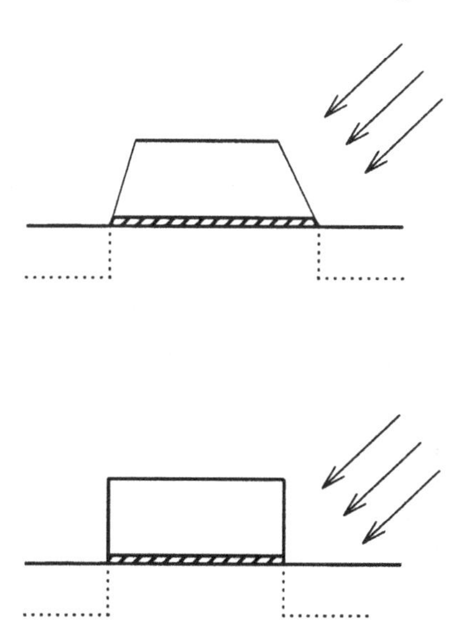

Fig. 6.22. A sloped etched gate sidewall profile can block the implant, causing different channel lengths.

of about 15–30 Å for advanced technologies. Extremely high etching selectivity (polysilicon vs. gate oxide) is required. Etching through gate oxide can lead to substrate damages and possible gate oxide undercut, and it can result in the degradation of the transistor performance. Fourth, the etched pattern must be free of polystringer; in particular, there should be no polysilicon residues at the foot of the polysilicon gate structure. These stringers cause gate-to-gate shortages. Finally, the gate oxide must be free from plasma damages after the gate etching. Gate oxide damages increase the local weak spots, which can result in device failure or chip reliability issues.

Figure 6.23 shows the etch rates of polysilicon and photo-resist with an ECR etcher with various halogen plasmas. The etch rates of fluorine plasma are faster than chlorine plasma, which in turn are faster than bromine plasma. These trends correlate well with the volatility trend of the etching products ($SiF_4 > SiCl_4 > SiBr_4$). The trend implies that for these etching reactions, the rates are limited by the product desorption step. Similar trends are observed for the photo-resist and silicon oxide etching rates. In terms of etching

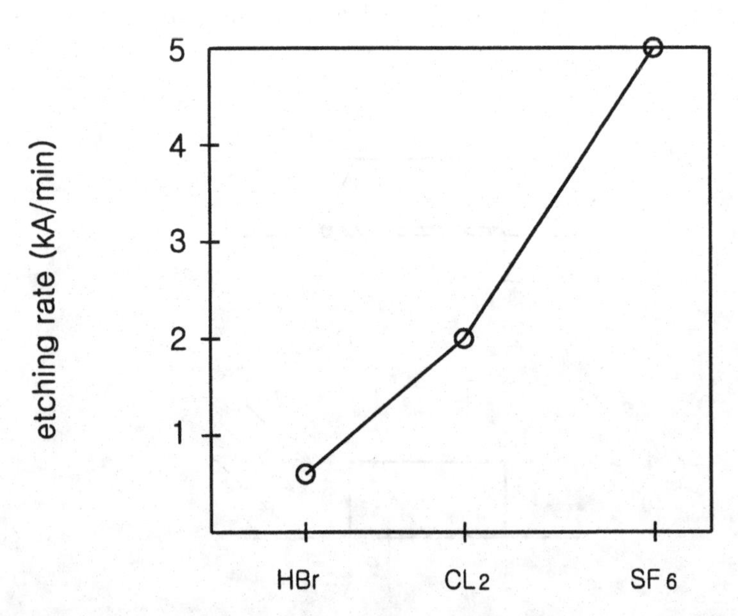

Fig. 6.23. Silicon etching rates for various halogen plasmas in an ECR with 2% oxygen flow rate ratio.

selectivity of polysilicon versus the oxide and the resist, HBr provides the highest selectivity, as illustrated in Fig. 6.24.

To discuss the etching profile, let us define the anisotropy as

$$\text{anisotropy} = 1 - \frac{\text{lateral etch rate}}{\text{vertical etch rate}} . \tag{6.12}$$

This is demonstrated in Fig. 6.25. Among the three, F-plasma has the largest lateral etching rate owing to its high reactivity, when compared to the other two halogens. However, once initial reactions are initiated, fluorocarbons and oxyfluorides are formed and deposited on the sidewalls; this reduces the lateral etching. For bromine plasmas, owing to the low reactivity of bromine with silicon, the lateral etching is minimal, and the anisotropy primarily results from the ion

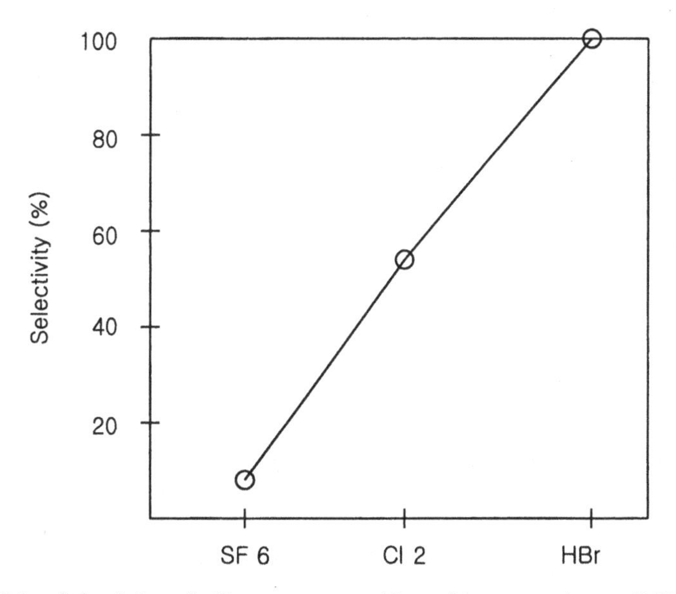

Fig. 6.24. Selectivity of silicon versus oxide etching rates in an ECR etcher with 2% oxygen ratio.

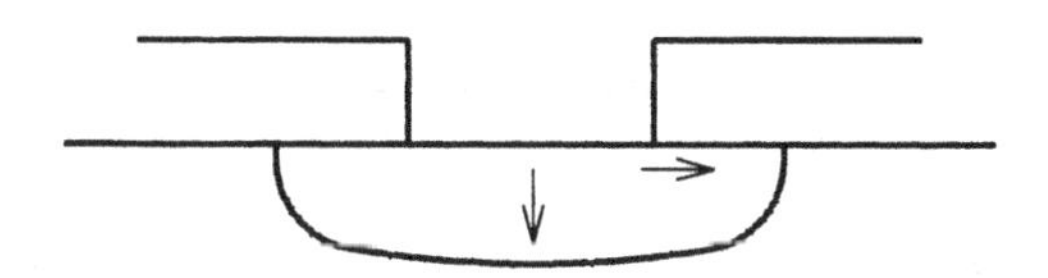

Fig. 6.25. The lateral and vertical etch rates.

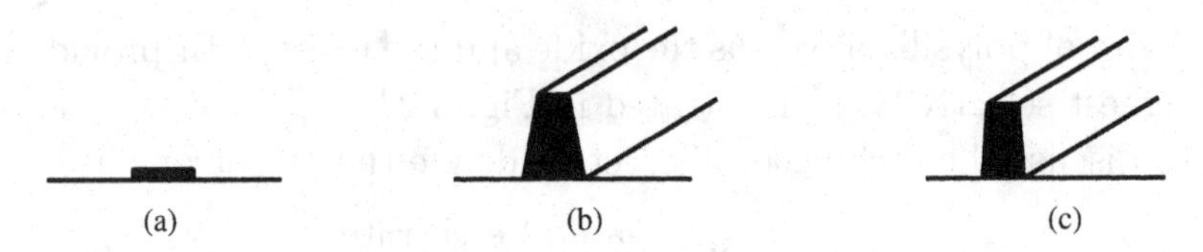

Fig. 6.26. Etched polysilicon gate profiles with (a) SF_6, (b) Cl_2, and (c) HBr.

bombardment in the plasma. Figure 6.26 shows the etching profiles of the three halogen plasmas. For SF_6/O_2, the polysilicon line can be totally etched and lifted off owing to its high lateral etching rates. The profile of Cl_2/O_2 is more tapered than that of HBr/O_2 because the former has a higher resist erosion rate. The polysilicon gate is often doped with phosphorous to form an n-type gate. The etch rates of the n-type polysilicon are much faster than the intrinsic polysilicon because the n-type polysilicon, which is rich in electrons, enhances the adsorption of halogen radicals. Therefore a higher etching rate results.

6.5.2. *Tungsten polycide etching*

A gate structure composed of a layer of tungsten silicide on polysilicon is named as a tungsten polycide gate structure. The polycide gate is a dual-layer structure. For its etching processes, apart from requiring a uniform CD uniformity, it is also important to have a vertical sidewall. Having either a smaller or larger silicide CD is not desired, as illustrated in Fig. 6.27, since they will cause ensuing process issues. For a larger silicide CD, the subsequent ILD planarization may not fill the sidewall, resulting in voids. For a smaller

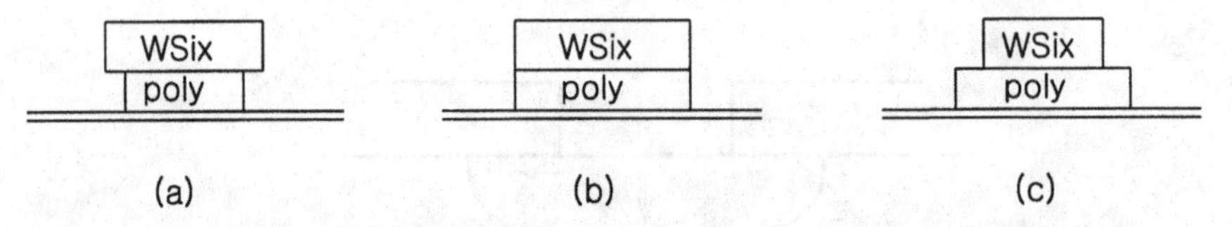

Fig. 6.27. Three polycide gates etching profiles. Case (b) gives the desired profile; profiles of (a) and (c) are not desired.

silicide CD, the resistance of the polysilicon lines will be increased. During etching, building up the sidewall passivation is essential to obtain a vertical and continuous sidewall profile. The polycide gate structure can be etched with a feed gas mixture of SF_6, Cl_2, and O_2. The substrate temperature is critical in affecting the etching rates and sidewall profile. Below 50°C, there is hardly any polycide etching due to the buildup of the tungsten etching products, which are not volatile in that temperature range. The sidewall profile as a function of temperature is illustrated in Fig. 6.28, and it can be used to optimize the sidewall profile. The addition of SF_6, even in small percentages, can have dramatic effects on eliminating the etching residues. A small percentage of SF_6 in the feed gas is effective in removing the tungsten polycide residues by converting nonvolatile tungsten chloride into tungsten chlorofluorides. On the other hand,

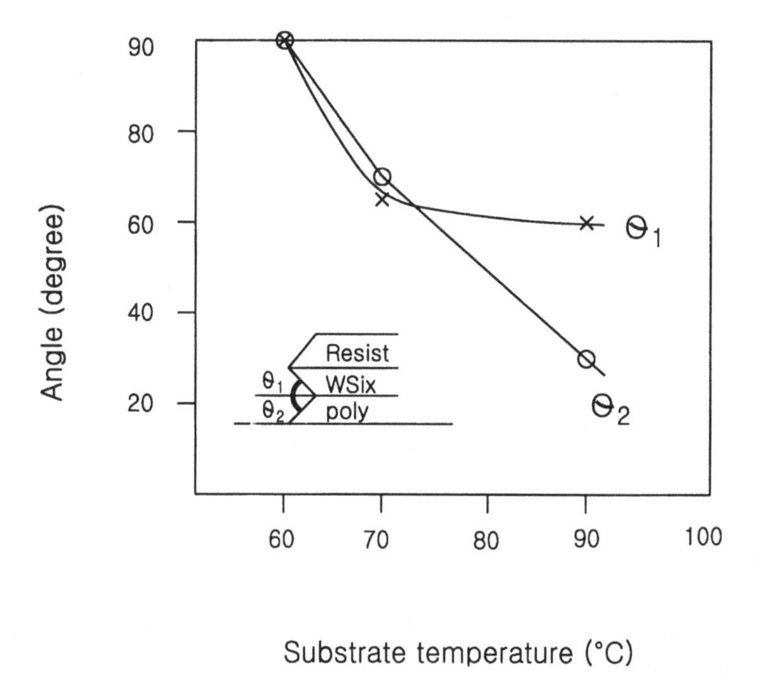

Fig. 6.28. Polycide etching profile as a function of substrate temperature.

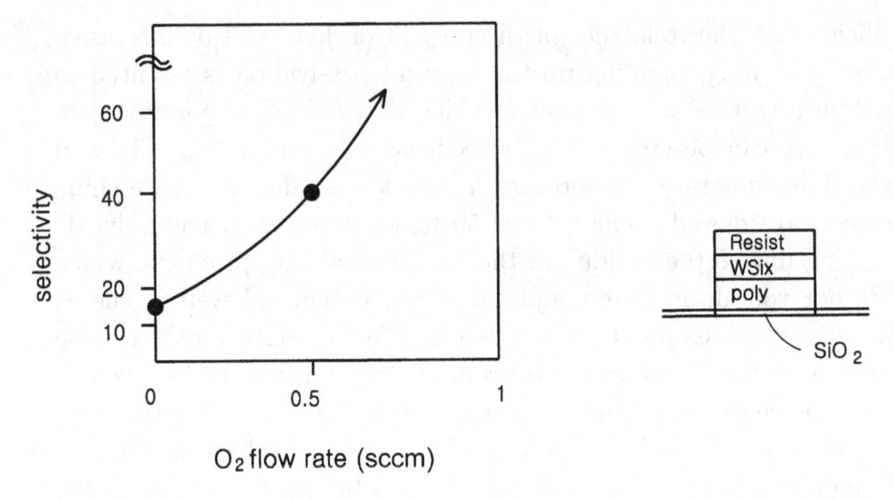

Fig. 6.29. Etching selectivity of W-polycide to the underlying oxide in an ECR etcher with Cl_2/SF_6 of 19/1 at 0.5 mTorr, 1 kW, and 60°C.

even a small amount of oxygen greatly improves the etching selectivity with respect to the gate oxide, as indicated in Fig. 6.29. The reasons are twofold. One reason is that the formation of the oxyhalogen deposit on the silicon oxide surface slows down its etching rates. The other reason is that the formation of oxyhalides reduces the amount of the silicon halides that are available for etching the oxide.

6.5.3. *Tungsten etching*

Tungsten has been widely used in device manufacturing as interconnect materials or contact plugs. The process requirements for the two are different. For interconnect applications, it is important to maintain good CD control, vertical sidewall profiles, and stringer-free etched patterns. For plug applications, it is critical to have residue-free surfaces in the field region after etching. Fluorine-containing gases such as CF_4, SF_6, or NF_3 with oxygen are popular choices for etching precursors; among them, SF_6 and O_2 are commonly used for tungsten etching in device manufacturing. With these etchants, various compounds, such as WO_xF_y, WS_2, SO_xF_y, WO_x, and WF_n, are observed on the surface. The reactions take place as follows:

$$\begin{aligned} \mathrm{W} + n\mathrm{F} &\longrightarrow \mathrm{WF}_n \uparrow, \\ \mathrm{WF}_n + \mathrm{O} &\longrightarrow \mathrm{WF_4O} \uparrow + \cdots, \\ \mathrm{W} + 2\mathrm{S} &\longrightarrow \mathrm{WS_2}(s). \end{aligned} \tag{6.13}$$

The tungsten fluoride and oxyfluoride are volatile products, while sulfur acts as the sidewall passivation agent, accounting for the vertical sidewall profiles. Oxygen radicals, on the other hand, can react and eliminate sulfur and tungsten by forming SO_xF_y and WO_xF_y. For improving tungsten adhesion on the substrate, a thin layer of titanium nitride (TiN) or titanium tungsten (TiW) is required to be used as the glue layer for tungsten deposition. Tungsten etching has good selectivity with respect to TiN but not with TiW. Consequently, when the TiN glue layer is used, the etching is often stopped on the TiN surface, while with TiW, it is stopped on the underlying silicon oxide. For plug and interconnect applications, it is important that the etching be finished with a step that is isotropic in nature to clear all the possible residues on the topograph.

6.5.4. *Dielectric layer etching*

The two most frequently used dielectric layers in semiconductor manufacturing are silicon oxide and silicon nitride films. PECVD oxide is often used in interlayer dielectrics (ILD) or intermetal dielectrics (IMD) planarization. Thermal silicon nitride is commonly used to define the active areas for field oxidation, while PECVD nitride is used for the passivation layer. For patterning active areas, the silicon nitride etching must have uniform CD across wafer and wafer to wafer. The dimension of the nitride directly influences the transistor width and thereby the circuit performance and timing, as illustrated in Fig. 6.30. The dual passivation layer etching is carried out through the underlying PECVD oxide and stops on the underlying metal pad. This is relatively straightforward because the fluorine can hardly attack the metal surface owing to the fact that AlF_x is not volatile.

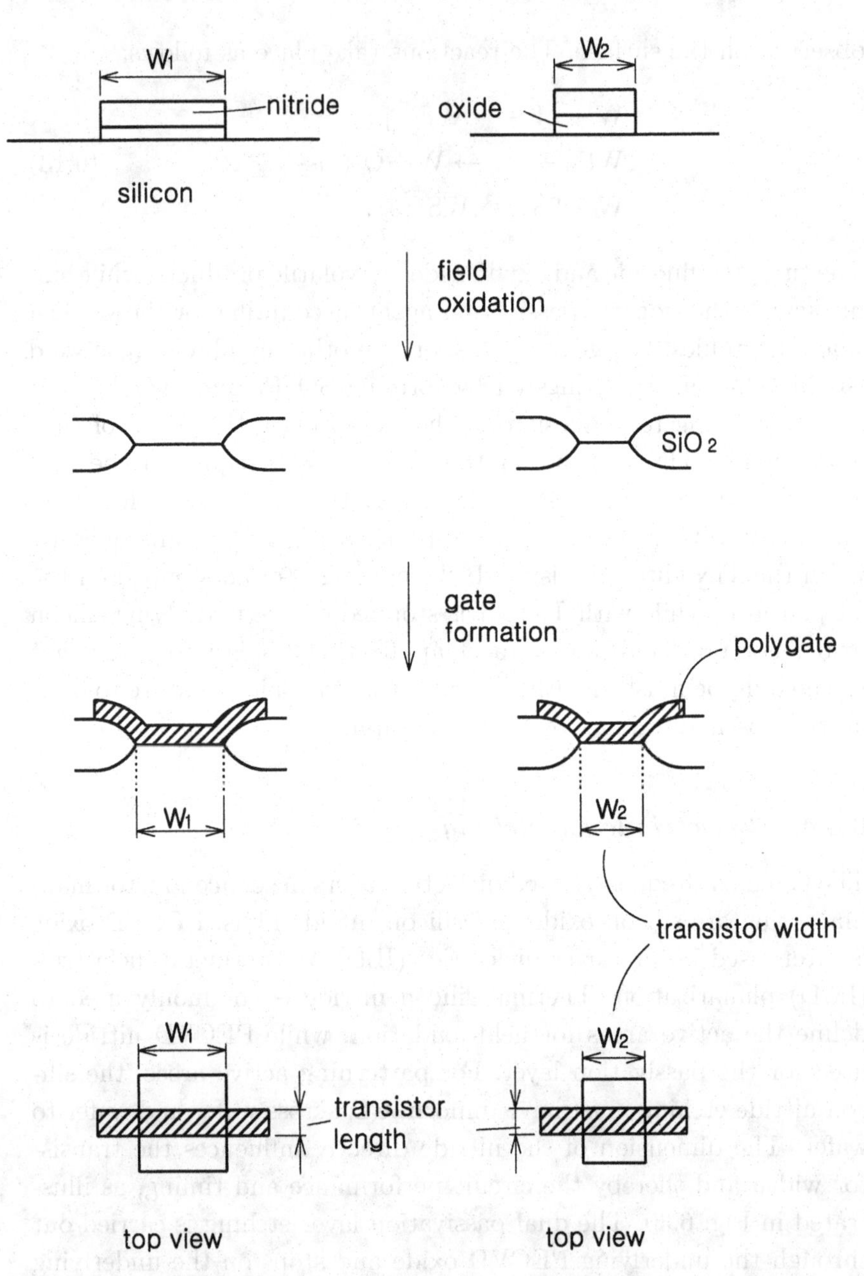

Fig. 6.30. Etching-induced silicon nitride (active area) CD variation can lead to transistor width difference.

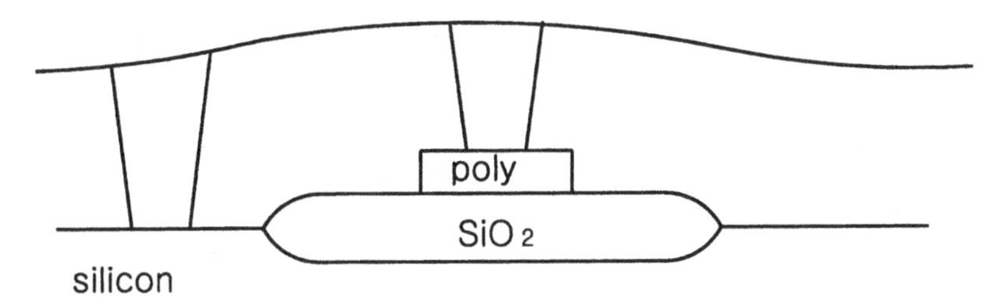

Fig. 6.31. Significant depth difference between the contact hole on silicon surface and that atop the polygate.

An important application of oxide etching is contact or via hole etching. The challenging aspect of this process is that the depths of these holes vary across the wafer owing to the topograph. There can be a difference in depth in the magnitude of several thousand angstroms. Figure 6.31 illustrates the contact holes on top of the silicon substrate and polysilicon on field oxide. The oxide in the deep contacts must be cleared, while the polysilicon in the shallow contacts cannot be overetched, especially when there is silicide on the top. A high etching selectivity of oxide to silicon is needed to meet the requirements. In an RIE etching system with CF_4/H_2, the fluorocarbons, CF_x, formed in the plasma tend to deposit on the silicon and inhibit the etching, whereas they react with oxygen in the oxide and evaporate. These characteristics lead to good oxide-to-silicon etching selectivity. In high-density plasma, such as magnetic enhanced reactive ion etching (MERIE), an extensive ionization exists, and thereby more reactive fluorine radicals are generated, which aggressively attack the silicon. To maintain high etching selectivity, a high carbon-to-fluorine ratio precursor, such as $C_2H_2F_4$, is used instead, which will render a selectivity of higher than 10.

Differently etched contact profiles and sidewall angles are required for different applications. For aluminum sputtering, sloped sidewalls of holes are needed to improve the metal step coverage. However, for advanced technologies, the design requirement for contact to gate edge is very small, and a vertical sidewall is needed, as illustrated in Fig. 6.32. The CD uniformity is always critical because

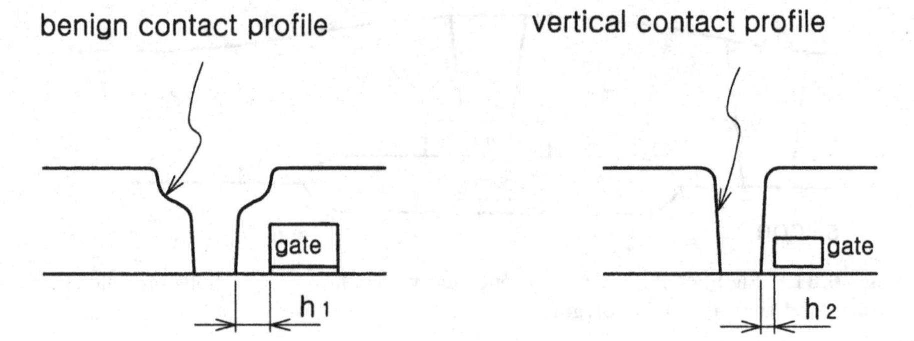

Fig. 6.32. Contact profiles required for Al-sputtering (left) and for W-plug (right).

different hole sizes lead to different contact resistances. This can result in circuit timing mismatching.

6.5.5. *Spacer etching*

One of the oxide etching applications is spacer etching. Oxide spacers are widely used in forming lightly doped drain (LDD) devices to avoid the hot carrier effect. A typical process flow is presented in Fig. 5.27. To avoid device asymmetry and nonuniformity across wafer and wafer to wafer, the lateral dimension and profile of the spacer must be kept constant. The spacer etching must be highly anisotropic so that it can remove the oxide on the horizontal surface without lateral oxide loss on the vertical wall. The anisotropy for this case can be achieved with strong ion bombardment and sidewall polymer passivation. The etching must be able to stop on the silicon surface with negligible loss of the surface silicon. The spacer width is mainly dictated by the oxide thickness.

6.5.6. *Silicided contact etching*

Titanium silicide is often used in advanced technologies to reduce source/drain sheet and contact resistances. In general, the resistances can be reduced by severalfold. For the silicided devices, the silicide must remain intact after the contact etching, or the device performance can be destroyed owing to a drastic increase in contact resistance. For silicide contact etching, a $CHF_3/CF_4/Ar$ gas mixture can be used. The gas mixture attacks the silicide by forming TiF_4. The etching selectivity, oxide versus $TiSi_x$, can be increased by a factor of 2 when the substrate temperature is decreased from 100°C to 70°C. This can be explained by the vapor pressure difference. The vapor pressure of SiF_4, is 760 mTorr at 90°C, while the vapor pressure of TiF_4 is 100 mTorr at 230°C. A decrease in temperature has a more drastic effect on TiF_4 than on SiF_4, which is already volatile. The addition of nitrogen, even by less than 10%, can double the etching selectivity. This is due to nitrogen-containing polymer being built up on the silicide surface, which hampers the silicide etching.

6.5.7. *Aluminum metal etching*

Aluminum interconnects are often doped with Si and Cu to improve the contact integrity and electromigration performance. Such a material can be etched with plasma of BCl_3 or HCl with Cl_2 and other additives. In normal ambient, aluminum is protected with a thin native aluminum oxide film. BCl_3 ion bombardment is often used to break through the oxide layer, followed by chlorine radical etching. A good metal etching process must possess a few characteristics. It should have good CD uniformity, vertical sidewall profiles, good etching selectivity with respect to the underlying oxide, and it should be residue- and corrosion-free. In general, metal etching recipes do not have a very high selectivity to resist. Significant resist loss can be observed after the metal etching is completed. The eroded resist material forms sidewall polymers to ensure good sidewall profiles. During the etching process, the underlying oxide thickness loss must be kept to a minimum since the oxide loss causes the aspect ratio between two neighboring metal lines to increase. Therefore it is more difficult to planarize at a later stage, as illustrated in Fig. 6.19. With Cu-doped aluminum lines, etching residues are often observed after etching, as shown in Fig. 6.33. The residues result from the copper precipitations that are converted to $CuCl_2$. The nonvolatile

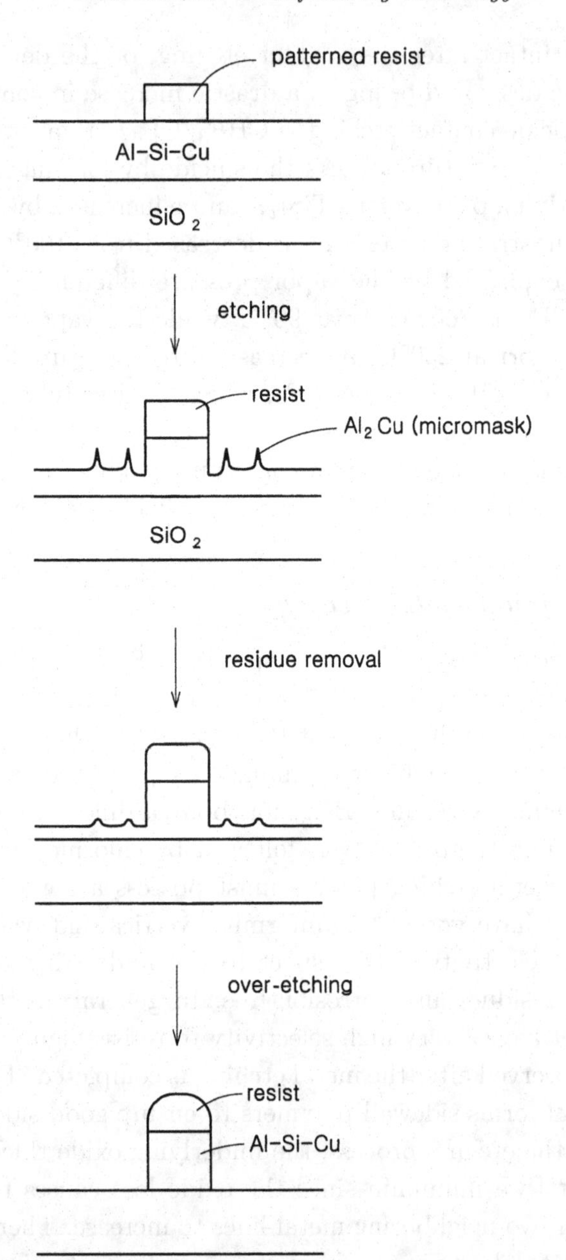

Fig. 6.33. Process sequence for eliminating micromasks in a metal etching step.

etching product, $CuCl_2$, forms micromasks and prevents the underlying material from being etched away. In such circumstances, the ion bombardment effect can be increased, or a recipe with a high lateral etching rate can be used to eliminate these micromasks and to remove the residues. Alternatively, the metal sputtering process can be optimized such that the Cu precipitation can be avoided.

A crucial issue in carrying out the metal etching is to avoid metal corrosion. As mentioned earlier, the eroded resist forms the sidewall polymer. Some of the etching products, such as $AlCl_3$, can be trapped in the polymer. When the etched wafer is exposed to the ambient moisture, the following reaction takes place:

$$AlCl_3 + H_2O \longrightarrow Al(OH)_3 + HCl\,. \qquad (6.14)$$

HCl then attacks the aluminum and causes metal corrosion. The corrosion is further accelerated by the galvanic coupling of Al and Cu. A few approaches are proposed to prevent metal corrosion. An *in situ* polymer stripping with H_2O/O_2 plasma can be used to avoid the effects of ambient moisture. Alternatively, the etched wafer can be exposed to fluorine plasma so that all the trapped chlorine is replaced with fluorine. Despite the fact that gross corrosion can be avoided, some minor corrosion, called mouse-bites, in which some portions of the metal width are lost due to corrosion, is observed sometimes. Once metal corrosion occurs, reliability is a concern.

6.5.8. *Chromium etching*

Chromium has been widely used as a light absorber on photomasks. Its thickness ranges from 500 to 1000 Å. To minimize the chromium surface reflectivity, it is often coated with a thin layer of chromium oxide. There are a few important requirements for the chromium etching. First, it should have a good CD uniformity across the mask because CD errors on the mask can be printed and amplified onto wafers due to mask error factors. The requirement on CD uniformity tightens as the technology migrates to smaller geometries. For example, in general, the CD uniformity requirement for 0.18-μm technology is 40 nm, for 0.13 μm, it is 22 nm, and for 0.09 μm, it is 18 nm. Second, it should have a vertical etched sidewall profile. The

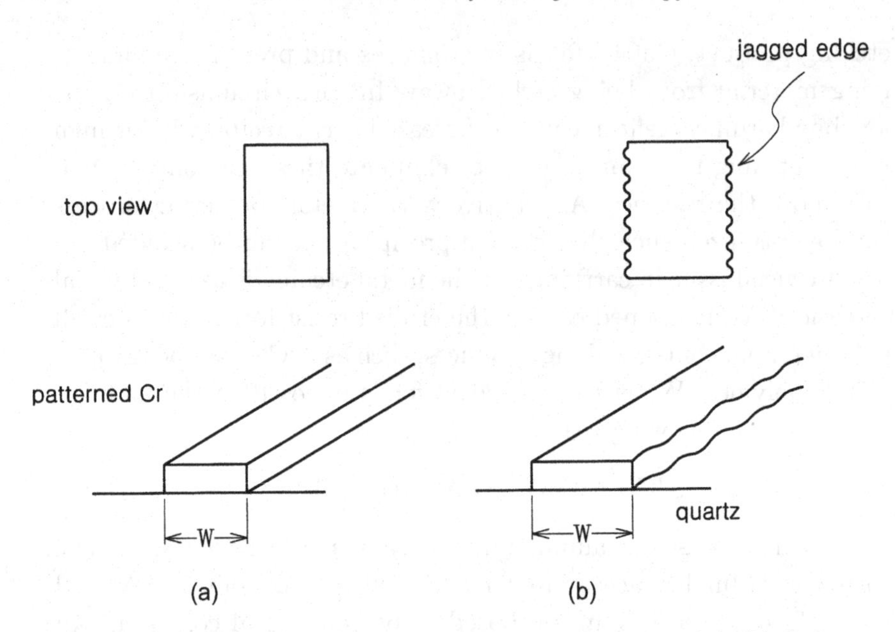

Fig. 6.34. (a) Patterned Cr has a desired smooth line edge while (b) Cr has a jagged line edge.

more vertical the sidewall is, the better the CD control will be. Third, the line edge should be smooth. A jagged edge or nonsmooth edge can result in different printed CDs on wafers, as indicated in Fig. 6.34. Chromium can be etched with chlorine and oxygen. Chromium is hardly etched with chlorine or oxygen plasma alone, mainly because their etching products, $CrCl_3$ and Cr_2O_3, are not volatile; also, their melting points are 1152°C and 2435°C, respectively. However, significant etch rates can be obtained if the plasma of a chlorine–oxygen mixture is used. Figure 6.35 shows that the Cr etching rate is very low with pure chlorine, but then it increases with the addition of oxygen. A maximum point can be reached, and then it starts to decrease again. The etching products are $CrClO$ and $CrCl_2O_2$ when plasma of a chlorine–oxygen mixture is used.

6.5.9. *Resist ashing*

Almost all etching processes end with the plasma ashing process, in which the photoresist (hydrocarbon) is removed with an oxygen

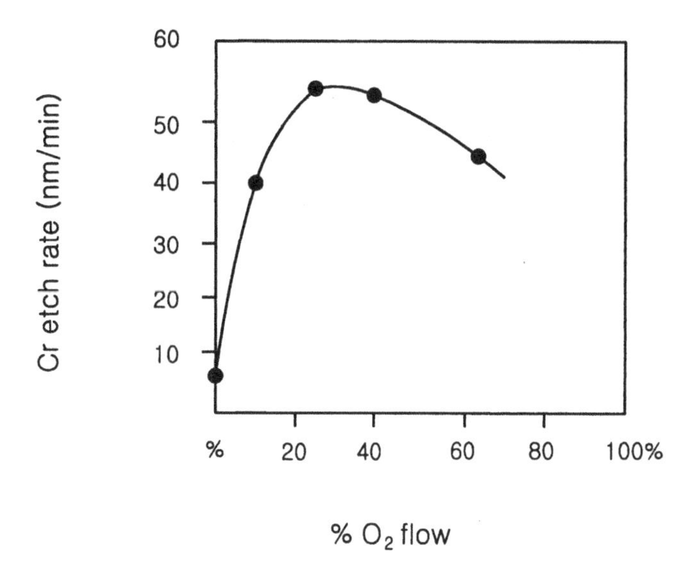

Fig. 6.35. Chromium etching rates plotted against oxygen concentration in feedstock (O_2 and Cl_2).

plasma:

$$C_nH_yO_z + O \longrightarrow CO + CO_2 + H_2O\,. \tag{6.15}$$

The resist ashing or stripping step does not require anisotropy as long as it removes the resist uniformly. Generally, for resist ashing, CF_4 is added to accelerate the resist etch rates. The effect of adding CF_4 is that the dissociated fluorine atoms can adsorb on the resist surface and modify its structure so that the etch rate is enhanced. Excessive CF_4 causes the etch rate to decrease because the abundant CF_2 radicals start to compete with oxygen radicals during the surface adsorption and reactions.

6.6. Plasma-Enhanced CVD and Etching Reactor Modeling

Plasma-enhanced etching and CVD often share the same reactor designs. Figure 6.36 shows a Reinburg-type plasma reactor, where the substrate holder is supported from a rotating shaft from the bottom, and the substrate table rotates during processing. Power is fed

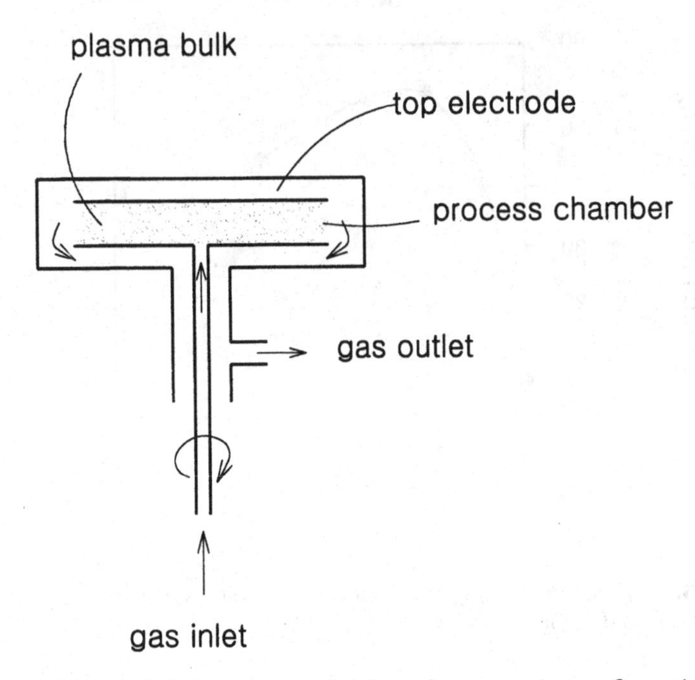

Fig. 6.36. A Reinburg-type radial flow plasma reactor configuration.

from the top electrode. During operation, the substrate table rotates to achieve good uniformity, and the gas feedstock can be fed from the center or from the periphery to form centrifugal and centripetal flows, respectively. Plasma is ignited in the process chamber as the gas feeds in. Another popular type of reactor is the barrel reactor, as shown in Fig. 6.37. Wafers are mounted on a tilted substrate holder. The tilted surface and chamber form a gas flow channel that decreases in size as gases flow toward the exit, so as to increase the mass transfer rates and to compensate for the depletion effects. The plasma is formed inductively from the coils around the reactor. The above two reactors belong to the batch type; in particular, a number of wafers can be processed at the same time. For advanced technology, the single-wafer reactor has been popular as a result of an ever-increasing demand for uniformity. Figure 6.38 displays a high-density plasma system, a magnetic-enhanced plasma reactor. The gas is fed through the top of the reactor. Microwave plasma is ignited with a 2.45-GHz source, as gases travel downward. The microwave frequency couples

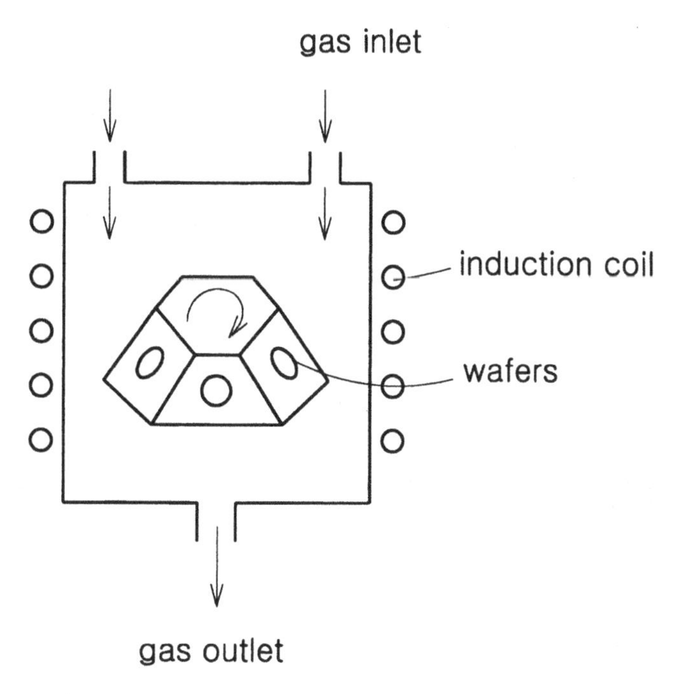

Fig. 6.37. A typical barrel-type plasma reactor with an induction coil for heating.

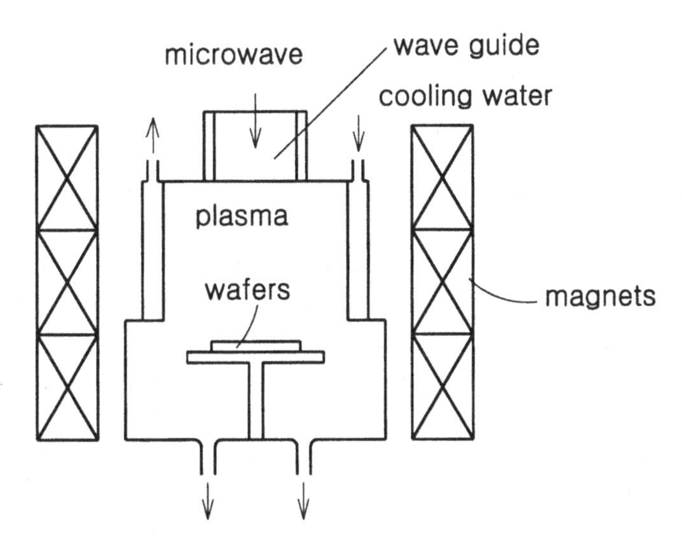

Fig. 6.38. A typical ECR plasma reactor.

with the magnetic field, and as a result, the plasma is enhanced in terms of plasma density. The wafer is placed under the plasma bulk.

Plasma-enhanced CVD and etching are processes with opposite goals. For deposition, a nonvolatile film is formed, while for etching, volatile reaction products are expected. There is no comprehensive model that allows the prediction of the deposition or etching behavior by knowing the system's operation conditions. This is mainly due to the fact that some fundamental aspects of these processes are still unknown. The modeling of the plasma-enhanced deposition or etching processes are very similar to those of thermal CVD processes, except for the involvement of the electron energy and density functions. A comprehensive model for these processes includes several balance equations. The Boltzmann equation accounts for electron density and energy distribution. The mass transfer equation provides the concentration profile in a reactor as well as the local deposition or etching rates. The momentum balance equation is required for obtaining the flow field. The energy balance provides the temperature profile in a reactor. Finally, a continuity equation and the surface reaction kinetics expressions must be added to complete the reactor model. Basically, the Boltzmann equation correlates the system electrical inputs, such as power and frequency, to the electron density and energy distribution. These two parameters provide the precursor gas dissociation, ionization, and excitation rate constants. The mass balance equation, energy balance equation, and momentum equations are coupled through physical parameters that are functions of temperature and gas composition. Although it is possible to tackle all the coupled equations at the same time, it is more practical to simplify the model. It is often assumed that one can decouple the Boltzmann equation from others and that the electron energy distribution is of Maxwellian or Druyvestyne type. Then, the electron energy and density can be expressed in terms of power input and system pressure, as discussed in Chapter 4. With the obtained electron energy and density, one can then estimate the species dissociation rate constants by knowing the related electron impact cross sections. One can further assume that the heat of reactions and the power input are negligible; in other words, the system is isothermal. This assumption leads to a simplified model, as follows:

$$\begin{aligned} &\nabla \bullet \rho\vec{V} = 0\,, \\ &\nabla \cdot \rho\vec{v}\vec{v} = -\nabla p - [\nabla \cdot \tau] + \rho\vec{g}\,, \\ &\nabla \cdot (\rho\vec{v}\omega_i) = \nabla \cdot (\rho D_{im}\nabla\omega_i) + \theta_{ij}r_j\,, \end{aligned} \tag{6.16}$$

where ∇ is the gradient operator, $\nabla\bullet$ is the divergence operator, $\vec{v}$ is the gas flow velocity, ρ is the gas density, τ is the shear force, p is the system pressure, D_{im} is the effective diffusivity of species i in the gas mixture, ω_i is the species fraction, θ_{ij} is the product of the stoichiometric coefficient and molecular weight ratio of a radical i to its precursor j, and r_j is the dissociation rate of precursor species j.

For gas flow, the nonslip boundary condition is assumed for the reactor wall; a fully developed flow profile can be used at the exit. For species balance, Danckwert's-type boundary conditions can be used for the inlet and outlet. However, for the reactor walls and the substrate surface, there are several approaches. First, assume that the surface etching or CVD reaction is species transport–limited. In other words, the species are consumed as soon as they arrive at the surface. Therefore the species surface concentration is zero, and the result is

$$\text{Surface boundary conditions: } \omega_i = 0.$$

Second, one can assume that the surface reaction rate equals the radical collision rate times a reaction probability:

$$-\rho D_{im}\frac{\partial\omega_i}{\partial y} = \gamma\sqrt{\frac{RT}{2\pi M_i}}\rho_i\omega_i\,, \tag{6.17}$$

where γ is the reaction probability of each radical that hits the surface and gets reacted. Third, one directly uses the surface reaction kinetics expressions in literature as the surface boundary condition. For example, if simple first-order reaction kinetics is assumed, then

$$-\rho D_{im}\frac{\partial\omega_i}{\partial y} = K\rho_i\omega_i\,.$$

After solving for the concentration profile, the surface deposition or etching rate can be expressed in terms of the concentration gradient

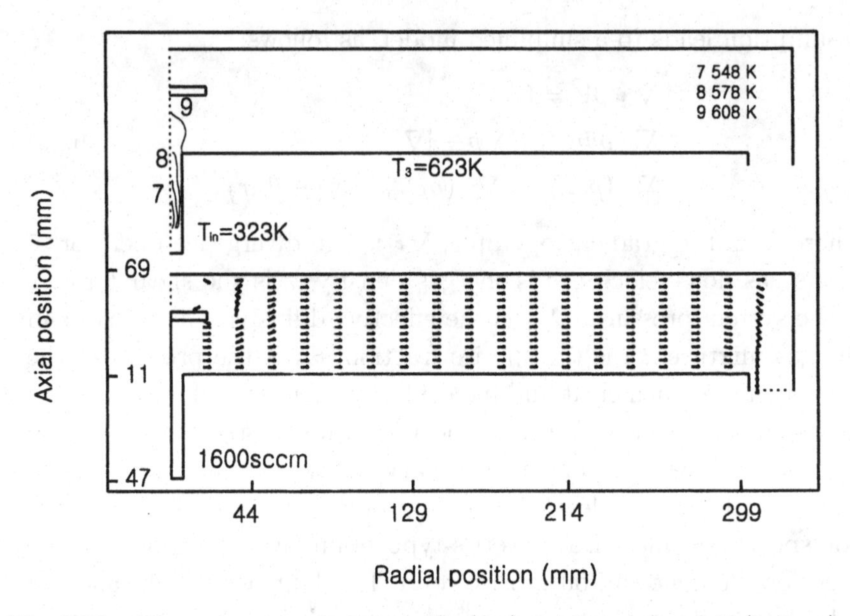

Fig. 6.39. The system temperature profile (top) and the gas flow field (bottom).

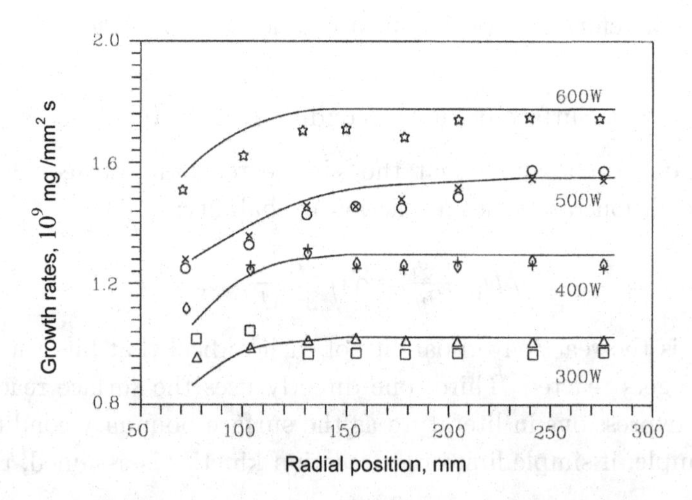

Fig. 6.40. Silicon nitride film growth rates at various radial positions. The marks indicate repeated experimental data; lines are model predictions with an electron temperature of 6.98 eV and K of 23.9 (Torr/k·W·cm^3).

at the surface:

$$-\rho D_{im}\frac{\partial \omega_i}{\partial y}\frac{1}{\rho_f} = \text{Deposition rate or etching rate}\,,$$

where ρ_f is the film density. The above model, assuming nonisothermal and Druveystine electron energy distribution, is applied to a Reinburg-type silicon nitride deposition system. Figure 6.39 shows the system temperature profile. It can be obscrvcd that the isothermal assumption is reasonable since the inlet gas is heated up to the substrate temperature level soon after passing the gas inlet. The velocity profile is smooth, except for the small region above the inlet gas baffle plate. Figure 6.40 illustrates that the actual deposition rates confirm the theory at different power inputs. In the above model, the effects of ion bombardment have been neglected.

Chapter 7

PATTERN TRANSFER: PHOTOLITHOGRAPHY

A good photolithography technology is the prerequisite for device manufacturing and technology migration since it enables circuit patterns to be transferred from masks to wafers. A photolithography process involves several steps, which are divided into three blocks: preexposure, exposure, and postexposure. Section 7.1 begins the chapter with an introduction to photolithography technology. Section 7.2 illustrates the HMDS treatment, resist coating, and soft-baking processes. Resist chemistry is the focus of the third section, 7.3, which explains resist chemistry and its evolution to meet requirements that have accompanied technological advances. Section 7.4 prepares readers with a few fundamental imaging-related principles for understanding the exposure system and printing requirements. It covers the fundamental principles of imaging such as Snell's law, interference, and other important principles in photolithography. It also illustrates resolution and depth of focus of an imaging system in terms of lens numerical aperture and wavelength. Section 7.5 is devoted to the introduction and evolution of industrial exposure systems and the fundamental requirements of photolithography in semiconductor manufacturing. Finally, Section 7.6 illustrates the last three steps in photolithography: post exposure bake, developing and hard bake.

7.1. Introduction

The key element of photolithography is the imaging step with an exposure system. With the exposure system, the mask pattern is imaged onto a resist-coated wafer surface as a latent image, similar to

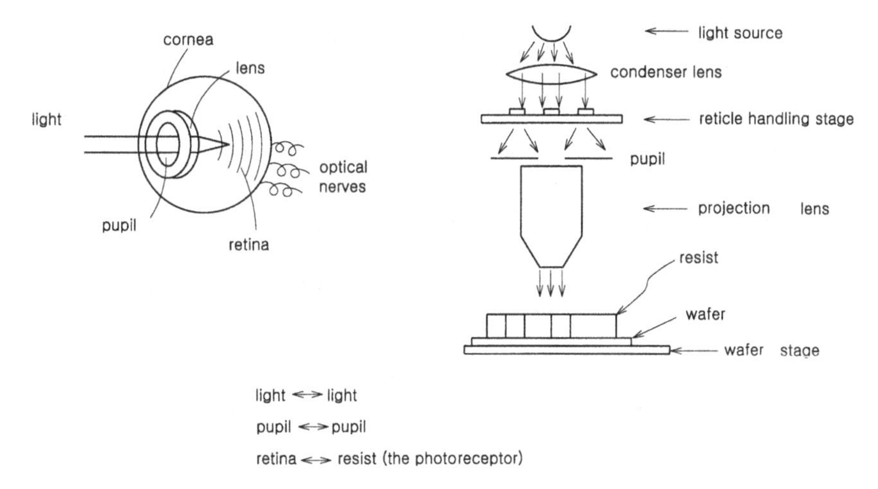

Fig. 7.1. The analogy between a human eyeball and an exposure system.

the way the human eyes work. This process is illustrated in Fig. 7.1. Photolithography is related to the semiconductor industry in the same way that gasoline is related to a car engine: a car cannot run without gas. The semiconductor industry cannot sail forward without photolithography technology. The capability or limitation of each technology node is essentially defined by photolithography.

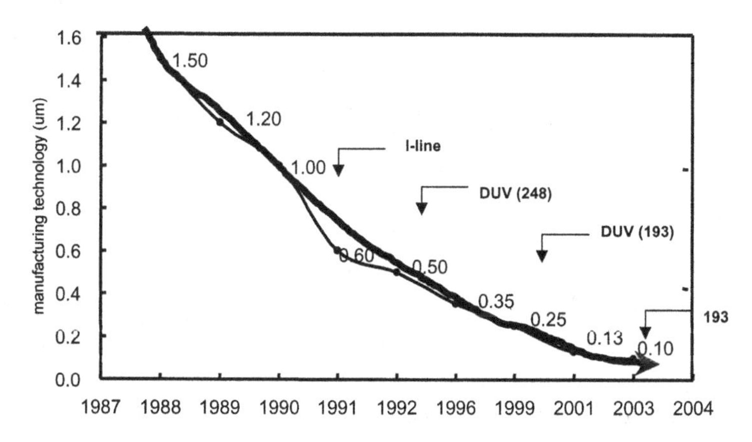

Fig. 7.2. The road map of photolithography technology and manufacturing technology. After introduction of each wavelength, there are several enhancements, such as an increase in NA or adding OPC or phase-shifting masks, etc., to reach each desired manufacturing technology.

Figure 7.2 shows the technology advances over time with various photolithography technologies. The fundamental and salient change is that the exposure wavelength decreases with advancing technology. The wavelength change brings about resist and pellicle material changes, among others. In semiconductor processing technologies, photolithography can best illustrate the interdisciplinary nature of the semiconductor industry. It heavily involves a knowledge background in optics, polymer material, and photochemistry. Photolithography and etching are often used in sequence. The former transfers a layer pattern from a mask to the photoresist on the wafer; the latter engraves the resist patterns into the underlying dielectric or conducting film on the wafer surface, as shown in Fig. 7.3. For example, such a process could be used to selectively remove some portions of an oxide film to form contact hole patterns, which connect a lower-level metal to an upper-level metal. Another example is to use the process to selectively remove some portions of a blanket aluminum film to form proper metal connections, as shown in Fig. 7.4.

7.2. Preexposure Steps

For the mask patterns to be transferred to the resist on the wafer surface, there are several preexposure steps needed to properly prepare the resist for the exposure as well as postexposure steps to develop the resist for the desired patterns. Figure 7.5 shows a process flow, including all the preexposure and postexposure steps.

7.2.1. *Priming*

In a normal manufacturing environment, a wafer would absorb moisture on its surface. Moisture can interfere with the adhesion between the wafer and the resist to be applied. A dehydration bake at around 100°C is intended to eliminate wafer surface moisture so as to enhance adhesion of the resist.

The silicon oxide surface tends to be hydrophilic. In some situations, water molecules readily adsorb on the surface; in others, silanol groups exist on the oxide surface, as shown in Fig. 7.6.

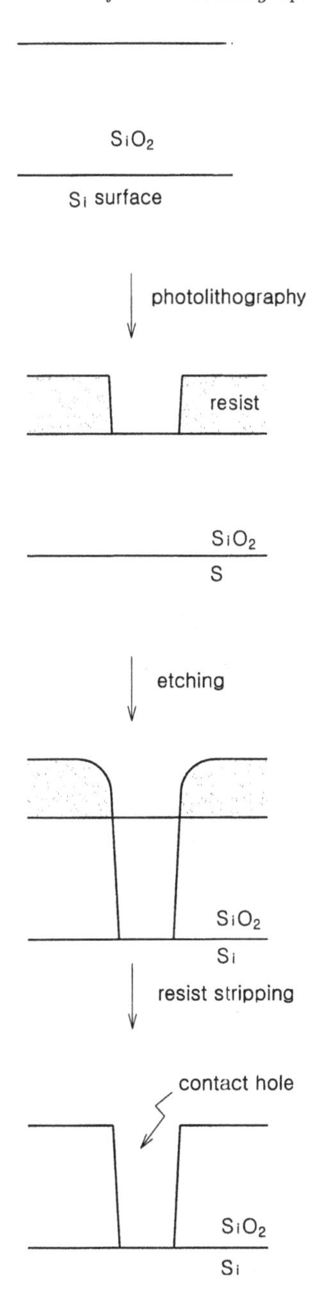

Fig. 7.3. Photolithography and etching are used in sequence to selectively remove oxide to form contact holes.

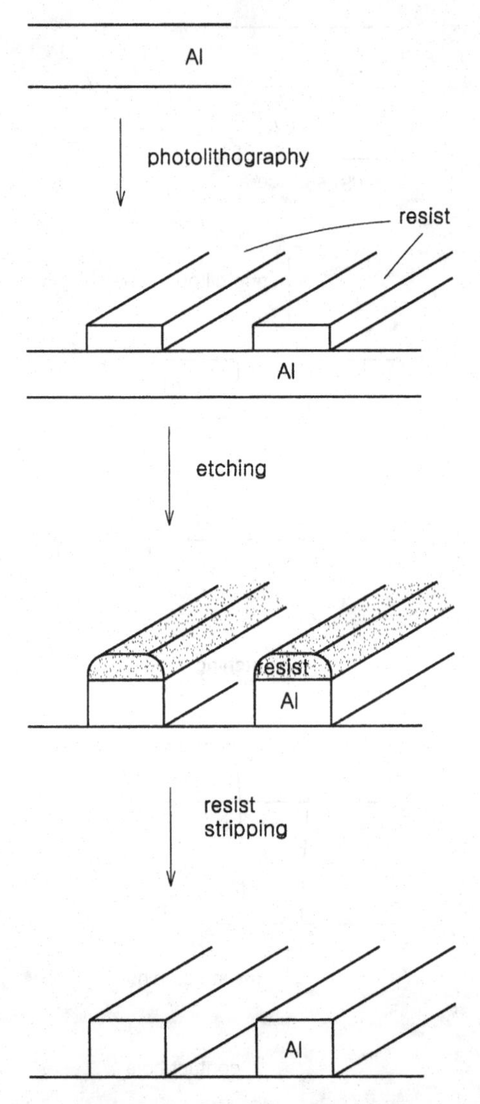

Fig. 7.4. Photolithography and etching used in sequence to form Al line/space patterns.

Priming prepares the dehydrated oxide surface for resist coating. The purpose of priming is to eliminate surface-adsorbed water and silanol groups, hence promoting resist adhesion. In a closed chamber, the vapor of hexamethyldisililazane (HMDS) is introduced onto

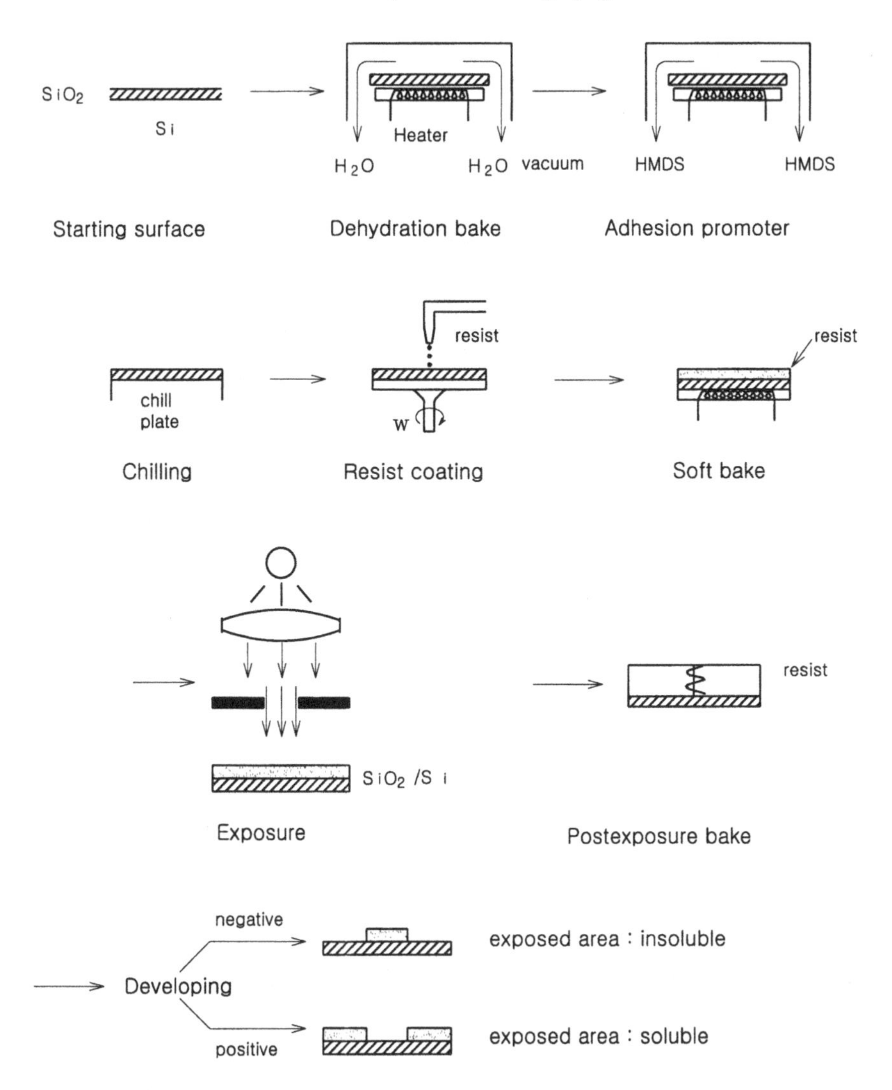

Fig. 7.5. Process steps involved in wafer photolithography.

the wafer surface. HMDS is an organosilane having several methyl groups attached to each molecule, as shown in Fig. 7.7. On reaching the oxide surface, HMDS reacts with the surface-adsorbed water or with the silanol groups, replacing them with trimethyl groups, as demonstrated in Fig. 7.8. The bulky trimethyl groups provide more Van der Waals bonding sites and repel moisture from the oxide-resist

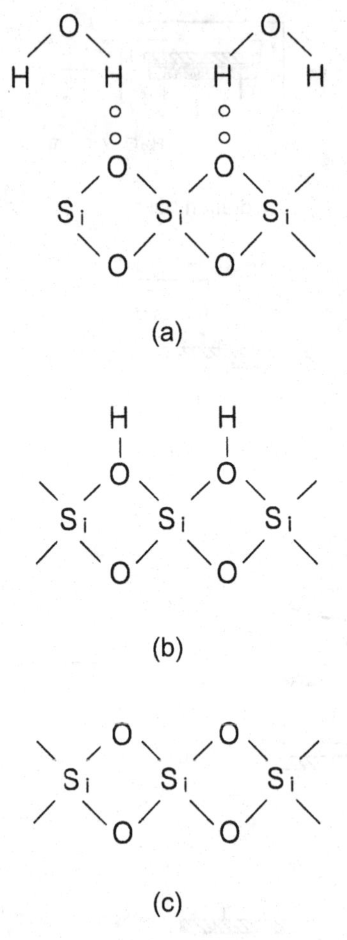

Fig. 7.6. The often seen silicon oxide surface: (a) with adsorbed water, (b) with silanol group (Si-OH), and (c) with siloxane.

```
        CH3        CH3
         |          |
H3C — Si — N -- Si — CH3
         |    |     |
        CH3   H    CH3
```

Fig. 7.7. The molecular structure of HMDS (hexamethyldisilazane).

(1) HMDS reacts and removes surface water

$$(CH_3)_3\text{-Si-NH-Si-}(CH_3)_3 + 2H_2O \longrightarrow 2\,\underbrace{(CH_3)_3\text{-SiOH}}_{\text{Trimethylsilanol}} + NH_3$$

(2) HMDS reacts with surfsce silanol group

$$\text{Si(-OH)(-OH)(-O-)}_2 \ldots \xrightarrow{\text{HMDS}}$$

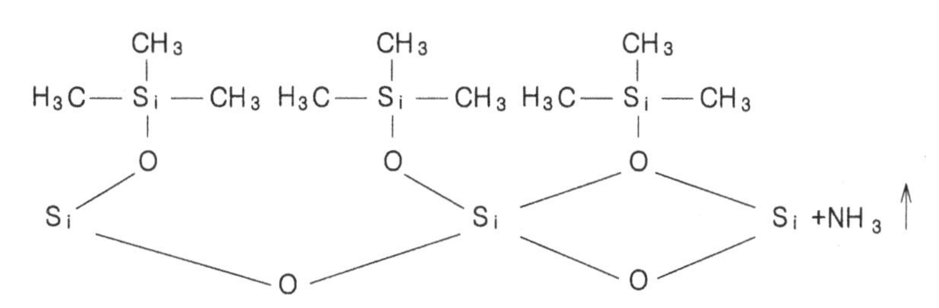

Fig. 7.8. HMDS reacts with various silicon surfaces, forming a large number for methyl ($-CH_3$) groups and enhancing resist adhesion.

interface; as a result, the adhesion of the resist to the oxide surface is promoted. Other substrate surfaces, such as metal or polysilicon, tend to be passivated with native oxides in magnitudes of tens of angstroms, and they also need the HMDS priming step to promote resist adhesion.

7.2.2. *Resist coating*

After HMDS priming, the resist is spin-coated on the oxide surface. To ensure good photolithography printing results, the resist coating must fulfill the following requirements:

(a) *Good thickness uniformity.* The effects of resist thickness nonuniformity on device processing are twofold. First, from a lithography point of view, the CD of lithography printing depends on the local resist thickness. Hence control of the uniformity of the resist thickness is extremely critical. Second, from a substrate protection viewpoint, in the etching process following the photolithography, one must have a uniform resist thickness across the wafer. Otherwise, when an extended etching is required, the substrate under the thin resist area may be damaged. Uniformity of the resist thickness strongly depends on such factors as air flow in the coater, spin speed, acceleration, ambient temperature, and humidity. There is no fixed formula to guarantee a uniform resist thickness, but an educated trial-and-error approach always works.

(b) *Free of pinhole.* The resist pinholes most likely result from the development of gas bubbles during coating. Once a pinhole is formed, image quality can be totally destroyed. Furthermore, the pinhole areas provide no protection at all for subsequent dry or wet etching. Thus they get transferred onto the underlying substrates.

(c) *Free of defects.* The postcoating defect count on the resist surface decides the success or failure of this photolithography process because any defects can show up on the printed wafers. The resist spray nozzles and piping must be maintained periodically. Defect control in the coating area is also very critical.

Resist coating constitutes a very complicated process of hydrodynamic modeling. As the wafer spins, the resist experiences a centrifugal force; in the meantime, there exists a shear force owing to the velocity gradient in the vertical direction. The higher the viscosity is, the higher the shear stress will be. Furthermore, because the solvent continues to evaporate, resist viscosity and density increase. This is a very complicated model of non-Newtonian fluid flow. The model solution gives the resist thickness uniformity across the wafer. Owing to the complexity of this microscopic hydrodynamic model, and the difficulties for obtaining an exact solution, a few empirical formulae have been proposed for estimating the average resist thickness for spin coating:

$$\delta = \frac{kS^2}{\sqrt{\omega}}\,. \tag{7.1}$$

In this formula, δ is the resulting averaged resist thickness, k is a constant related to resist viscosity and coater system, S is the solid content, and ω is the angular velocity of the coater. This macroscopic model sheds no light on the uniformity of the resist, but it does give clues on how to tune the process for a desired thickness. It is obvious that once a resist is chosen, the variable that can be altered to obtain a desired resist thickness is the coater's spin speed. On the other hand, if the resist performance is acceptable, viscosity can be changed so as to achieve the desired thickness within a range of operable spin speeds.

Coater spin speed ranges from a few hundred to a few thousand rpm. At such speeds, some of the resist applied onto the wafer surface is actually wasted. The resist coating is essentially a very expensive step. The desired resist thicknesses are actually determined by a few factors. A thick resist is one way of assuring that the wafer is pinhole-free. On a severe topography, a thicker resist is often required to cover all extruded areas. The resist thickness must be thick enough to endure an etching process step. Furthermore, a thick resist is needed for an implant application, where the resist is supposed to block ions in areas where implantation is not intended. Nonetheless, for resolution considerations, thin resists are desired, as resolution decreases with resist thickness. Normally, the resist thickness range is first chosen by applications and then fine-tuned with swing curves to the ultimate thickness.

7.2.3. *Soft baking*

After the resist is coated on wafers, a soft baking is carried out to drive out most of the solvent to render resist photosensitivity. Normally, less than 3% of the solvent is left in the resist after soft baking. For advanced device manufacturing lines, soft baking is conducted on a hot plate at around 85°C to 90°C for about 90 to 120 s. Soft baking is a very critical step as it determines not only the exposure conditions, but also resist adhesion on the wafer surface and development rates.

The development rate of the exposed resist depends highly on the soft baking conditions. An underbaked resist is more prone to developer attack. It is therefore often misinterpreted as high photospeed, but in fact, the CD uniformity can be severely degraded. On the other hand, it has been shown that among many factors, such as the HMDS priming, soft baking, and postexposure baking, soft baking is considered to be the most significant factor in terms of affecting resist adhesion. Resist adhesion, as shown in Fig. 7.9, is indicated by the ratio of the lateral etching rate to the vertical etching rate. The higher the ratio is, the worse the adhesion will be. This is because lateral etching is manifested by the etchant penetrating through the resist and oxide interface, if the adhesion is poor. Soft baking must be optimized in terms of process latitude and the resist adhesion. On the basis of the optimized soft baking conditions, the exposure conditions can then be set.

7.3. Resist Chemistry

Resist is one of the key elements that dictate wafer printing quality. Resists are polymers that consist of three major components: resin, photoactive compound (PAC), and solvent. There are two types of resist: positive and negative. The major difference lies in the solubility difference between the exposed and the unexposed portions of the

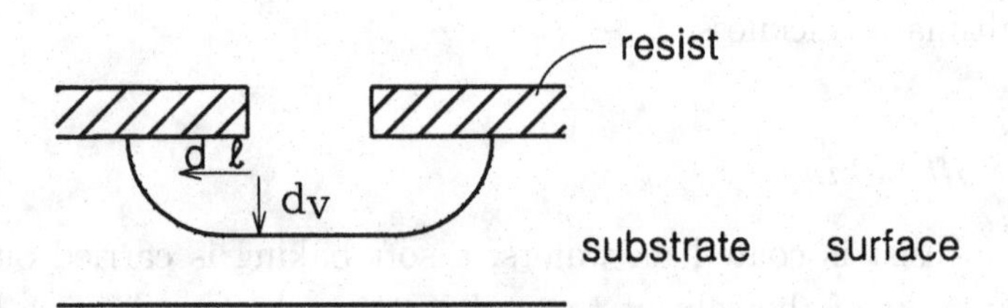

Fig. 7.9. Evaluation of resist adhesion with *dl/dv* (lateral to vertical etching ratio) ratio.

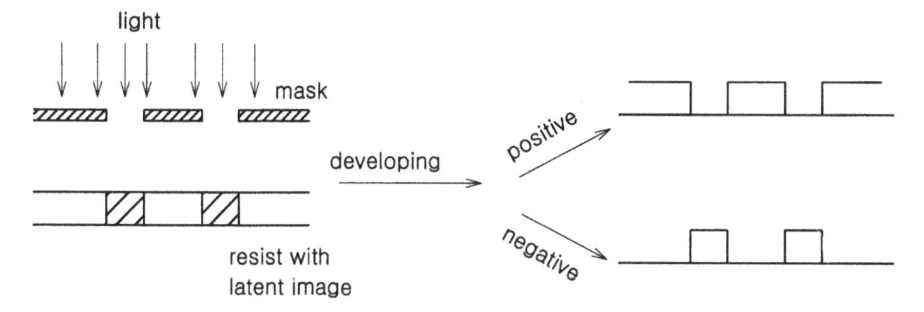

Fig. 7.10. The comparison between positive and negative resists.

resist, as shown in Fig. 7.10. An exposed positive resist turns soluble in a base solution, whereas the unexposed portion of a negative resist is soluble in a developing solution, and the molecules in the exposed areas cross link to form larger molecules and hence become insoluble in developing solutions. In general, positive resists give better resolution and CD control. Positive resists prevail in wafer processing, while negative resists are widely used in high-end (beyond 0.13 μm) mask manufacturing, where the exposed area is around 20% to 30% of the total area. A good photoresist should possess the following characteristics:

(a) *High absorption at the exposure wavelength.* One of the approaches to improve photolithography resolution is to use a shorter-wavelength light source, as will be discussed in later sections. The resist material used must be matched to the ever-decreasing wavelength. Its absorption peak must coincide with the emission characteristic wavelengths of the light source to ensure a high-throughput exposure step.

(b) *Good resolution.* A resist material has its own generic resolution at a specific wavelength. Resist resolution is strongly related to its contrast value, which is defined as the slope of the curve representing the remaining thickness versus dosage, as illustrated in Fig. 7.11. The larger the slope, the higher the contrast, and the better the resolution. One good approach to evaluate resist resolution is to check its printing on hole patterns. The patterns consist of various hole sizes, ranging from above to below the target design rule. After wafer

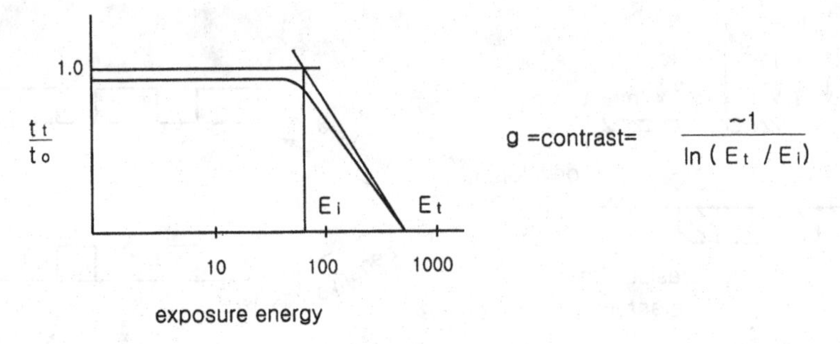

Fig. 7.11. Definition of resist contrast (t_o, t_t: initial and final thickness).

printing, measure the hole size and check on the corner rounding, and then plot the dimensions on the mask versus those on the wafer. The resulting correlation should be a straight line and close to the diagonal line, as shown in Fig. 7.12. The point at which the wafer printing size (point B) starts to deviate from the diagonal indicates the resolution limit. This approach can be used to compare the resolution limits of different resists as a part of the selection procedure.

(c) *Good photosensitivity.* In evaluating the sensitivity of resists, a fixed thickness of resist is often used. The exposure dosage is then varied to measure the remaining resist thickness after developing.

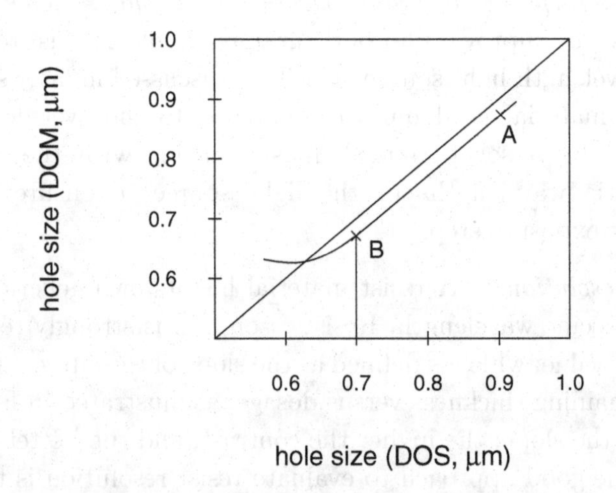

Fig. 7.12. An evaluation of resist resolution limit by plotting hole size of DOM (dimensions on mask) versus DOS (dimensions on silicon).

Sensitivity is defined as the dosage needed to remove a fixed thickness of the resist after developing. The photosensitivity is directly related to photolithography throughput. A high-sensitivity resist requires a low dosage to clear out the exposed area; hence a short exposure time is needed.

(d) *Good thermal stability and mechanical strength.* Resist thermal stability and mechanical strength are dominated by characteristics of the resin material. Good thermal stability is critical in the baking step and in plasma etching. During baking, the temperature is around 100°C for tens of seconds. In plasma etching, the resist could be under severe ion bombardment, and wafer temperature can go up significantly. A poor thermal stability could result in resist profile deformation, as shown in Fig. 7.13, which causes CD variations across the wafer with respect to the target CD. Resist is often used for blocking ion implantation. The mechanical strength of the resist must be strong to endure implantation without being destroyed or sputtered off, especially for high voltage and high current implantation.

(e) *Good dry etching resistance.* In plasma etching, it is essential to define etching selectivity, i.e., the ratio of the substrate etching rate to the resist etching rate. To ensure a faithful pattern transfer from the resist to the underlying substrate, a high-selectivity etching process is needed, meaning that the resist does not have significant erosion after etching. A poor resist profile evolution with etching time is illustrated in Fig. 7.14. Resist erosion due to etching can

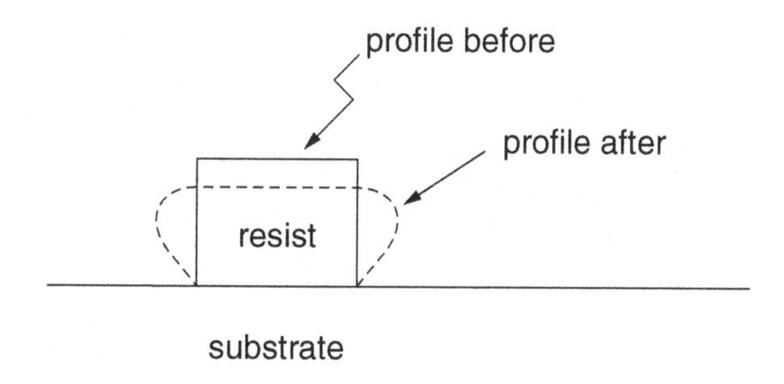

Fig. 7.13. The resist profiles before and after excessive high-temperature treatment.

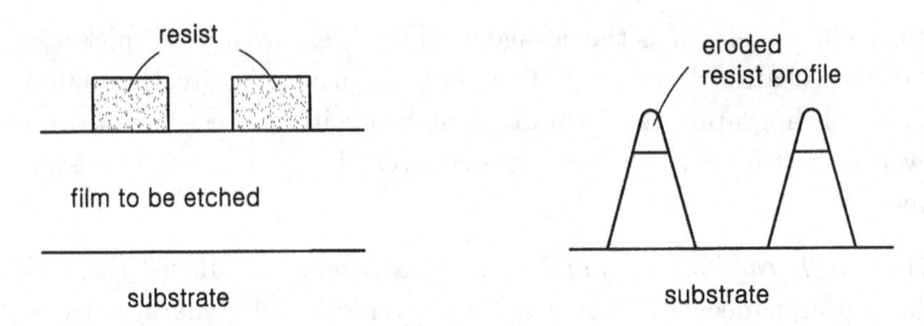

Fig. 7.14. An eroding resist leads to a poor after-etching film profile.

certainly affect the resulting CD values. A good dry etching resistance is essential to ensure a faithful pattern transfer.

Nowadays, advanced wafer manufacturing predominantly uses positive resists. G-line or I-line process resists use novolac resin. These resins provide the resist with etching resistance, mechanical strength, and thermal stability. Together with solvent, the resin renders film-forming characteristics. Novolac resin is shown in Fig. 7.15. The resin is mainly polymeric, with molecular weights ranging from 1000 to 10,000. The resin molecular weights are designed to be relatively low to ensure good solubility in developer in the subsequent developing step. More important, the light absorption of the resin should be as low as possible at the desired exposure wavelength, leaving photons to be absorbed by the PAC, so that the light can reach the bottom of the resist. The photoactive compound is diazonaphthoquinone (DNQ) or its derivatives, some of which are shown in Fig. 7.16. The compound's main feature is the C=N functional

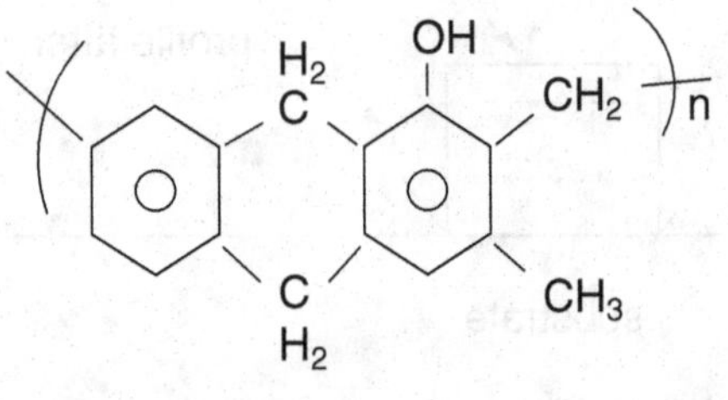

Fig. 7.15. The novolac resin as used in G- and L-line resists.

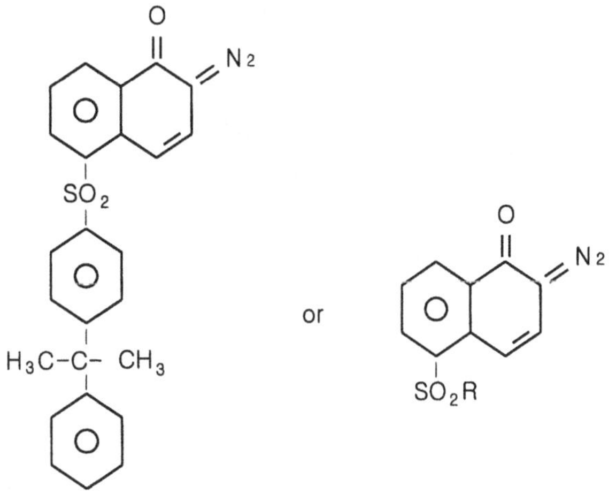

Fig. 7.16. The molecular structure of the PAC, diazoquinone. The photoactive portion is above the SO_2.

group. On receiving photons, it releases nitrogen molecules and converts to ketene, which then absorbs moisture and converts into carboxylic acids, as demonstrated in Fig. 7.17. The acid is soluble in an alkaline solution.

The fact that an exposed positive resist is soluble and the unexposed resist is insoluble can be explained as follows: Before the PAC is dismantled by photons, although resin is soluble in alkaline solution, its dissolution rate is significantly inhibited or blocked by the PAC, which is highly hydrophobic. One of the inhibition mechanisms is that the sulfonate group on the unexposed DNQ provides hydrogen bonding sites to the phenolic hydroxyl groups on novolac resins. The greater the hydrogen bonding, the lower the solubility will be. When the resist is exposed to light, the DNQ turns into carboxylic acid, losing its hydrophobicity and rendering the resist, as a whole, soluble.

As photolithography technology advances, the exposure wavelength must be decreased to keep up with the ever-increasing high resolution demand. As the wavelength shortens, the first hurdle for photolithography to overcome is the search for an appropriate resist.

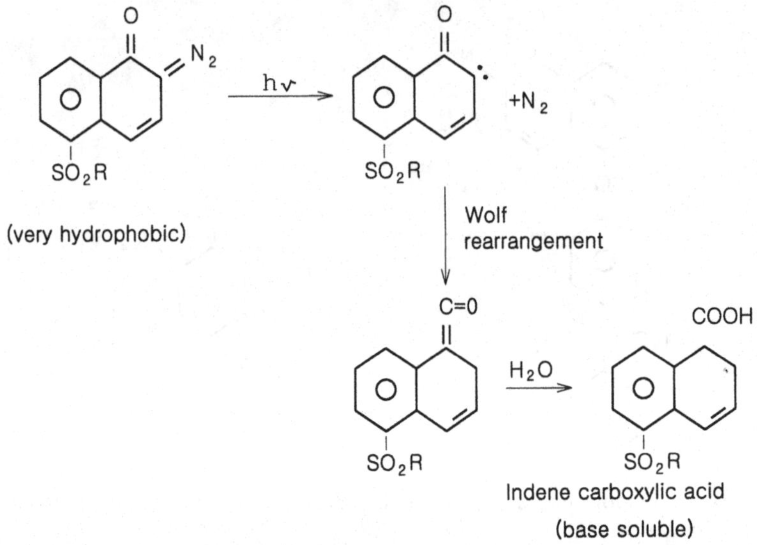

Fig. 7.17. PAC is converted to carbene radicals and releases nitrogen after absorbing photons.

For PAC, adding different substituents or changing the functional group bonding sites can modify the resist absorbance at different wavelengths. Quantum efficiency is often used to evaluate the PAC. It is defined as the number of PAC molecules converted by each photon received:

$$\eta = \frac{N_{cv}}{N_p}, \tag{7.2}$$

where N_{cv} is the number of PAC molecules converted and N_p is the number of photons received. For G- and I-line resists, the quantum efficiency ranges from 0.3 to 0.8; for chemical amplified resists, quantum efficiency can be much larger than 1.0 due to acid catalysis of the photochemical reactions.

Resins are derivatives of aromatic compounds. The organic solvents in the resist, such as xylene and n-butyl acetate, should be able to mix the resin and PAC into a homogenous phase. Solvents give the resist film-forming characterization and viscosity. Most of the solvents will be evaporated during soft baking.

Negative resists prevailed in the early stage of the semiconductor industry in the 1960s and 1970s, when photoresolution requirements were not as stringent as they are today. The resin in the exposed area cross links, reducing its solubility in the developer solution; meanwhile, the resin in the unexposed area readily dissolves in the developer. The photoactive compound is an arylbisazide; on receiving photons, will decompose into nitrene, which then releases nitrogen molecules. The highly reactive nitrenes cross link to form polyisoprene, as demonstrated in Fig. 7.18. The cross link, or polymerization, causes the molecular weights to increase, rendering the compound insoluble. Aromatic compounds are often used as solvents in negative resists.

As technology evolution reaches beyond 0.25 μm, a deep UV (DUV) light source ($\lambda < 300$ nm) is required to print the quarter micron features. The novolac-type resins and DNQ photoactive compounds no longer meet the requirements. The resin is excessively light absorbing, and the DNQ takes several photons to generate one useful product, resulting in low photosensitivity. The industry then turns to look for chemically amplified resists. In DUV resists, photons initiate reactions in photoacid generators (PAG) to release acid in the exposed area. In the subsequent bake, the acid reacts with resins catalytically, changing their solubility and regenerating more acids. A large number of chemical reactions can result from the absorption of a photon, as illustrated in Fig. 7.19. This catalytic reaction gives rise to very high quantum efficiency and photospeed resist materials. Figure 7.20 shows an example of the tBOC DUV resist, in which the protons are generated from PAC as a result of receiving photons. Then, the protons cleave the pendant group in the hydrophobic or base-insoluble PBOCST, converting it into a base-soluble or hydrophilic poly-(4-hydroxystyrene) (PHOST). Another proton is regenerated as a result. The reactions carry on catalytically. Various examples of DUV resists with different acid generators and resin structures are shown in Fig. 7.21. When migrating into 193-nm technology, the polymer has to be changed from aromatic to aliphatic (without benzene rings) for better 193-nm transparency.

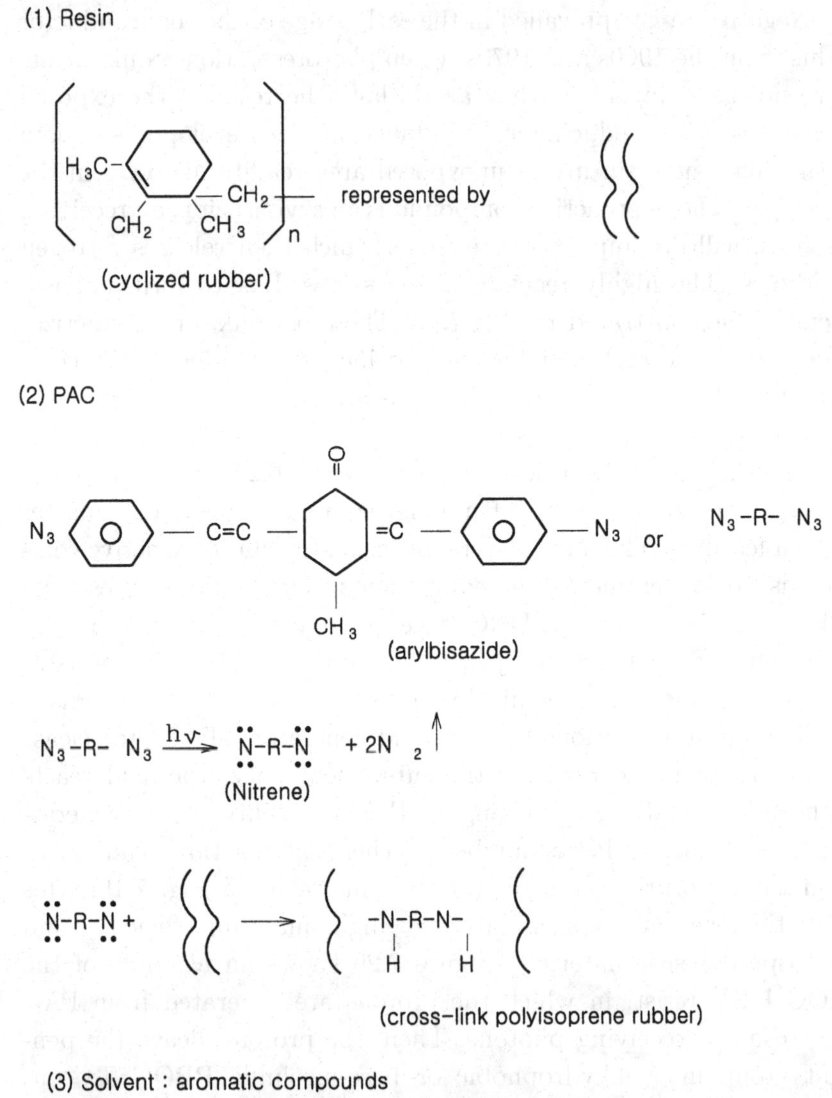

(1) Resin

H_3C- ... $-CH_2-$ ... CH_2 CH_3 n represented by

(cyclized rubber)

(2) PAC

N_3 — C=C — (O, CH_3) =C — — N_3 or N_3-R-N_3

(arylbisazide)

$$N_3-R-N_3 \xrightarrow{h\nu} \ddot{N}-R-\ddot{N} + 2N_2\uparrow$$

(Nitrene)

$$\ddot{N}-R-\ddot{N} + \;\rightarrow\; -N(H)-R-N(H)-$$

(cross-link polyisoprene rubber)

(3) Solvent : aromatic compounds

Fig. 7.18. Molecular structures of the resin and PAC for a typical negative resist.

Chemically amplified resists (CARs) have several advantages such as high contrast, high quantum efficiency (photospeed), and good resolution. They also pose a few challenges in production, however. Because of their catalytic reaction nature, if the catalytic

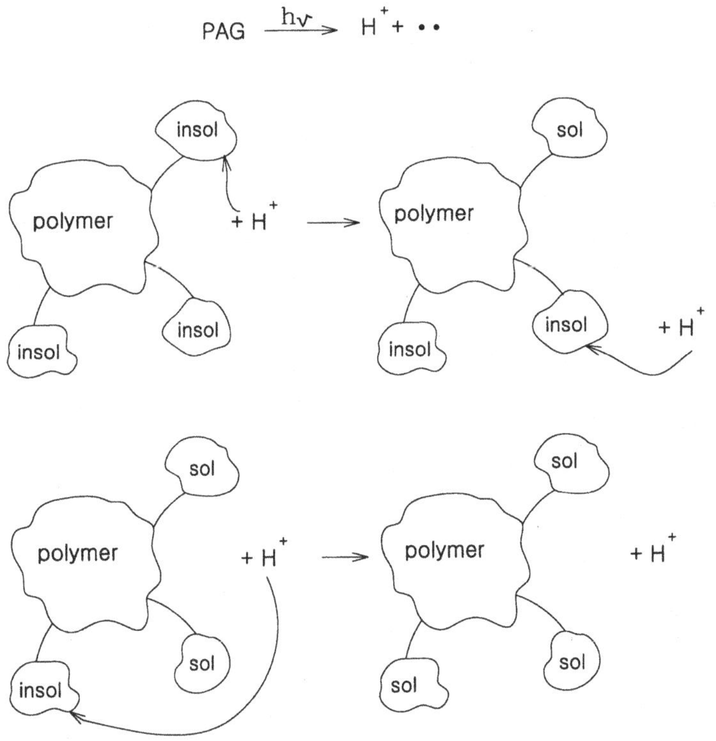

$$\text{PAG} \xrightarrow{h\nu} \text{H}^+ + \bullet\bullet$$

Fig. 7.19. Schematic explanation of the acid generation and regeneration.

reaction is interrupted unexpectedly, all lithographically important reactions would not take place, and the resist would totally fail.

The two most challenging issues associated with the chemically amplified resists are ambient base concentration control and baking temperature control. It has been shown that storing CAR in normal fabrication environments can result in poor CD uniformity as well as CD target drifts. Airborne ammonia can fatally affect CAR performance, mainly because ammonia neutralizes the acids in the photochemical reactions. Several basic molecular contamination sources have been identified. These include human bodies, ambient air, exhaust from other manufacturing lines, and some clean room construction materials (ceiling material and HEPA).

The impact of base concentration varies with the sensitivity of the CAR and feature sizes. Smaller features are more sensitive to

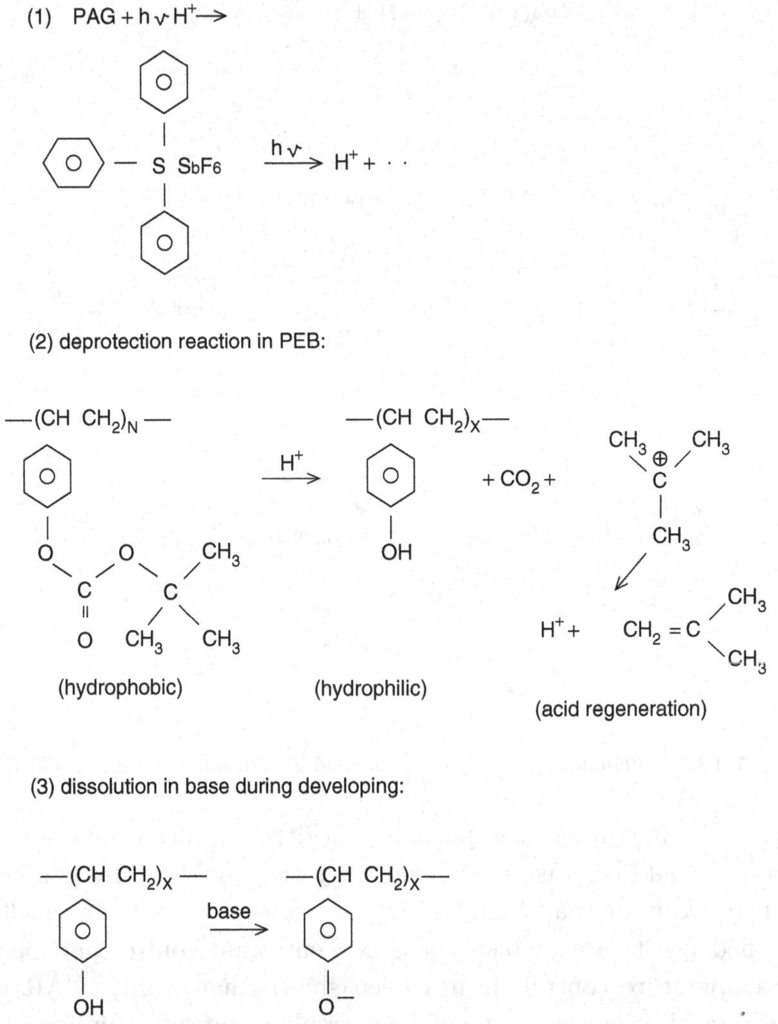

(1) $PAG + h\nu \cdot H^+ \rightarrow$

$S\,SbF_6$ $\xrightarrow{h\nu}$ $H^+ + \cdots$

(2) deprotection reaction in PEB:

$—(CH\ CH_2)_N—$ $\xrightarrow{H^+}$ $—(CH\ CH_2)_X—$ $+ CO_2 +$

CH_3 CH_3 $\oplus$ C CH_3

O O CH_3 C C O CH_3 CH_3 OH

$H^+ +$ $CH_2 = C$ CH_3 CH_3

(hydrophobic) (hydrophilic) (acid regeneration)

(3) dissolution in base during developing:

$—(CH\ CH_2)_X—$ $\xrightarrow{base}$ $—(CH\ CH_2)_X—$

OH O^-

Fig. 7.20. CAR resist (tBOC) chemistry in exposure, PEB, and developing steps.

ambient base concentrations. In a 0.13-μm technology regime, the base concentration should be kept as low as a few ppb. To meet such stringent requirements, the scanner and the track are normally isolated, with individual hoods equipped with ammonia chemical filters.

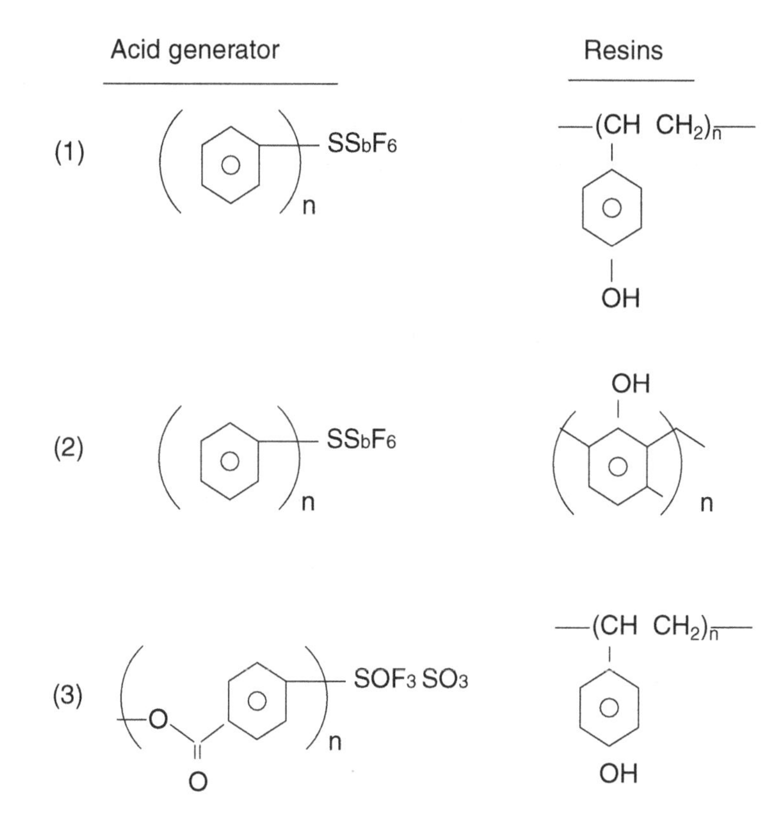

Fig. 7.21. Examples of DUV resists with various PAG and resins.

7.4. Fundamentals of Image Formation

Before examining exposure systems, we should first review some of the important imaging principles and terms that are often used in photolithography. After the resist is coated on a wafer and soft baked, the next step is to transfer the circuit patterns from a mask to the wafer surface. This is done with an exposure system. An exposure system consists mainly of a light source, a lens system, a mask handling mechanism, and a wafer stage, as shown in Fig. 7.1. The light source gives off the desired monochromatic light, which is collimated with a condenser lens to shine on a mask. The mask contains all the circuit information in a pattern of binary tones (either clear or opaque) and is placed under the exposure light with the pattern side facing the wafer stage during exposure. The light rays get diffracted after passing through the mask pattern and are collected with a projection lens system and focused onto the wafer surface, forming latent images in the resist layer. To form an image of the mask pattern information, optics demand that at least two diffracted rays be collected and interfere with each other. The projection lens, therefore, must be of a reasonable size to collect the necessary rays. State-of-the-art steppers or scanners are of the projection type. The capability of an exposure system is measured mainly with two factors: resolution and depth of focus. The resolution is defined as the minimum feature size that an exposure system can resolve with a decent depth of focus. The depth of focus refers to the vertical distance, away from the best focal point, within which the imaging quality is considered acceptable. These two factors are related to the numerical aperture and wavelength of an exposure system, as follows:

$$\text{Resolution limit } R = k_1 \frac{\lambda}{\text{NA}} \tag{7.3}$$

$$\text{Depth of focus DOF} = k_2 \frac{\lambda}{(\text{NA})^2}\,, \tag{7.4}$$

where λ is the exposure wavelength and NA is the numerical aperture of the lens, which increases with lens diameter. The proportional constant k_1 is related to photolithographic process characteristics. Better resolution resists, or resolution enhancement techniques such as phase-shifting masks or optical proximity correction, would give smaller k_1 values. On the other hand, k_2 is close to unity. A good exposure system is characterized by having a small resolution limit and a large depth of focus so as to have large photoprocess latitude. According to Eqs. (7.3) and (7.4), it is obvious that decreasing wavelength and increasing numerical aperture are the two major approaches to achieve better resolution limits, and these approaches have been the theme of photolithography technology evolution. By using these approaches, one obtains better resolution but faces a diminishing DOF, which can be a viable photoprocess if the wafer topograph is minimized with appropriate planarization techniques.

With exposure wavelength changes, a lot of related materials have to be changed or modified accordingly. To name a few, the materials of the scanner lens, mask blanks, and pellicles have to be changed to match the shorter wavelength so that they have minimum absorption of the wavelength. Excessive light absorption can induce material degradation or distortion, in addition to light intensity loss. Obviously, the resist system also needs to be changed with exposure wavelength to gain good resolution. With the resist systems changed, all the photosteps shown in Fig. 7.5 need to be reoptimized to have the best photolithographic performance.

7.4.1. *Exposure wavelength*

The light for exposure systems comes mainly from high-pressure mercury lamps for wavelengths of a wide range. The underlying principle is that the mercury atoms are excited by the energetic electrons in the mercury plasma; as some of the excited mercury atoms return to their original low-energy states, they emit photons. The intensity of the emission is not uniform with respect to wavelengths. Some high-intensity emissions are observed at wavelengths of G-, H-, and I-lines, which are widely used in lithography exposure systems. The wavelength of G-line is 428 nm; that of I-line is 365 nm; and that of DUV is 248 nm. A good exposure source should have monochromatic emission with high intensity and a long lifetime. The monochromatic wavelength eliminates chromatic aberration. A high-intensity exposure light boosts the wafer throughputs. A long lightbulb lifetime is always desired for cost saving and low maintenance efforts.

For high-intensity exposure light sources at DUV (248 nm) and 193 nm wavelengths and beyond, excimer lasers are employed owing to their high intensity, monochromaticity, and availability. *Excimer* stands for "excited dimmers." In other words, excimer lasers employ excited halogen molecules to react with inert gases:

$$\mathrm{Kr} + \mathrm{F}_2^* \rightarrow \mathrm{KrF} + h\upsilon \quad \text{for 248 nm wavelength} \tag{7.5}$$

$$\mathrm{Ar} + \mathrm{F}_2^* \rightarrow \mathrm{ArF} + h\upsilon \quad \text{for 193 nm wavelength} \tag{7.6}$$

$$\mathrm{F}_2^* \rightarrow \mathrm{F}_2 + h\upsilon \quad \text{for 157 nm wavelength} \tag{7.7}$$

The excess energy of the reaction is given off in the form of a laser at a specific wavelength. For example, the reaction of krypton with a fluorine-excited dimmer gives off 248 nm of light; argon reacting with the dimmer gives off 193 nm of light. The relaxation of an excited fluorine dimmer gives off 157 nm of light. Beyond the 157-nm exposure light source, the technologies are referred to as next generation lithography (NGL).

7.4.2. *Lens imperfections*

The lens system is one of the key elements of an exposure system. Lens quality determines, for the large part, the circuit imaging quality on wafers such as critical dimension uniformity (CDU). The CDU is essential to the device yield. For a perfect lens system, all light rays from an object converge onto a corresponding image point, forming a clear image. Practical lens have never been perfect. The behavior of a real lens system differs from that of a perfect one; the deviation is called aberration, as demonstrated in Fig. 7.22. Lens aberration takes several forms such as spherical aberration, chromatic aberration, coma, oblique astigmatism, field curvature, and distortion. These aberrations cause the diffracted wavefronts to deviate from the ideal ones; as a results, the image blurs. For a photolithography exposure system, aberrations result in printed dimension nonuniformity or unacceptable resist profiles.

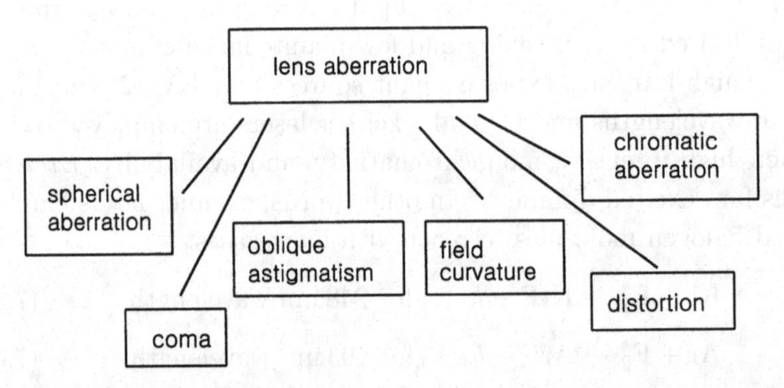

Fig. 7.22. Types of lens aberration that degrade image quality.

A fan of rays, each at a different distance away from the optical axis, fails to converge to the same point, as shown in Fig. 7.23. This type of aberration is called spherical aberration. Spherical aberration can be corrected by bending the lens. For a lens system, the spherical aberration of one lens can be corrected by another. Chromatic aberration results from nonmonochromatic light passing through a lens and failing to converge to the same focal point, as indicated in Fig. 7.24. Coma is an off-axis aberration resulting from the lens having different magnifications at different zones from its center. Figure 7.25 shows the image of a point source with an increasing extent of coma. Astigmatism, which results from different lens curvatures in different planes, causes the tangential and sagittal rays to focus on different points, as shown in Fig. 7.26. Field curvature is the type of aberration that causes the image of a planar object to be curved, as illustrated in Fig. 7.27. In other words, the images move toward the lens for off-axis light rays. Unlike other aberrations due to lens imperfection, distortion is a type of aberration that may also result from

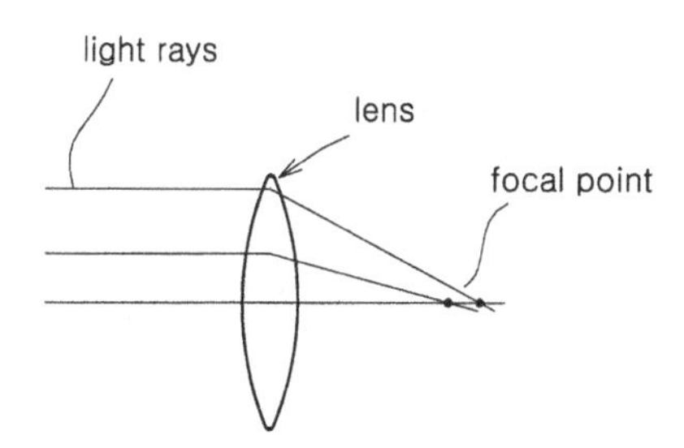

Fig. 7.23. Spherical aberration: parallel light rays fail to converge to the same focal point.

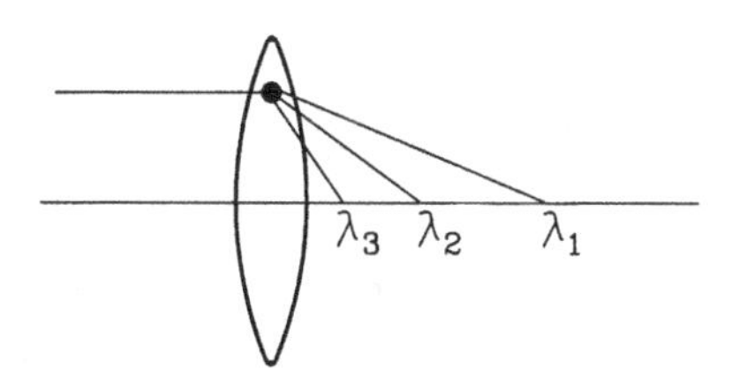

Fig. 7.24. Chromatic aberration: a beam of light with different wavelengths fails to converge to the same focal point.

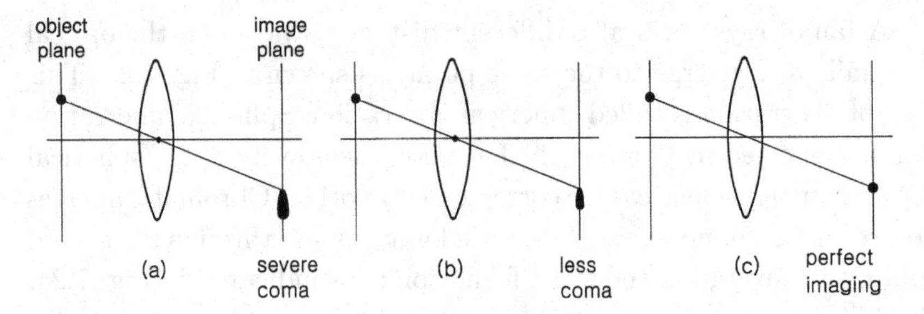

Fig. 7.25. Coma: an off-axis point on the object plane is imaged as a comet-like blur, trailing away from the optical axis.

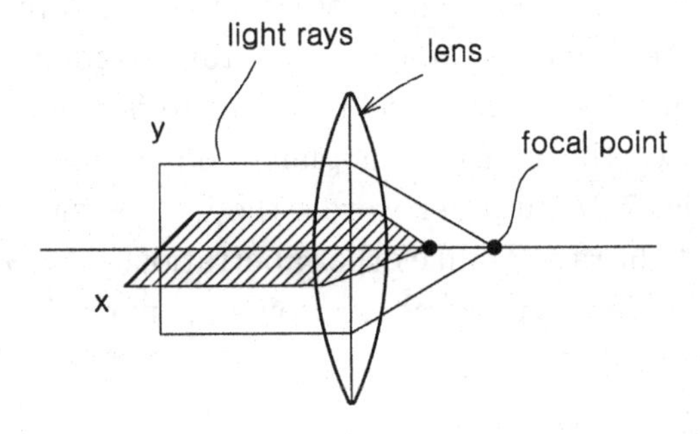

Fig. 7.26. Astigmatism refers to the tangential and saggital light rays being focused onto different points.

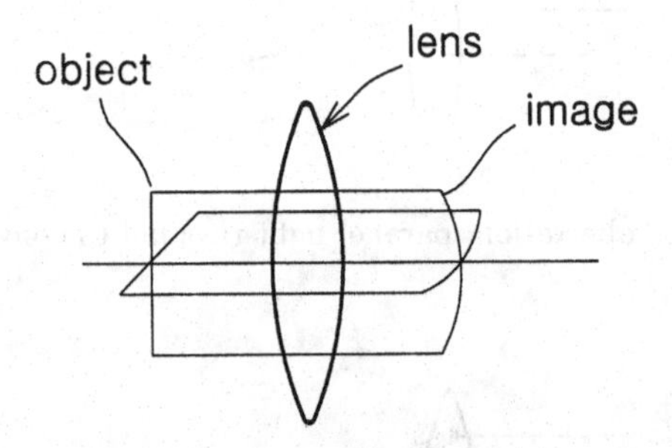

Fig. 7.27. Field curvature causes a distorted image, moving toward the mirror.

lens geometry. Distortion makes the image of an object distorted, as demonstrated in Fig. 7.28. All these imperfections of the lens system deteriorate the photoperformance of an exposure system in terms of printed dimension nonuniformity for both inter- and intrafield. The

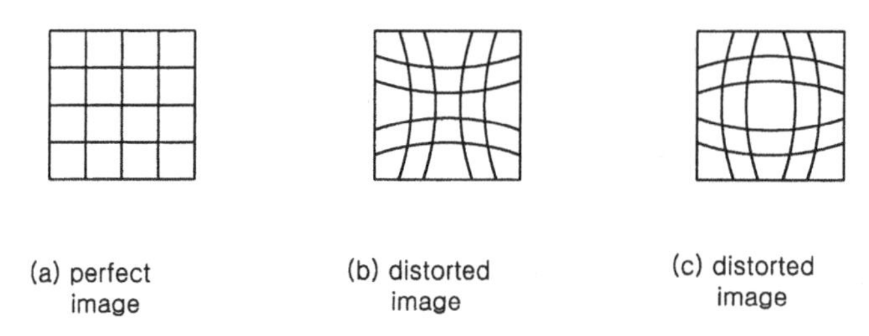

Fig. 7.28. Lens distortion results in distorted images on imaging planes (b and c).

field is defined as the image of a mask on the wafer. It could be a single-die or multidie image, depending on the chip size.

If a lens designer is allowed to design a lens system from scratch, he may be able to come up with a perfect lens, but this can be formidably expensive. The most common lens system design approach is to combine a number of lenses with different curvatures or shape factors so that the aberration of one lens can be compensated by the other ones. Nanometer-range device manufacturing technology requires the lens system to be almost perfect to have a wide process latitude. As a result, the lens system of a stepper or scanner often consists of tens of lenses with different curvatures and shape factors to compensate for each other's imperfection.

7.4.3. *Interference*

Light travels in a vacuum at its speed; it slows down when it travels in a medium of refractive index n:

$$C_m = \frac{C}{n} \quad \text{or} \quad n = \frac{C}{C_m}\,. \tag{7.8}$$

In other words, refractive indices of media are always larger than unity. In our daily life, light rays travel from one medium to the other, for example, from air to a glass of water or from air to a glass panel. When this occurs, the direction of the traveling light rays is refracted, as illustrated in Fig. 7.29. The direction change across the media follows Snell's law:

$$n_1 \sin\theta_1 = n_2 \sin\theta_2\,, \tag{7.9}$$

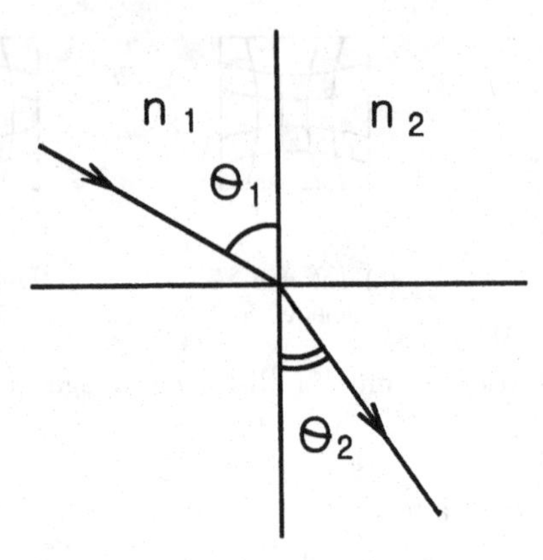

Fig. 7.29. Light travels across the boundary of two media and is refracted according to Snell's law.

where n is the refractive index of the medium and θ_1 and θ_2 are the angles of incidence and refraction, respectively. Snell's law explains image formation through a simple lens. A simple image formation of an object through a lens can be worked out through ray tracings, as shown in Fig. 7.30. In the real world, we do see deviations from this approximation, especially when the light rays travel through an aperture of a size close to the light wavelength. The smaller the aperture as compared to the wavelength, the more pronounced the diffraction phenomena.

Huygen's principle states that every point on a wavefront is a source of secondary wavelets, which radiate in all directions from their centers with the propagation speed of the original wave. The

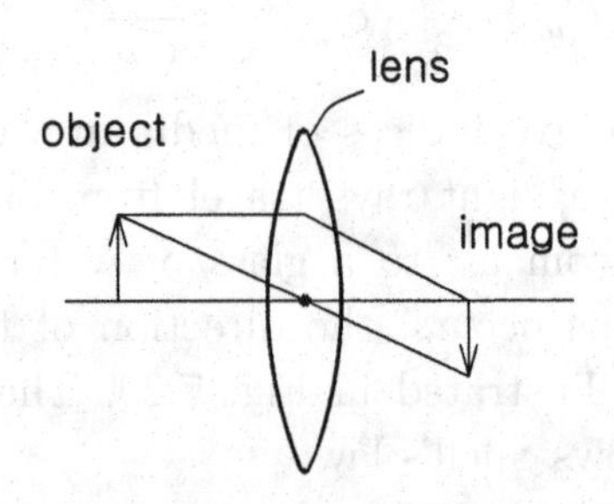

Fig. 7.30. The image formation through ray tracing.

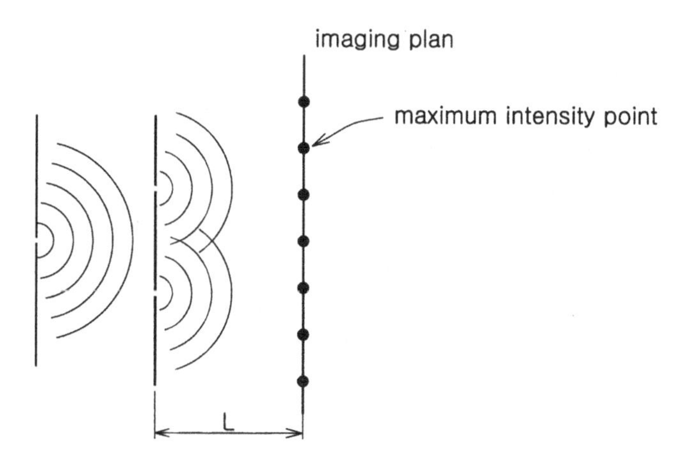

Fig. 7.31. The interference through a two-slit screen.

new wavefront is the envelope of secondary wavelets. In a homogeneous isotropic medium, the wavefronts from a point source are spherical. Employing Huygen's principle, one can see from Fig. 7.31 that the wavelets from the apertures would propagate in all directions and interfere with each other to form a shadow image (the interference pattern) on the screen. One would observe a bright and dark pattern due to constructive and destructive superposition of the waves. The optical path difference, ΔS, between two light rays radiated from the slits is

$$\Delta S = nd \sin\theta \,, \tag{7.10}$$

where d is the spacing between the slits. It then follows that the constructive superimposition of the two rays gives a maximum; a destructive superimposition gives a minimum:

$$\Delta S = nd \sin\theta_n = N\lambda \,, \quad \text{for the } N\text{th maximum} \tag{7.11}$$

$$\Delta S = nd \sin\theta_{n+1/2} = (N + 1/2)\lambda \,, \quad \text{for the } N\text{th minimum} \,, \tag{7.12}$$

where n is the ambient refractive index, d is the space of the slits, θ is the angle between the direction of diffracted light and the horizontal axis, and λ is the wavelength. Furthermore, for a fixed value of wavelength, the Nth maximum or minimum would appear further away from the optical axis as d or pitch of the grating decreases, indicating a more pronounced diffraction effect. The larger the θ is, the higher the diffraction angle will be. Now, imagine if one were to collect all the diffracted lights and image them onto the screen. The more the diffracted lights are collected, the better the image quality will be. Hence, to have faithful imaging, one would have to have an infinitely large lens to collect all diffracted light. In reality, the lens is of finite size. Thus, in real imaging systems, some information is lost due to the use of a limited size lens. As a result, the intensity modulation of the light rays right after passing through the grating (a mask pattern) is degraded on being reconstructed with a lens, as demonstrated in Fig. 7.32.

In optics, the size of a lens with numerical aperture is often defined as

$$\mathrm{NA} = n \sin\theta\,, \tag{7.13}$$

where n represents the ambient refractive index and θ represents the angle subtended by the two outermost diffracted rays. Physically, the larger the lens diameter, the larger the NA will be. The lens NA indicates the capability of collecting diffracted light. We now rewrite

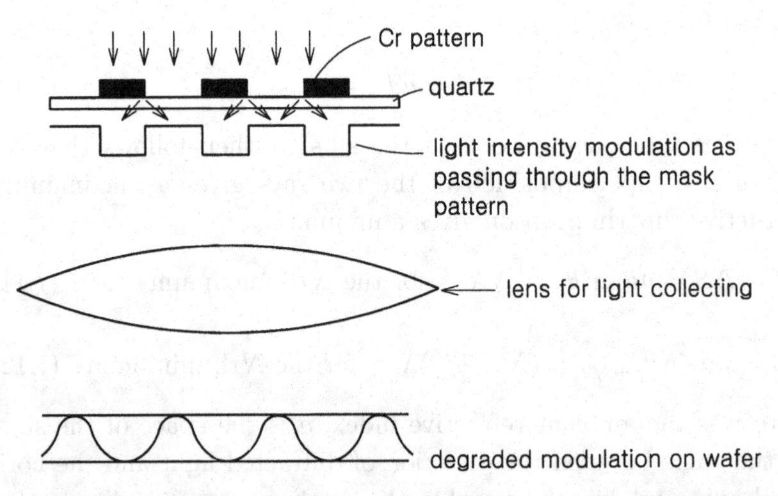

Fig. 7.32. The light intensity modulation is degraded, as seen on the wafer surface.

Eq. (7.11) as

$$pn \sin\theta = \lambda \tag{7.14}$$

(assuming N is equal to unity, the first diffracted order), and p is the one-half pitch of the grating aperture. Alternatively, with Eq. (7.14), one has

$$p = R = \frac{\lambda}{\mathrm{NA}}, \tag{7.15}$$

where R is the minimum feature size that a lens of NA can resolve. A criterion was proposed by Rayleigh that two neighboring slits or features are considered resolved if the interference pattern is such that the first maximum superimposes with the first minimum of the other. Rayleigh's resolution criterion has been widely used in photolithography and is conventionally expressed as

$$R = k_1 \frac{\lambda}{\mathrm{NA}}, \tag{7.16}$$

where k_1 factor is used and is a function of photolithography process parameters. Comparing Eq. (7.15) with Eq. (7.16), one can realize that NA of a lens corresponds to the minimum lens size that is capable of capturing the zero-order and first-order diffracted light rays of its minimum resolvable feature size. The zero-order diffraction carries the background light intensity, while the first-order diffraction provides the image information through interference.

7.4.4. *Depth of focus*

The other important performance index of an optical system is the depth of focus (DOF). Along the optical axis, by moving the imaging plane by a certain distance, x, either toward or away from the screen, the images formed still meet acceptable quality requirements. The distance x is referred to as DOF, expressed as $\pm x$. Surely, on the focal plane, the image is the sharpest; as the imaging plane moves away from the best-focused position, the image blurs. However, the true criteria of so-called acceptable image qualities are vague or somewhat arbitrary as long as we can see. In the field of imaging science,

Rayleigh's criterion has been widely accepted to define the criteria of acceptable image qualities. It states that the optical path difference (OPD) between the best-focused and defocused wavefronts should be less than one quarter of the operating wavelength to have acceptable image quality. A perfect lens should deliver at its exit pupil a spherical wavefront centering at the origin of the imaging plane, as shown in Fig. 7.33. The equation of the wavefront, based on the pupil coordinates, is expressed as

$$x^2 + y^2 + (z - R)^2 = R^2 , \tag{7.17}$$

where R is the radius of wavefront curvature. Expanding the third term of the left-hand side and assuming that the exit pupil is small compared to the radius of curvature of the wavefront (x and y are small as compared to R), one can then neglect the z^2 term, arriving at

$$z = \frac{x^2 + y^2}{2R} = w(x, y) . \tag{7.18}$$

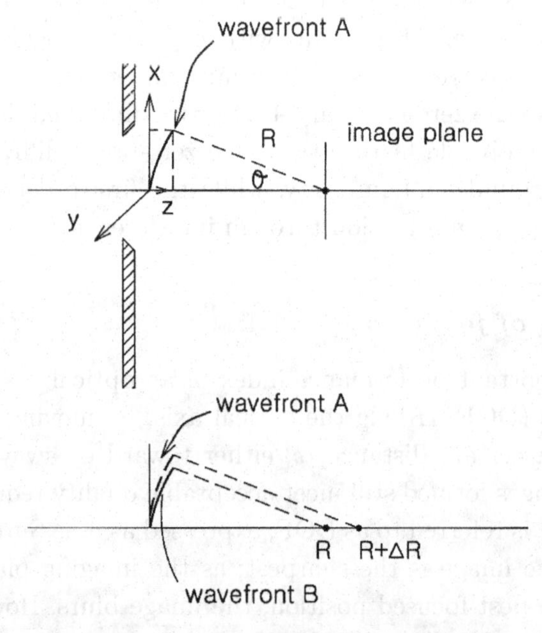

Fig. 7.33. Two wavefronts A and B, having different radii of curvature, converge at different imaging points along the Z-axis.

Now, if one looks at the other wavefront that centers at $R + \Delta R$, then the OPD, Δw, between the two wavefronts would be

$$\Delta w = -\frac{(x^2 + y^2)}{2R^2}\Delta R \tag{7.19}$$

$$\frac{dW}{dR} = -\frac{(x^2 + y^2)}{2R^2}. \tag{7.20}$$

Applying the Rayleigh criteria, $\Delta w \leq \lambda/4$, one has

$$\Delta w = \frac{-(x^2 + y^2)}{2R^2}\Delta R = \frac{\lambda}{4}. \tag{7.21}$$

It then follows that

$$\Delta R = \text{DOF} = \frac{\lambda}{2\sin^2\theta} = \frac{\lambda}{2\text{NA}^2}. \tag{7.22}$$

Here we reach a conclusion that DOF of an optical system is proportional to its operating wavelength and inversely proportional to the square of the numerical aperture of its lens.

In photolithography, the criterion of acceptable image quality is defined by the sharpness of the resist edges and critical dimensions, rather than by Rayleigh's criterion. Therefore DOF is often expressed as

$$\text{DOF} = k_2\frac{\lambda}{\text{NA}^2}, \tag{7.23}$$

where the proportional constant, k_2, is a function of photoprocess parameters. Its value ranges from approximately 0.8 to 1.0. For a lens with its minimum resolvable feature, DOF can also be interpreted as the distance between the point and the best focal point within which the OPD between the zero and the first diffracted light rays is equal to one quarter of wavelength, as illustrated in Fig. 7.34. It can be observed that the DOF decreases quadratically with numerical aperture or lens diameter.

7.4.5. *Coherence*

Coherence and incoherence are terms that describe the correlation between propagation waves. They are further defined in two terms: temporal and spatial coherence. If two propagating waves have a

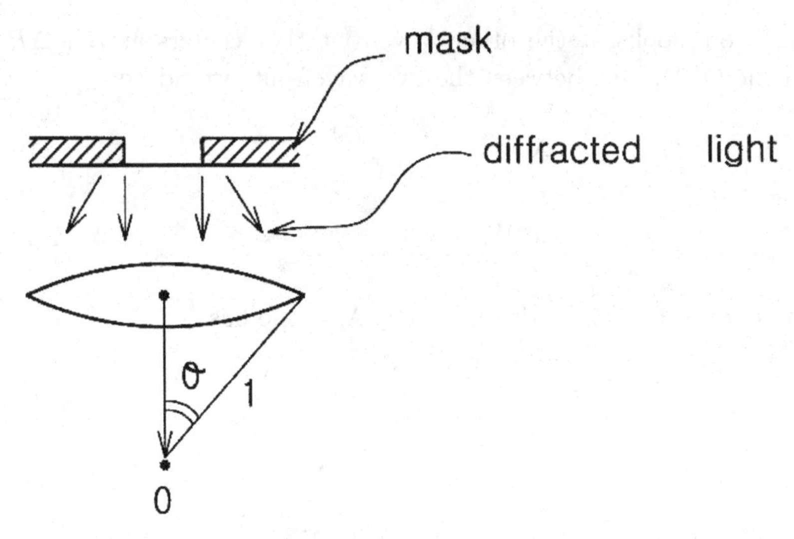

Fig. 7.34. The optical path difference between the zeroth and first-order diffracted light rays.

fixed phase difference at all times, they are said to be temporally coherent. Temporal coherence between two waves is characterized by the monochromaticity of the waves. The narrower the frequency bandwidth ($\Delta\omega$) of a monochromatic light, the greater the temporal coherence will be. On the other hand, if two waves travel in parallel (collimated), they are said to be spatially coherent. In general, unless specified, coherence implies both temporal and spatial coherence. Ideally, two monochromatic waves of the same wavelength that travel in parallel are coherent waves. For example, a so-called ideal highly collimated 193-nm laser light source used in photolithography is a coherent light source. If two light waves travel in completely random directions and have no fixed relative phase relationship, they are said to be incoherent, for example, the waves from two independent lightbulbs. In fact, all scenarios in the real world lie in between coherence and incoherence and are termed partial coherence. Figure 7.35 illustrates that coherent light beams through a lens give a sharp single focal point, while incoherent light beams give blurred images, as each beam focuses onto different points.

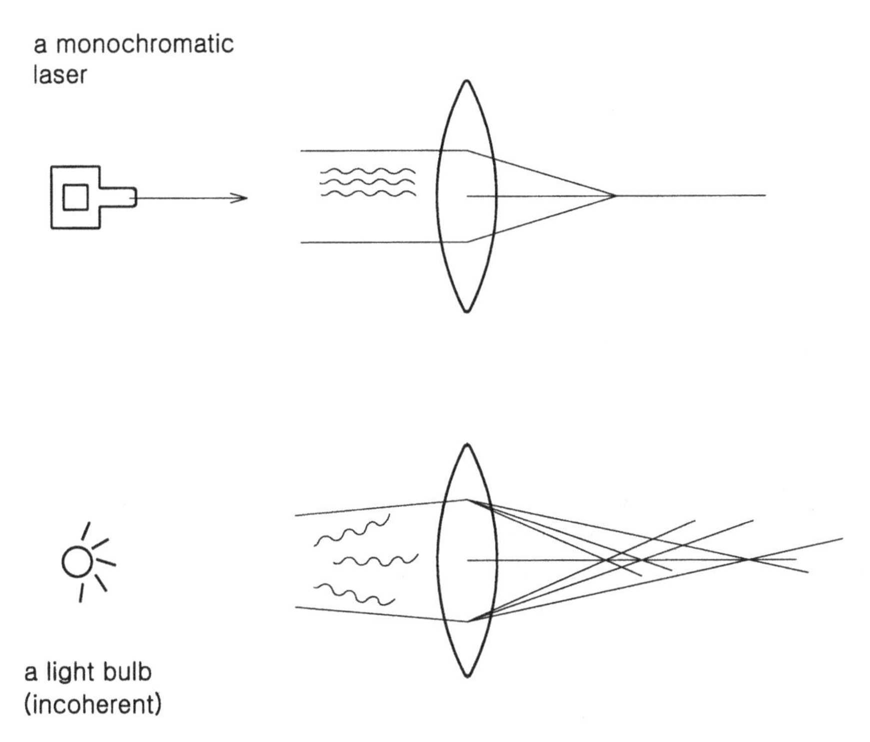

Fig. 7.35. Difference in imaging for coherent and incoherent light.

For the best photolithography performance, the exposure light source has to be monochromatic, with a bandwidth that is as narrow as possible and highly collimated. Furthermore, exposure tools are designed such that the condenser lens is smaller than that of the objective lens:

$$\text{Coherence} = \frac{\text{NA}_{\text{condenser}}}{\text{NA}_{\text{objective}}}\,. \tag{7.24}$$

Ideally, when the coherence value is zero, the light is coherent; at infinity, it is incoherent.

Photolithographically, it has been proved experimentally that a partially coherent light gives better results than does a coherent one. The incoherence optimization procedure is a very tedious step as incoherence is one of the many photoprocess parameters that may affect the final critical dimension uniformity as well as the DOF latitude.

7.4.6. *Optical transfer function*

The optical transfer function (OTF) is a function that relates the input of an optical system to the output, or vice versa, in the frequency domain. This approach has been widely used in the field of electronic communications and control. The underlying idea is that if the input to a system is sinusoidal or cosinusoidal (in fact, sine and cosine have the same functional form but differ in phase by $\pi/2$), the output is also sinusoidal. For the lens system, the sinusoidal input would be the sine wave grating of the mask; the output would be its sinusoidal image (remember that all functions can be expressed in terms of sinusoidal functions). The amplitude of the input and output is related by a transfer function:

$$O_{\text{out}}(s) = T(s)I_{\text{in}}(s)\,, \tag{7.25}$$

and the phase is related by

$$\phi_{\text{out}}(s) = \phi_{\text{in}}(s) + \varphi(s)\,. \tag{7.26}$$

The imaging equivalents of Eqs. (7.25) and (7.26) are as follows:

For the contrast

$$C_{\text{out}}(s) = \text{MTF}(s)C_{\text{in}}(s) \tag{7.27}$$

For the phase

$$\phi_{\text{out}}(s) = \text{PTF}(s) + \phi_{\text{in}}(s)\,. \tag{7.28}$$

The OTF concept can be applied to imaging systems owing to the contribution of the vintage Fourier transform, invented by a French mathematician in the 18th century. The theory states that any function can be expressed in terms of a summation of a series of cosine and sine functions. Discussion of the theory can be found in advanced engineering mathematics textbooks. Mask patterns are also periodic to some extent such as the one illustrated in Fig. 7.36.

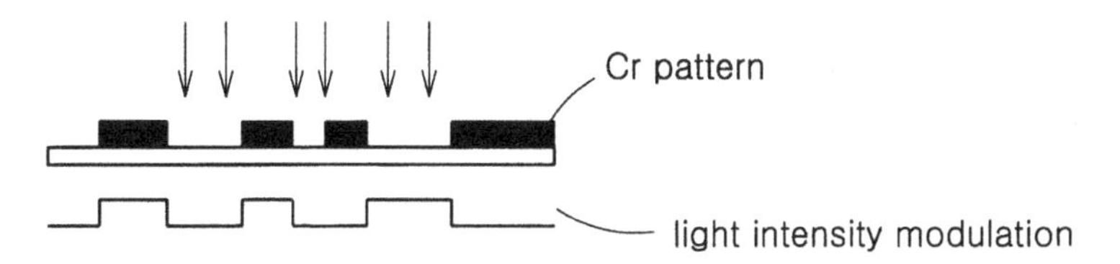

Fig. 7.36. The light intensity modulates after passing through a mask pattern.

The main advantage of using OTF is that the product of the OTF of each individual optical element equals the OTF of the overall system:

$$\mathrm{OTF} = \mathrm{OTF}_1 \times \mathrm{OTF}_2 \times \cdots . \tag{7.29}$$

As in Eq. (7.29), the transfer function has two parts. One deals with the amplitude, the other with phase; namely,

$$\mathrm{MTF} = \mathrm{MTF}_1 \times \mathrm{MTF}_2 \times \cdots \tag{7.30}$$

$$\mathrm{PTF} = \mathrm{PTF}_1 + \mathrm{PTF}_2 + \cdots . \tag{7.31}$$

Assuming that the phase remains unchanged by going through the imaging system, one has OTF equals MTF, which is defined as

$$\mathrm{MTF} = \frac{C_{\mathrm{out}}}{C_{\mathrm{in}}} = \frac{\left[\dfrac{I_{\mathrm{max}} - I_{\mathrm{min}}}{I_{\mathrm{max}} + I_{\mathrm{min}}}\right]_{\mathrm{out}}}{\left[\dfrac{I_{\mathrm{max}} - I_{\mathrm{min}}}{I_{\mathrm{max}} + I_{\mathrm{min}}}\right]_{\mathrm{in}}} . \tag{7.32}$$

For a photolithography exposure system, the mask pattern is transferred to wafer surfaces through the lens and resist transfer functions:

$$\mathrm{MTF} = \mathrm{MTF}_{\mathrm{lens}} \times \mathrm{MTF}_{\mathrm{resist}} . \tag{7.33}$$

MTF of a system tends to decrease with the spatial frequency of the mask patterns to be printed, as illustrated in Fig. 7.37. Obviously, this is because the higher the spatial frequency, the larger the light

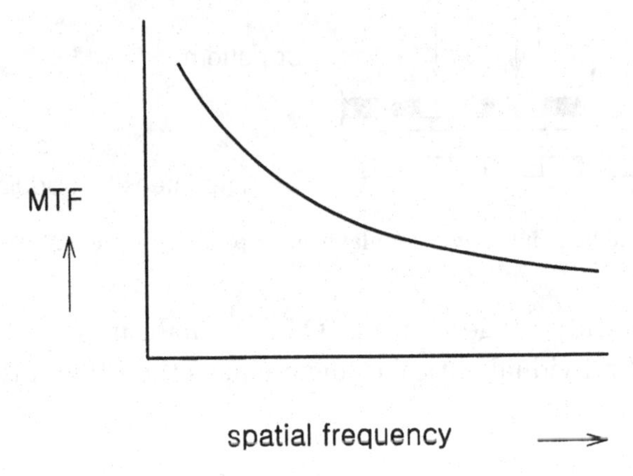

Fig. 7.37. The MTF of an optical imaging system decreases with spatial frequency (or 1/pitch).

diffraction angles, and hence more information is lost. The MTF of the resist is related to its contrast value.

7.5. The Exposure System Evolution and Photoprocess Variations

With the background covered in the last section, we are now ready to explore the exposure system. Looking at the exposure system evolution in semiconductor manufacturing, there are basically four major categories, as shown in Fig. 7.38. The contact printing system is the

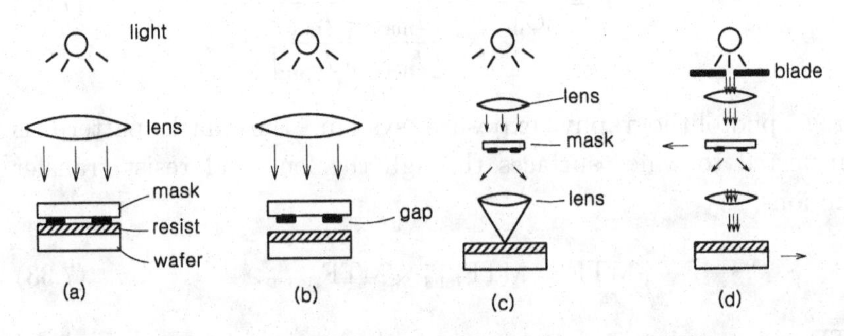

Fig. 7.38. Evolution of photolithography exposure tools: (a) contact printing; (b) proximity printing; (c) projection printing; and (d) scanner (a slit of the mask is exposed at a time).

earliest exposure system. It essentially has a mask directly pressed against the resist-coated wafer surface. This type of system has the advantage of having the best resolution — below 0.5 μm is possible — as the diffraction effect is minimized. But owing to the direct contact of the resist surface and the mask, resist residues can become a contamination source.

As technology advanced to a point where contaminations became printable defects, the sizes of which started to affect device yields, the industry turned to proximity printing, in which the wafer surface and the mask surface are pulled apart to a distance of roughly tens of microns, to avoid contamination due to direct contact. By doing so, the resolution of the system is significantly degraded, as diffraction effects come into play. The resolution limit of proximity printing is governed by the Fresnel diffraction as

$$\text{Resolution limit } R = k\sqrt{\lambda d}\,, \tag{7.34}$$

where R is the resolution limit, λ is the operating wavelength, and d is the gap between the wafer surface and the mask surface. For example, if the k factor takes a value of 0.9, the wavelength 400 nm, and the gap 15 μm, the resolution limit is 2.2 μm. The resolution can be improved by decreasing the gap or the operating wavelength.

As device technology advances into the regime of a few microns, the projection system dominates. The projection system has several advantages. First, it has the best resolution, down to a nanometer regime. Second, it avoids the contamination due to direct wafer–mask contacts. Third, it has a demagnification factor, such as 4$\times$ or 5$\times$. In other words, the patterns on the mask are printed onto wafer surfaces with size reduction ratios of 1/4 or 1/5. It makes the projection printing process more forgiving in terms of tolerating particles and CD errors on masks as they are all reduced by a factor of 4 or 5. The resolution limit of a projection printing system is shown in Eq. (7.3). With projection systems, the semiconductor industry still tries to drive down the resolution by increasing the NA or decreasing the wavelength. As NA gets larger, it becomes extremely difficult to manufacture a large lens with the uniform quality needed. Hence the printing strategy is changed from the step-and-repeat (stepper) to the step-and-scan (scanner). For steppers, the system utilizes the whole mask area as a printing field. After one shot is done, it steps the wafer onto the next field, and so on. The field is placed purposely at the center of the lens system so as to obtain the best quality. For scanners, only a stripe of mask and lens is used during scanning. This ensures good uniformity across the stripe area. The mask and wafer move in opposite directions to complete the scan on a whole wafer.

The ultimate goal of a photolithography process is to transfer circuit patterns on masks to wafer surfaces with high fidelity. The performance of the process is often measured in terms of uniformity of critical dimensions and pattern fidelity with wide enough process latitude, as well as good overlay accuracy. The uniformity of critical dimensions means that the dimensions of features of the same size should be equal or close to equal across-wafer and from wafer to wafer. This is very important; for example, in the case of a polysilicon gate definition, the gate dimension dictates the transistor speed. If the gate dimensions vary with position, transistors at some locations may be too fast, while at others, they may be too slow. The whole chip speed can be degraded. The timing can be mismatched, leading to circuit malfunctioning. Pattern fidelity refers to the two-dimensional profile matching between the features on a mask to those on a wafer. Figure 7.39 illustrates a pattern that is not faithfully transferred from the mask to the wafer. Such a scenario can induce a transistor failure if the line end of the polysilicon gate is so severely rounded that it does not stretch across the active area.

Overlay refers to the alignment performance between layers. This is important because a complete chip circuit is defined into a set of N-layer masks, as discussed earlier. The wafer process is done layer

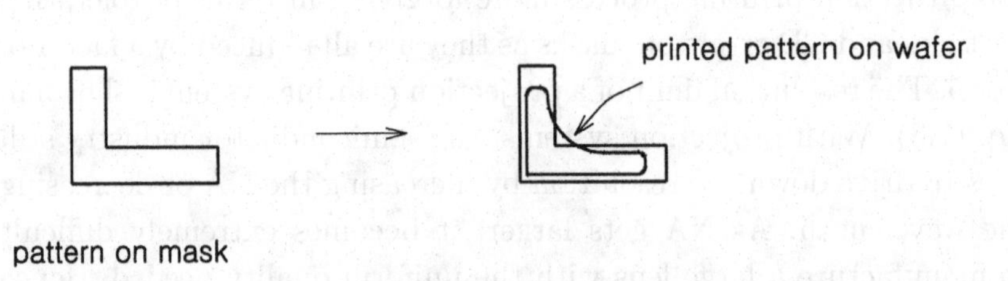

Fig. 7.39. Poor pattern transfer from the mask to wafer.

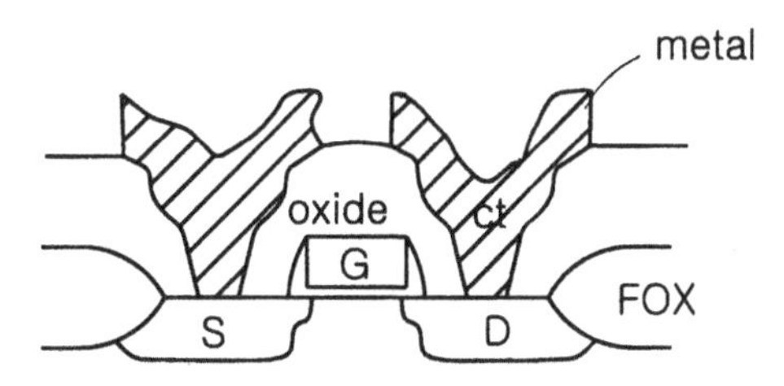

(a) perfectly aligned

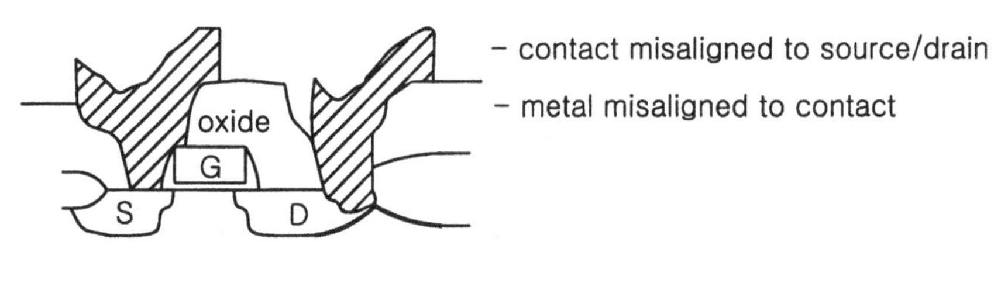

(b) misaligned

Fig. 7.40. A demonstration of a perfectly aligned MOS versus a misaligned MOS device.

by layer. For the circuit to function properly, a basic requirement is to have all layers aligned. Figure 7.40 demonstrates a situation where all related layers of a MOS structure are perfectly aligned versus one that is not. As one can imagine, the misaligned structure would have excessive leakages among layers, and it is hard to imagine that the circuit will work. In the real world, there exist certain process variations that cause nonuniform CD, corner rounding, and overlay errors.

There are several phenomena that can degrade CD uniformity (CDU) across a wafer such as standing wave, swing, and optical proximity effects. A wave propagates from a medium into the other; the wave can partly reflect from the interface and partly pass through the interface. As light shines on a resist-coated wafer surface, the

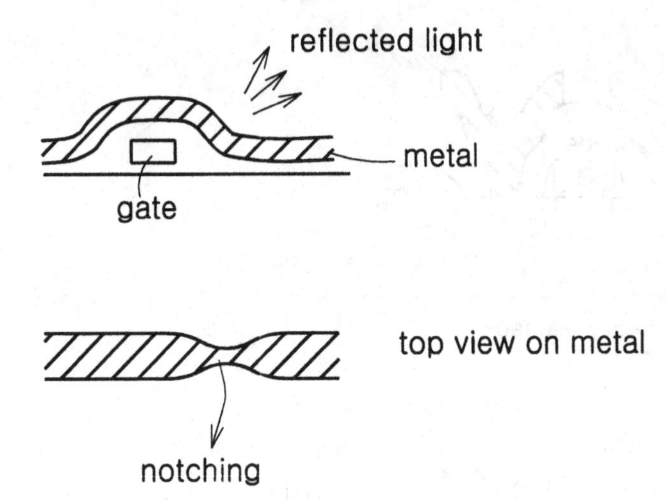

Fig. 7.41. As a metal layer crosses over the topograph, the sloped metal surface acts like a mirror, causing a notching.

reflected light and incident light would interfere and result in a standing wave. The standing wave effect results in a wiggling resist sidewall, as discussed and shown in Fig. 7.46. This can degrade the CDU when there is slight resist thickness variation. A strong reflective substrate can cause notching, as illustrated in Fig. 7.41. The light incident on a shining sloped surface bounces back and exposes the resist bulk that is not supposed to be exposed. This obviously degrades the CDU.

The other effect that would degrade the CDU is the swing effect, as shown in Fig. 7.42. The exposure energy needed to clear off a certain thickness of resist increases with the resist thickness. The trend is not monolithic, but sinusoidal. With this effect, any resist thickness variation across a wafer, either due to coating nonuniformity or topography, leads to CDU degradation. This is often seen in memory circuits, in which the memory array is much denser than peripheral layouts, giving rise to thinner resist in the array than in peripheral circuits. If, unfortunately, one falls at the maximum and the other at the minimum, the CD could be way off. A straightforward and effective approach to mitigate the interference-induced notching, standing wave, and swing effects is to use an antireflective coating (ARC), either organic or inorganic. The

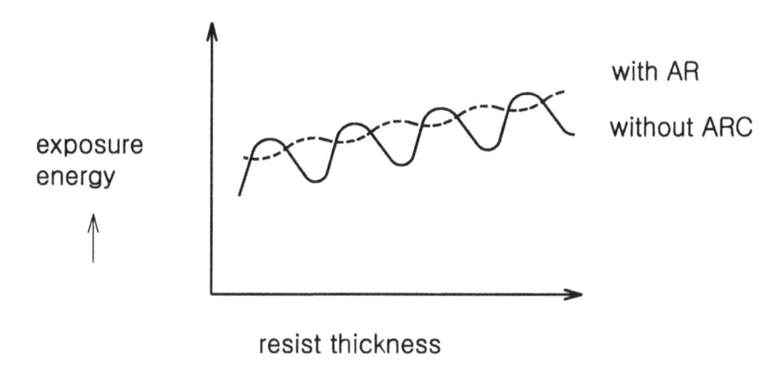

Fig. 7.42. The swing effect: resist thickness variation results in CD variation.

organic ARC is highly absorbing, and therefore it greatly reduces the reflected light intensity. For the inorganic ARC, the thickness is optimized such that the reflected light from the resist–ARC and ARC–substrate interfaces cancel out. Figure 7.42 also indicates that with the use of ARC, the substrate reflectivity is significantly reduced; so is the swing effect. The ARC thickness is structure-dependent. It must be optimized in terms of the underlying layer structures and thicknesses as well as the exposure wavelengths.

Advanced technologies drive for ever-decreasing pitches — the width plus the spacing of features. Given the same size of NA of the projection lens, the exposure light will get a larger extent of diffraction. This means that more information is lost during imaging. This causes a poor pattern transfer fidelity issue. The corners are rounded off, and the line ends are shortened. The extent of shrinking and rounding aggravates with ever-shrinking feature sizes. The use of optical proximity correction (OPC), as will be discussed in Chapter 8, can be employed to improve pattern fidelity on wafers.

Process latitude refers to the tolerance of process parameter variations within which one can still maintain acceptable photolithography performance. Taking into account the abovementioned phenomena, the exposure energy range and DOF are often considered as the process latitude or process window for CD uniformity control. Figure 7.43 shows a window of energy and focus variations in which the CD variations of a specific target meet the

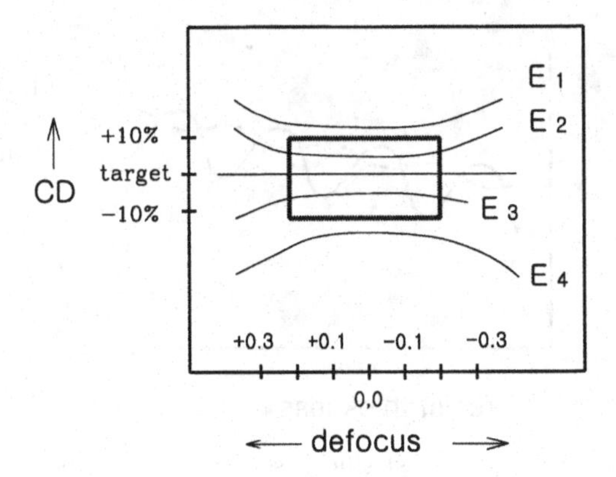

Fig. 7.43. CD variation of a specific target with exposure energy and focus values. The rectangle indicates the process window.

requirement of ± 10% of target CD values. With the optimal energy and focus, the resist image is the sharpest. Moving away from the center energy, CD values increase with decreasing energy. As the focus offset values change away from the center, the images smear; the sidewall angles degrade, which affects the etching uniformity. The photolithography parameter settings must be further refined by considering the area of overlapping windows in energy focus plots for both isolated and dense patterns. To be production worthy, a large process window is required. Of course, the cost of manufacturing always sets a limit on the search for a larger process window.

Layer-to-layer alignment errors originate from two sources: the mask-to-mask misalignment during mask making and the wafer-to-wafer misalignment during wafer exposures. In semiconductor manufacturing, there are two different overlay approaches. One is to align all layers to a reference layer, while the other is to align the current layer to a related previous layer. The overlay error measurements are done with a specifically designed structure placed on each of the related layers. As shown in Fig. 7.44, a previous layer is embedded with a square frame box of 20 μm, and then the current layer is embedded with a square bullet of 10 μm. After both layers are processed, the top-view SEM can be taken, or errors can be read automatically with metrology tools. Factors such as thermally induced or

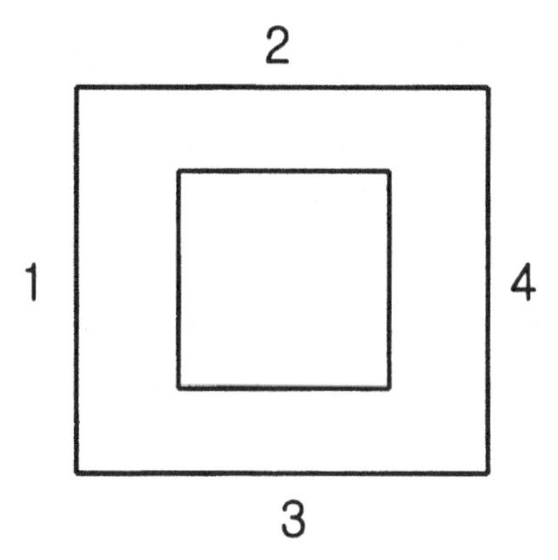

Fig. 7.44. Overlay boxes provide the overlay reading after both layers are processed.

stress-induced wafer distortion or lens distortion can cause misalignment. These misalignment sources are supposed to be automatically calibrated or corrected on exposure machines; however, there is still misalignment seen from layer to layer. These misalignments will certainly cause device malfunctioning and hence yield loss. Figure 7.45 illustrates an example of a misaligned polysilicon gate with respect to the active layer. A misalignment causes the gate not to be able to stretch across the active layer. From the top and side views of the

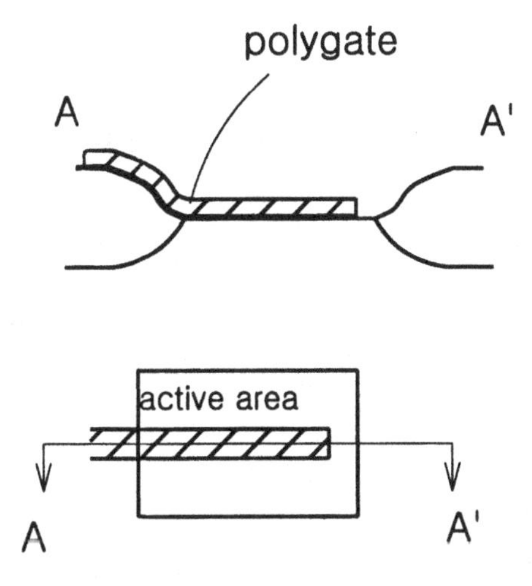

Fig. 7.45. Schematic of misalignment between gate and active area.

transistor structure, one can see that the transistor will not work as the source and drain are short. For each layer, there must be an overlay tolerance specified with respect to a related previously processed layer. The values dictate a part of the design rules. For example, let us look at the metal 1 extension to contact design rule setting:

$$\begin{aligned}&\text{(the design rule of metal 1 to contact extension)}\\ &\quad \geq \text{(CD variation of contact)} + \text{(CD variation of metal 1)}\\ &\qquad + \text{(variation of metal 1 to contact misalignment)}\,. \end{aligned} \tag{7.35}$$

One can see that if both process variation and misalignment can be controlled and minimized, the design rule can be set to a smaller value. If this applies to all layers of a device, the whole chip can be made more compact. Hence one obtains more dice/wafer, and hence lower cost can be achieved.

7.6. Postexposure Steps

7.6.1. *Postexposure bake*

As shown in Fig. 7.5 previously, the postexposure steps include postexposure bake, developing, and hard bake. For novolac resists, the high temperatures of PEB allow the carboxylic acid to diffuse out to mitigate the standing wave effects, as shown in Fig. 7.46. For CARs, PEB is even more critical as the acids generated during exposure are supposed to undergo deprotection reactions, as discussed previously. The temperature uniformity in PEB is extremely important for CARs in terms of determining the ultimate CD uniformity.

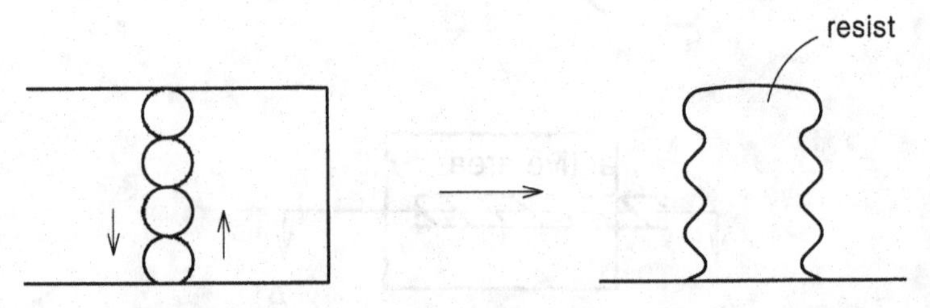

Fig. 7.46. Standing wave effect during exposure (left) results in a wiggling resist edge profile after development (right).

7.6.2. *Developing and hard bake*

Developing is a step that uses a base solution to dissolve the exposed areas (acid-containing areas) of positive resists. After developing, the mask pattern is imagined onto the resist. With etching, the resist patterns are transferred to the underlying substrate. To ensure good pattern transfer fidelity, the resist must withstand the harsh wet chemical or plasma etching ambient without being lifted or eroded. With implantation, the ions are implanted into the desired pre-defined areas to form the proper device structure. The resist must not outgas under high vacuum conditions. Furthermore, the resist must be strong enough to withstand the highly energetic impinging ions without deformation or burning. To achieve these goals, a hard bake at around 110°C to 120°C is conducted.

During hard bake, residual solvent is further driven out to prevent the residual solvent from bursting under high vacuum conditions such as ion implantation. The resist volume contracts as a result of hard baking; the film shows a tensile stress, which can be relaxed after a period of time. The hard bake also enhances the adhesion between resist and the substrate interface. The optimal temperature for the best adhesion can vary with different interfaces. It should be noted that an overbaked resist could be very difficult to remove, even after the etching or implantation steps have been completed.

Chapter 8

PATTERN GENERATION

This chapter is intended to give readers a fundamental understanding of the mask making, or pattern generation, process. In addition to that, the chapter will introduce a few resolution enhancement techniques and principles. We start out in Section 8.1 with an introduction of the overall mask-making process flow and the requirements for masks. Further explanations of the steps are given in Sections 8.2 and 8.3. Section 8.2 covers the front-end processes, including e-beam writing and resist chemistry. Section 8.3 covers the back-end processes, including inspection, repair, cleaning, and pellicle mounting. The last section, 8.4, explains the principles of a few types of resolution enhancement techniques RET such as phase-shift masks, optical proximity corrections, off-axis illumination, and subresolution assisting features. The use of these RETs do affect mask making. The implications are illustrated.

8.1. Introduction

Pattern generation or mask making is a process in which a layer of the IC circuit pattern is engraved onto an absorber-on-quartz blank; this results in a photomask. The photomask is then repeatedly used for photolithography in wafer manufacturing. In contrast to the pattern generation, photolithography is a pattern transfer step, in which the pattern on a mask is transferred, or printed, onto wafer surfaces. Comparing to wafer processing, mask making is done on quartz substrates with a single layer of absorber. There are no topography issues. However, because this is the mother pattern that will

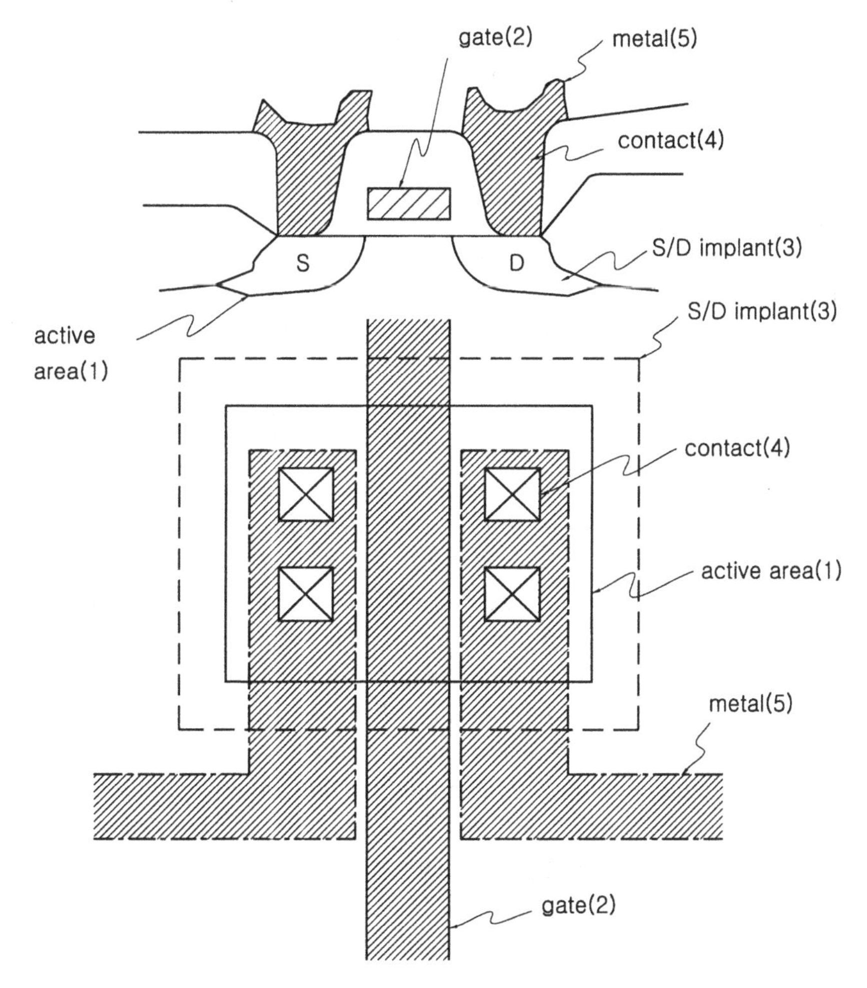

Fig. 8.1. A schematic MOS transistor structure with five masking layers. The numbers in the parentheses indicate the masking sequence.

be repeatedly printed on wafers, the quality must be impeccable, in the sense that no printable defect is allowed. Otherwise, any defect will be printed on every wafer.

An IC product design, be it a consumer product or a memory product, starts from a market survey and consolidation and a product definition. The product is then designed and verified. Once the product design is complete, the database is sent to a mask shop. The mask shop receives the database on which the circuit is presented in the form of physical representations composed of polygons. The circuit is also split into a number of layers; each layer is drawn with a distinctive color or pattern of stripes. Figure 8.1 shows an example of a simple MOS structure with its physical layouts for each layer. Each of these layers corresponds to a mask. It is customary in the industry to refer to a complete chip on a mask or a wafer as a die. On each mask, there could be one die or a number of dies, depending on the die size and the image field size of the exposure tools. Figure 8.2 shows a single-die mask and a multidie mask. For the former, one exposure of the mask prints a die on a wafer; for the latter, one exposure prints two dies on a wafer. A complete device structure is then processed from the bottom to the top. Each mask corresponds to a photolithography step, followed by an etching or implantation step. A typical 0.25-μm logic product has about 25 layers of masks.

A typical mask-making process with a quality checking procedure is shown in Fig. 8.3. Quartz is chosen because of its low thermal expansion coefficient and high transmittance for the exposure light. The blank is about one quarter of an inch thick, to avoid deformation due to gravity. The blank is cleaned and then sputter-coated with absorber; the most commonly used material is chromium. The thickness of the Cr is around 500–1200 Å, depending on the technology node. On top of the Cr layer, a thin layer of chromium oxide is

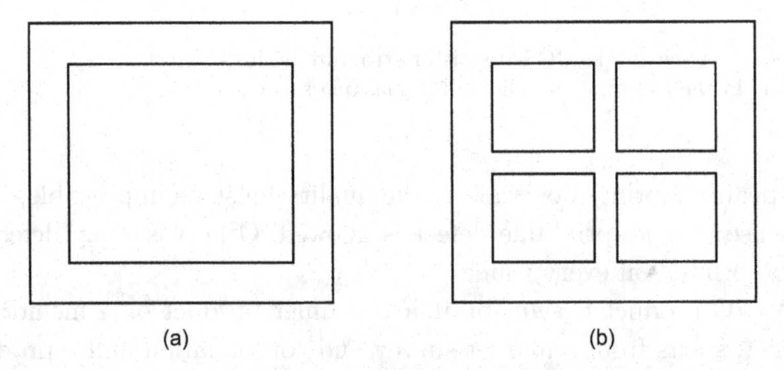

Fig. 8.2. A (a) single-die versus a (b) multidie mask: the first prints one die on wafers with one exposure, while the second prints four dies.

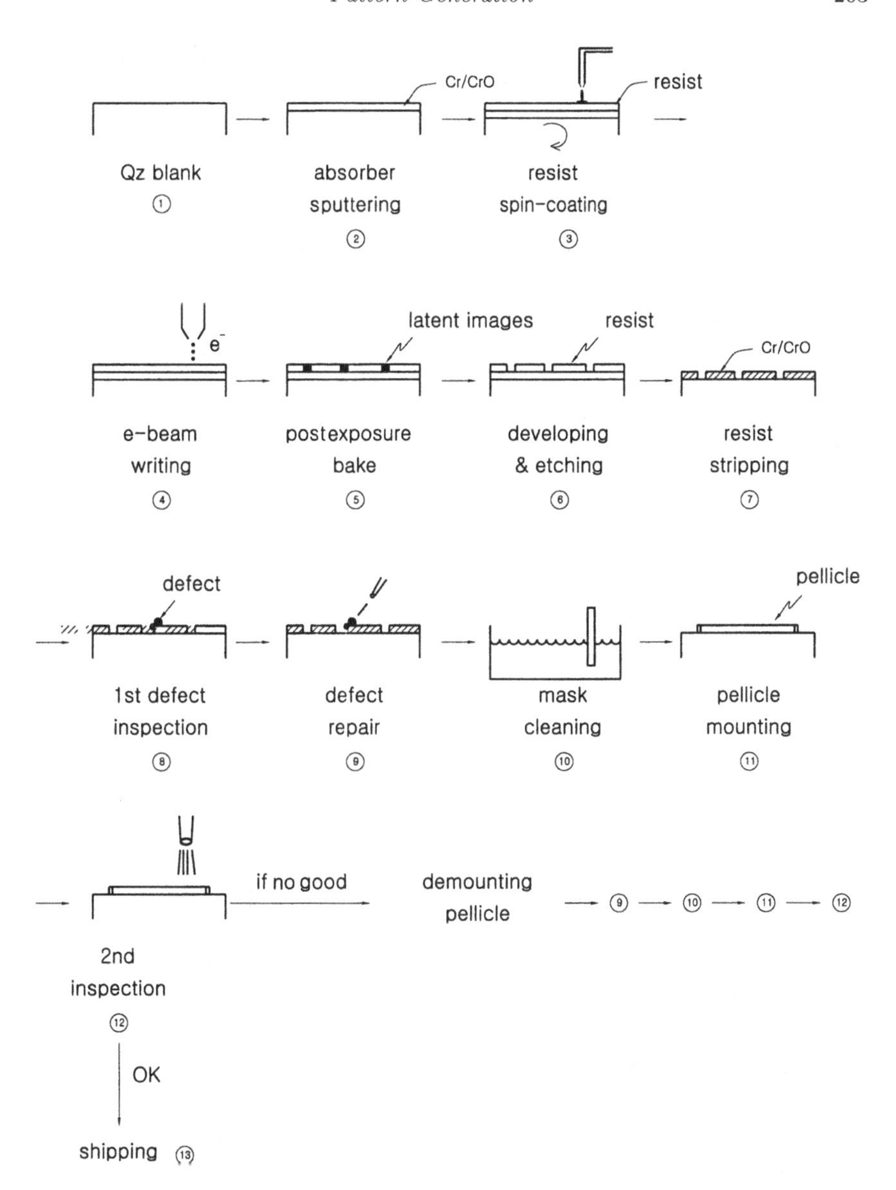

Fig. 8.3. A typical mask-making process with quality checking flow.

deposited to lower the Cr surface reflectivity, which could induce stray light, degrading wafer printing quality. The flatness of a mask blank surface is around 0.5–2 μm. There are two types of blanks: one for binary masks and the other for phase-shifting masks. For binary masks, Cr is used as the absorber; the tone is either clear (without Cr left, light transmitting) or dark (covered with Cr, opaque). For a phase-shifting mask, a phase-shifting material (mostly $MoSi_xNO$) is used to shift the phase angle of the incident light so as to enhance the image contrast. Further details will be discussed later in this chapter. After the absorber is deposited, photoresist is then spin-coated onto the square substrate.

Unlike wafer manufacturers, most mask shops start their mask-making process from resist-coated blanks provided by external suppliers; that is, they leave the quartz blank polishing, cleaning, Cr sputtering, and resist coating to the suppliers. There are a couple of reasons for this. One is because the volume is often too low to be economical for an individual mask shop; the other is associated with the difficulty of resist coating on square substrates. A mask shop starts the mask-making process with e-beam writing on the substrate, using mask layer data transferred from computers. The circuit layer data must be converted to a format that is readable to the e-beam writer at this stage. After the exposure, chemical reactions take place in the bulk of the resist. The blank is often placed in a hot plate for post-exposure baking to allow the reactions to continue in a controlled ambient environment. The latent image is then developed after post-exposure baking. After that, the pattern on the resist is transferred onto the absorber, either Cr or shifter, using plasma or wet chemical etchings. Resist is then stripped off after the etching is complete.

In contrast to wafer processing, the mask has to be defect-free; otherwise, the defects get printed on every single wafer that goes through the exposure. Therefore an inspection procedure is needed to capture the possible defects. There are two approaches to inspection. One is the die-to-die approach, comparing the manufactured patterns of two neighboring dies on the same mask. The other is the die-to-database approach, comparing the manufactured mask patterns to the database. Any differences found in the comparison are considered as defects. The defect coordinates are transferred to a repair machine. The repair machine removes the extrapattern defects with an ion beam or a laser beam and fills the void defects with an ion-beam-induced or laser-induced microdeposition of carbon.

A cleaning procedure is required after the repair to clean off residues and particles left on the mask. Just as with the passivation at the end of wafer manufacturing, in mask making, a pellicle is needed to protect the mask pattern from being scratched. More important, the pellicle prevents particles from directly landing on the quartz surface. Particles on the pellicle surface are not printable on wafers as they are off-focus. A second pattern inspection is required to ensure that no particles are incorporated during mounting. Once the second inspection result is acceptable, the mask is ready for shipping.

The ultimate requirements for mask making are threefold. The first requirement is having good critical dimension uniformity (CDU) across the mask and the mean CD-to-target difference. Figure 8.4 shows the trend for CDU requirements for advanced technologies. Unlike wafer measurements, mask CDUs are often expressed in a range (maximum CD–minimum CD) instead of 3σ, mainly due to the relatively small number of measurement points. These normally range from 20 to 100 points per mask made. Nowadays, there is a demagnification factor, 4× or 5×, for photolithography technology.

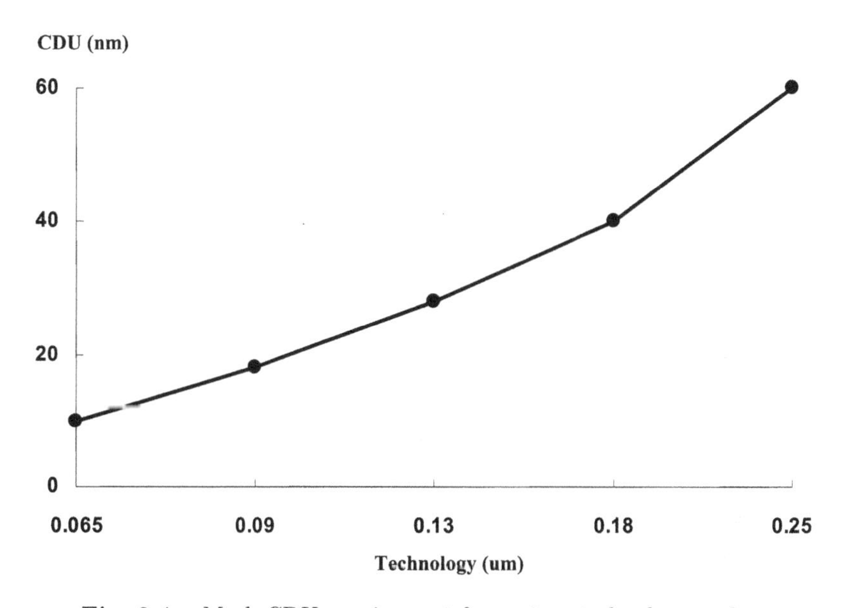

Fig. 8.4. Mask CDU requirement for various technology nodes.

This allows more forgiving specifications for mask making. However, as photolithography technology advances to a point where the to-be-printed feature sizes are close to the operating wavelength, there emerges an error factor, the mask error enhancement factor (MEEF):

$$\text{MEEF} = \frac{\partial(\text{CD}_{\text{wafer}})}{\partial(\text{CD}_{\text{mask}})/4}. \tag{8.1}$$

The perfect MEEF is unity, but a severe diffraction effect tends to enlarge it. In general, the MEEF increases with decreasing geometry and increasing pattern density. The MEEF is also related to the photolithographic process parameters and the resist system used. Figure 8.5 demonstrates that the MEEF for an advanced mask tends to increase with decreasing design pitches. One can see that the CD error on a mask can cause the same size of wafer CD error if MEEF is close to 4. This seems quite common in nanometer technologies. It is this factor that pushes mask CDU specifications. The difference between the mean CD to target is another important mask CD specification. If the CD were off target, it would be very difficult for the wafer exposure tool to shoot the correct CD on the wafer, which in turn would significantly affect device performance.

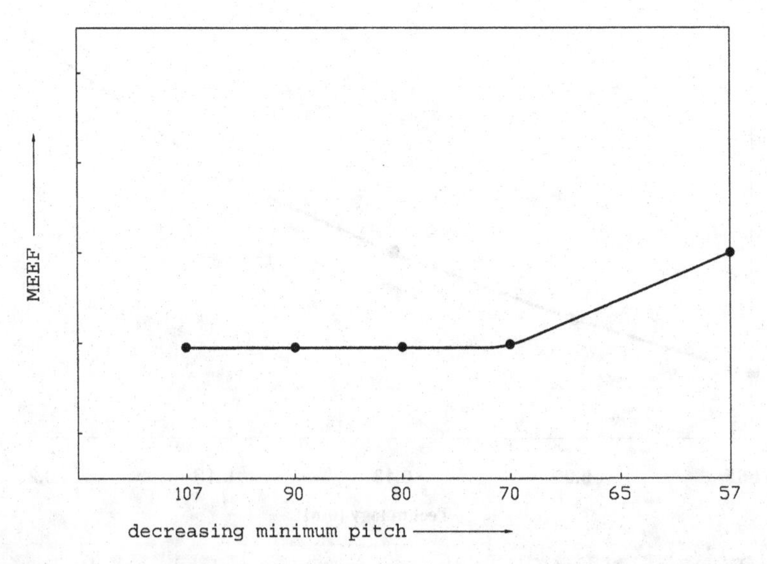

Fig. 8.5. MEEFs tend to increase with decreasing design pitches.

The second requirement for mask making is that all the masks of a whole set must be aligned with each other. All mask layers must be aligned to a reference file, which is composed of the coordinates of a number of alignment marks. Once a reference file is set up on an overlay measurement tool, it is considered the reference grid. In mask manufacturing, each mask is embedded with the same number of alignment marks, evenly located on the edges of the mask patterns. For each mask made, the positions of these marks must be measured and compared to the reference coordinates. The maximum deviation values of the x- and y-directions are considered the overlay errors. Again, the mask overlay readings are reduced by the demagnification factor of a wafer exposure system, $4\times$ or $5\times$; that is, the mask-induced wafer alignment error equals the mask overlay error divided by the demagnification factor. As long as the mask overlay errors are known, one can set up the wafer exposure so as to partly compensate for the errors.

The third requirement for mask making is that each mask must be free of printable defects. The printability of defects depends on the resolution of the exposure tool. If defects are too small to be resolved under the exposure tool, they are considered nonprintable; there is no need to repair them.

8.2. Electron Beam Writing and Resists

8.2.1. *The e-beam system*

Electron beams are often used to generate circuit patterns on masks, mainly because of their high resolution and accuracy. Electron beam lithography systems have evolved from SEM systems. Figure 8.6 shows a schematic of an electron beam system. It includes the e-gun, the column, the stage, and the control computers. The electron beam emitted from the e-gun passes downward through the column, which focuses and deflects the beam toward the substrate. The substrate is placed and fixed on a stage, the movement of which is controlled with respect to the beam position so as to control placement accuracy. The stage is mounted on an antivibration table that absorbs all

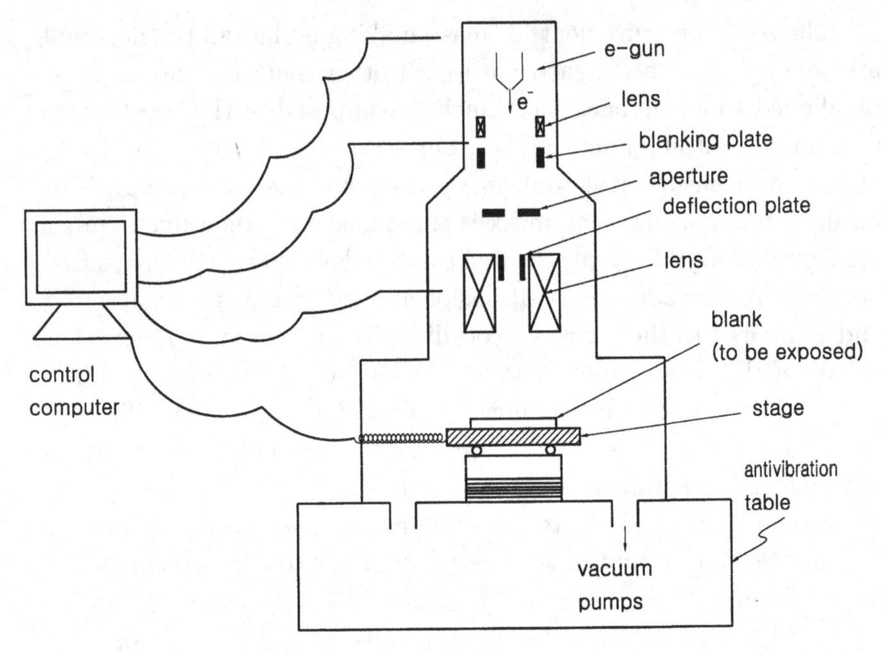

Fig. 8.6. A schematic of an e-beam system, composed of an e-gun, column, stage, and control computers.

vibrations from the environment. The control computer is the brain of the system, controlling the beam and stage movement as well as data transfer.

An e-gun is used to deliver an electron beam with a narrow electron energy range. There are two ways to extract electrons from a conducting solid surface. One is thermo-ionic emission, as shown in Fig. 8.7. This method resistively heats the solid to a very high temperature so that electrons in the solid material essentially evaporate from the solid surface into the surrounding space. The other method is field emission, as shown in Fig. 8.8. This method applies a high electrical field ($\sim 10^9$ V/m) to accelerate electrons to escape from the solid surface. For e-beam mask writing applications, the lanthanum hexaboride crystal is a commonly used electron beam source because of its relatively high brightness, low electron energy spread, and long lifetime as compared to a tungsten filament. Lanthanum hexaboride

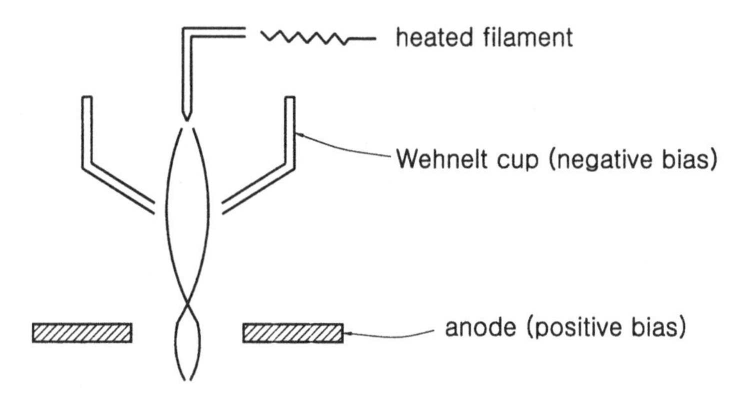

Fig. 8.7. Thermo ionic e-gun: electrons are emitted from a heated filament tip, moving toward the anode.

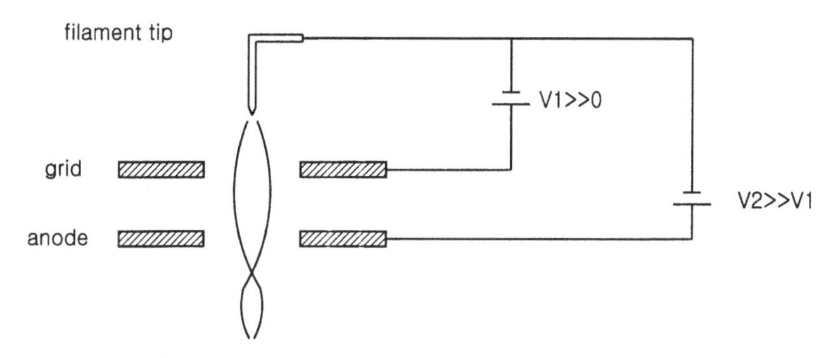

Fig. 8.8. Field emission e-gun: a strong electrical field (> 10^9 V/m) is applied to extract electrons from the filament tip.

can be grown in a Tammann vertical furnace by mixing La, B, and high-purity LaB_6. The grown crystal is then further polished and machined into a tip shape, as needed.

The tip is mounted on a cathode. As the cathode is heated up, an electron stream is accelerated down the column by the positive potential of the anode plate. Electrons exit through the small hole of the Wehnelt cup (a grid). The negative potential of the cap pushes the electron stream toward the optical axis of the column, forming a cross-over point below the grid, which is the virtual source size of the e-gun. The emission current increases with the power input to the filament; so does the brightness. The higher brightness represents a higher dosage that the beam can deliver to the exposure substrate per unit time. In particular, the higher the brightness is, the higher the exposure throughput will be. Unfortunately, as the beam current increases, the Coulomb effect (interactions) among the electrons in the beam increases. This results in electron energy spread (aberrations), which degrades the beam resolution. Refocusing through the magnetic lenses can improve the edge resolution.

The positively biased anode aperture extracts the electron beam, only allowing electrons to pass through its center hole. The voltage of this anode plate determines the energy of the electron that lands on the substrate. The electron beam can be focused or deflected electromagnetically or electrostatically, just like a conventional lens that can be used to refract the light, as illustrated in Fig. 8.9. A magnetic field, if properly set up for an electron beam, can send moving electrons into a spiral motion, hence focusing the beam. A magnetic lens is basically a copper coil embedded in an iron pole. The opening of the pole gives the magnetic field. By adjusting the current in the copper coil, one can alter the magnetic field strength.

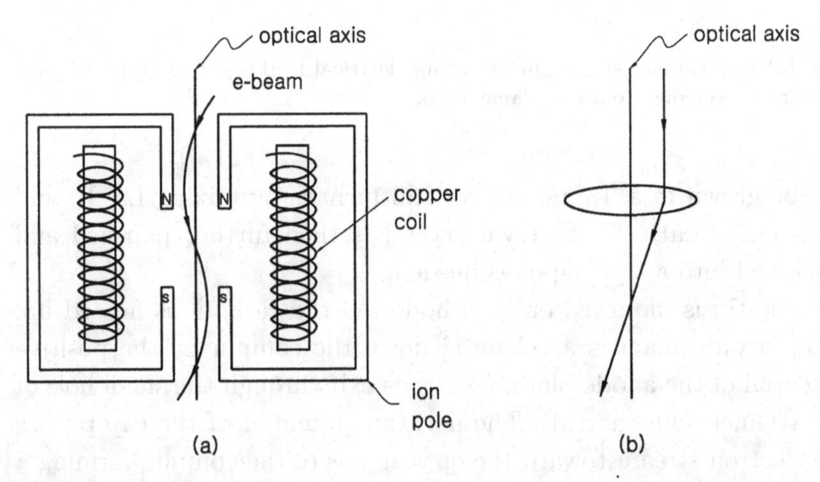

Fig. 8.9. A magnetic lens (a) versus an optical lens (b); both are capable of bending the incoming rays.

The electrical field, on the other hand, can directly deflect the beam. Important characteristics of an electron beam source include source size, brightness, and energy spread. The blanker can basically turn the beam on and off. The aperture lets the portion of the beam pass through the aperture, shaping the beam cross section.

There are various mask writing (exposure) strategies for e-beam mask writers. Each has its own pros and cons. The writing strategy basically involves varying the electron beam shapes and the ways the beam moves during writing. There are three common types of electron beams. The Gaussian beam is essentially a focused beam spot, that is, the cross-sectional area of the beam at the cross-over spot right after the Wehnelt cup. The beam size is fixed. The Gaussian beam is most often coupled with raster scan, in which the beam rasters through every location of the blank. As the beam rasters, the stage moves perpendicular to the direction of beam movement, as shown in Fig. 8.10. The beam is blanked off in areas where exposures are not intended.

A fixed shaped beam uses a large spot beam shaped by an aperture. It exposes large features such as polygons or parts of a polygon. An even more flexible and larger size of beam is the variable-shaped beam, the beam size and shape of which can be varied according to

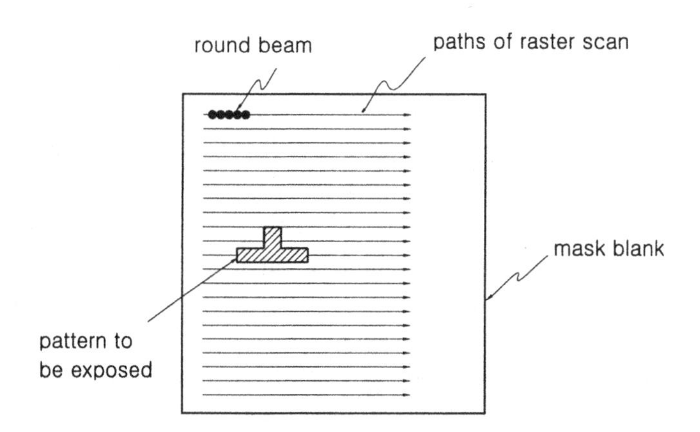

Fig. 8.10. A typical raster scan e-beam writing strategy. A fixed Gaussian round beam rasters through every location of the blank.

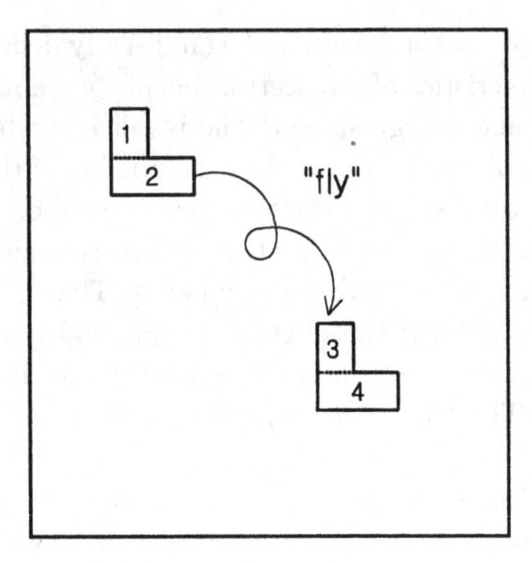

Fig. 8.11. A typical variable-shaped vector scan e-beam writing strategy. The beam shape changes when arriving at each small figure.

the figures that need to be exposed. This approach greatly improves the exposure throughput if coupled with a vector scan. A vector scan differs from a raster scan in that it does not sequentially go through every location of the blank; instead, it finishes exposing one local area and then flies to the next. Figure 8.11 shows a vector scan with variable-shaped beams. The obvious disadvantage of the vector scan is that as it flies to the next exposure area, a beam settling time is needed before the subsequent exposure resumes. This settling time is not needed for the raster scan. The variable-shaped beam coupled with a vector scan is used in most advanced e-beam mask writers. This approach saves a great deal of writing time when applied on sparse patterns such as hole patterns. As pattern density increases for dense layers, such as high-end gate layers or active layer masks, the vector scan system requires a significant increase in writing time. This is not true for the raster scan because it rasters through every location anyway.

When they impinge on the resist and substrate, charged particles, such as electrons and ions, see the resist and substrate as arrays of nuclei surrounded with electron clouds. This is different from the situation for optical waves. As a result, for e-beam lithography, the incident electrons can be forward-scattered or backward-scattered, depending on the incident angles and the atomic number of the target atom. Owing to their light weight, electrons lose an insignificant amount of energy on scattering. When the electrons are scattered back to the resist bulk, they expose the resist in exactly the same way as the primary incident electrons from the e-beam source, as illustrated in Fig. 8.12. These scattered electrons affect (add on to) the exposure dosage of the neighboring patterns, causing the CDs of a specific feature to vary with neighboring pattern density. This is called the proximity effect. For the same target CDs in positive resists, features located next to a blanket exposure tend to be smaller than those located next to an unexposed area or dense patterns. The range of influence of the backscattered electrons tends to decrease with increased electron beam voltage; the higher the electron voltage, the smaller the affected range. On the other hand, thinner resist is less affected by the scattered electrons. As a result, the minimum achievable feature size decreases with increasing e-beam voltage and decreasing resist thickness. To obtain uniform CD across the whole

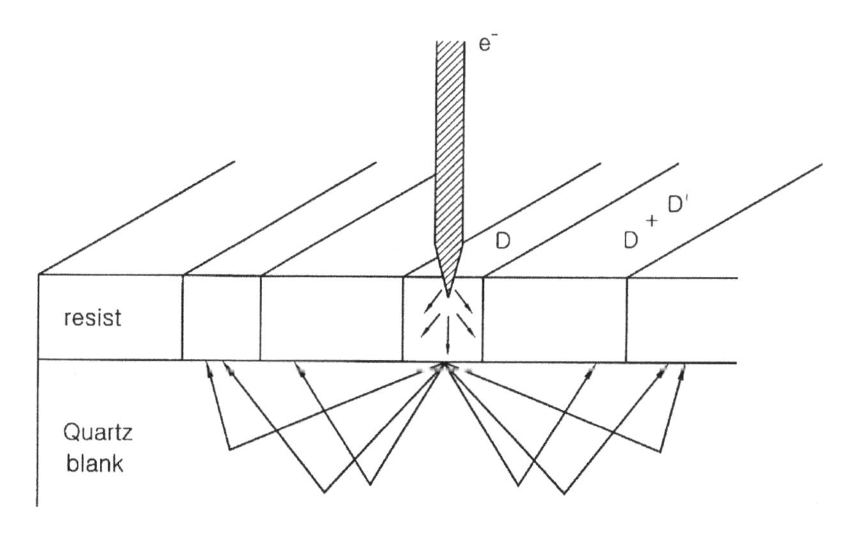

Fig. 8.12. The electron beam backscattering phenomenon.

mask, extra dosage compensation, that is, optical proximity correction, must be implemented. There are two approaches for achieving this purpose: one is to adjust the exposure energy according to adjacent pattern densities, and the other is to compensate for the dosage of the backscattered electrons with a defocused predosage.

8.2.2. *The e-beam resist*

In most resists, functional groups can be cleaved with an electron beam. The cleaved molecules then undergo various reactions that lead to differential solubility during the developing stage. If the exposed area turns out to be insoluble in developer, the resist is called a negative resist; conversely, if the exposed area is soluble in developer, the resist is a positive resist.

Polymethylmethacrylate (PMMA), as shown in Fig. 8.13, is one of the early successful resists developed for DUV and e-beam lithography processes. In the 1970s and 1980s, extensive research activities were concentrated on modifying polymers to suit various needs. The electron energy causes molecular chain scissions, which result in polymers with lower averaged molecular weight. The averaged molecular weights decrease with increasing electron dosage or energy density.

The developer solution is typically a mixture of methylisobutylketone (MIBK) and isopropane alcohol (IPA). PMMA has good resolution and wide process latitude but lacks dry etching resistance and requires high dosages (50–100 $\mu c/cm^2$ at 20 keV). The high dosage

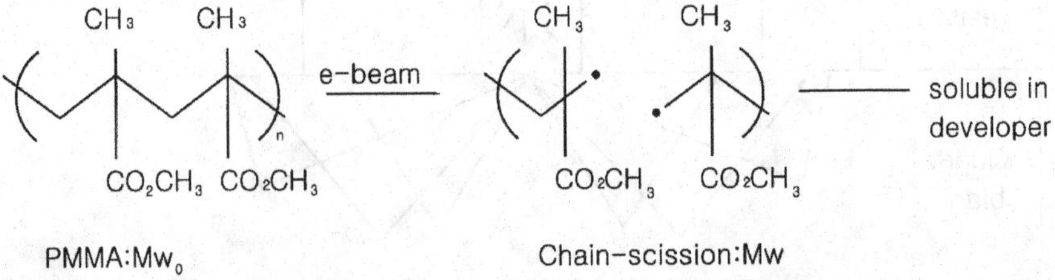

Fig. 8.13. PMMA undergoes e-beam-induced chain scission reaction.

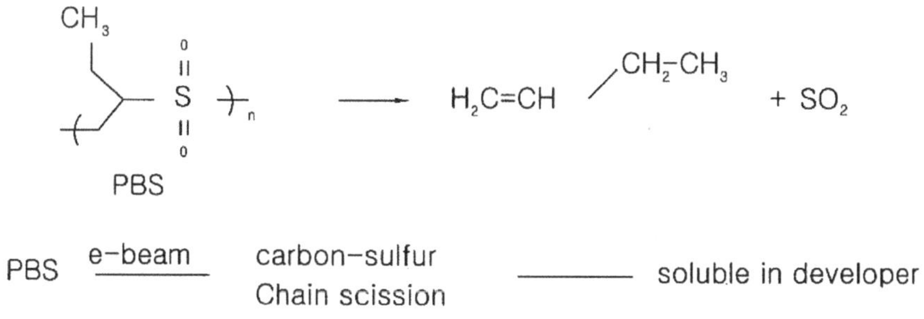

$$PBS \longrightarrow H_2C{=}CH{-}CH_2{-}CH_3 + SO_2$$

PBS $\xrightarrow{\text{e-beam}}$ carbon–sulfur Chain scission ——— soluble in developer

Fig. 8.14. PBS resist structure.

means low throughput and high cost in production. As a result, PMMA is rarely used in production.

Poly-butene-1-sulfone (PBS) has emerged as the alternative to PMMA in terms of sensitivity. The PBS structure is shown in Fig. 8.14. The electron energy mainly dissociates the C–S bonds, giving rise to smaller fragments of sulfur-containing polymers. Further scissions occur if the resist is overheated to above 65°C or so, and sulfur dioxide is released as a result. This outgassing should be eliminated to avoid exposure chamber contamination. The required dosage for PBS is about 3–5 $\mu c/cm^2$ at 30 keV, which is roughly an order of magnitude smaller than that required for PMMA. This translates into a much higher production throughput. The drawback of using PBS is that the high sensitivity may lead to narrow CD control latitude. Although PBS is good for wet etching processes, applicable down to a 0.25-μm technology node, it cannot withstand dry etching plasma.

Poly-methyl-α-chloroacrylate-co-α-methylstyrene (ZEP) has appeared more recently as a good alternative for low dosage requirements and good plasma etching resistance. The molecular structure of ZEP is illustrated in Fig. 8.15. The electron-energy-induced carbon–carbon scission is the mechanism leading to a lower averaged molecular weight polymer in the exposed area, and hence a differential solubility in developer solution. The applicable dosage for ZEP is around 25 $\mu c/cm^2$ at 30 keV. ZEP demonstrates excellent plasma etching resistance. It has been used for mask-making technologies from 0.25 to 0.13 μm. As technology migrates beyond

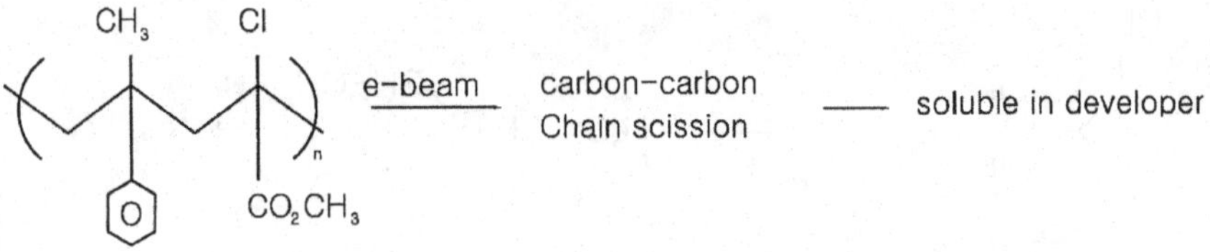

CH_3 Cl

e-beam → carbon-carbon Chain scission → soluble in developer

CO_2CH_3

Fig. 8.15. ZEP resist structure.

0.18 μm, the CD control budget tightens significantly. Advantages offered by ZEP tend to lag behind advanced technology needs.

Chemically amplified resists (CARs) technology, as developed for 248- and 193-nm photolithography, are also adopted for e-beam lithography. CARs provide several advantages over the above mentioned e-beam resists. CARs have very good resolution and CD control and require a low dosage, around 8 $\mu c/cm^2$ at 50 keV. CARs also have very good plasma etching resistance. They are abundantly available with stable quality, mainly due to the fact that they are being used for wafer photolithography production. Figure 8.16 demonstrates a typical acid catalyzed deprotection reaction of the tBOC CAR resist. Use of negative-tone CAR is gaining momentum in high-end mask production. Apart from its excellent resolution, high throughput, and plasma etching resistance, it is obviously advantageous for high pattern density layers such as high-end poly-gate or active layers. For these layers, the e-beam exposing area is about 60% to 80% for positive-tone resists. These translate into 40% to 20% for negative-tone resists, hence gaining throughput. Furthermore, the negative resist has much less proximity effect for isolated features than does the positive resist. For this application, a negative resist gives the potential for better CD uniformity control.

Despite the above mentioned advantages, chemically amplified resists have some lingering issues that remain to be resolved. First, owing to the action of the acid catalytic reactions, CARs are extremely sensitive to airborne bases such as ammonia. The base neutralizes the generated acids, poisoning the catalytic reaction and leading to resist scums, CDU degradation, or profile deterioration.

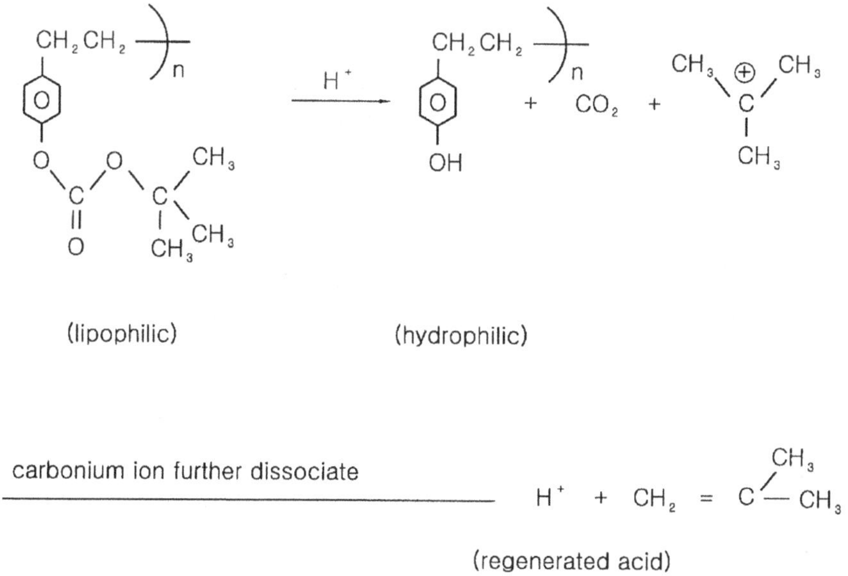
CH2CH2
n
H+
CH2CH2
n
+ CO2 +
CH3 ⊕ CH3
C
CH3
O O CH3
C C
O CH3 CH3
OH
(lipophilic)
(hydrophilic)
carbonium ion further dissociate
CH3
H+ + CH2 = C — CH3
(regenerated acid)

Fig. 8.16. Acid-catalyzed deprotection reaction of a positive CAR resist.

Second, CD performance tends to be affected by various delay times such as post coating delay (PCD), postexposure delay (PED), and post-PEB delay (PPD). The three delays are somewhat related to the ambient conditions. With nitrogen ambient, the CD drifts of PCD and PED are reduced, while the PPD shows marginal difference. It has also been shown that PED seems to dominate the CD drift among the three steps, even with nitrogen ambient. This is related to the fact that the generated acids are prone to base neutralization. In view of the resist stability and CD control, it is very important to keep ammonia concentration as low as possible. It can be achieved by adding chemical filters in clean room airflow paths.

8.3. Back-End Processing for Mask Making

The back-end processing, the process segment that occurs after pattern definition is done and resist is stripped, includes mask inspection, repair, cleaning, and pellicle mounting. In mask manufacturing, there are two types of defects: soft and hard. Defects that can be removed by the mask cleaning process are called soft defects. Conversely, defects that cannot be removed with cleaning are called hard defects, and they must be repaired (removed).

8.3.1. *The mask inspection*

A mask inspection system, as shown in Fig. 8.17, is typically equipped with powerful computers that handle data preparation, illumination, light sources and optics, the mask handling mechanism, and the image processing and detection mechanism. The light source illuminates a reflected mirror toward the condenser lens through the mask pattern, where the light is either reflected or transmitted. Both the reflected and transmitted lights are collected and processed by computers.

There are two types of inspections: pattern inspection and particle inspection. Pattern inspection is further classified into die-to-die and die-to-database inspections. Die-to-die inspection compares

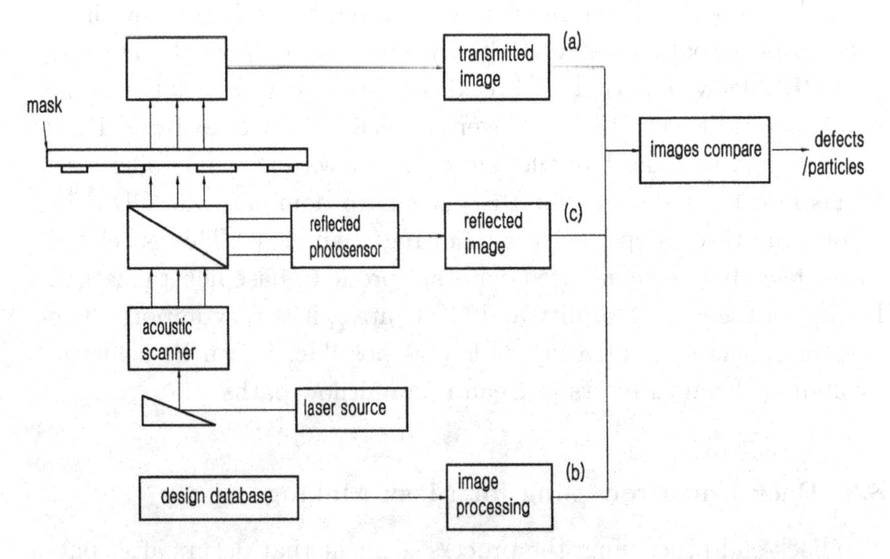

Fig. 8.17. A representative setup for a mask inspection system. (a) a, transmitted image of two neighboring die: die-to-die; (b) $a + b$, transmitted image versus processed database: die-to-database; and (c) $a + c$, transmitted image versus reflected image: particle inspection.

the transmitted images of two neighboring dies. On the other hand, die-to-database inspection compares the transmitted image of a die to its own processed image. Particle inspection (soft defect) results are revealed by the sum of the reflected and the transmitted light intensities, as demonstrated in Fig. 8.18. For a chromium surface, the reflected light accounts for 100% of the incoming light intensity. For quartz, the transmitted light is 100%. In the case of a particle on the surface, some of the light rays are refracted, and the sum of the transmitted and reflected light does not equal the incoming light intensity. The existence of the particle is therefore detected.

For die-to-die inspection, as illustrated in Fig. 8.19, the images of two neighboring dies are compared. The probability of having two

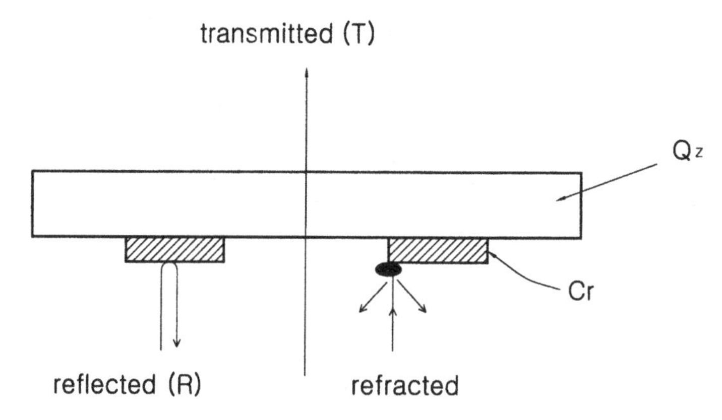

Fig. 8.18. The concept of particle inspection. In a particle that causes light refraction, $\mathrm{T} + \mathrm{R} < 100\%$, a defect is detected.

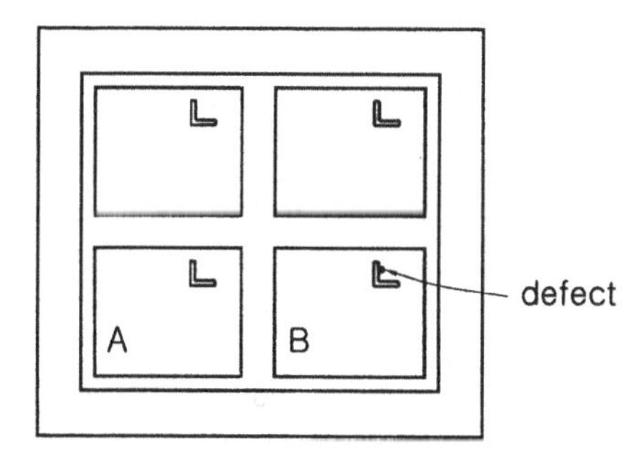

Fig. 8.19. A four-die reticle with a defect lands on the right lower corner die. Comparing A with B die, the defect can be detected.

defects at the same coordinate of two neighboring dies is nearly zero. Hence the image differences between the two dies are classified as defects. With die-to-database inspection, the transmitted light forms a die pattern image, which is then compared with the database. The database is not exactly the CAD database, but the so-called processed database, which looks very similar to aerial images of the database, as shown Fig. 8.20. The processed database images are calibrated to mask processing characteristics. The comparison shows the difference as defects. It is obvious that the defect count of pattern inspections is a result of the inspection sensitivity settings. Which differences are regarded as defects and which are regarded as false counts must be ultimately correlated to the wafer printing results. A tight sensitivity setting gives a high defect count, but many of the counted defects may be false. Appropriate inspection sensitivity often comes with experience and machine stability; however, the best approach would be determining the sensitivity with wafer printability.

Figure 8.21 demonstrates various types of mask defects. Depending on the defect types and sizes, some can be captured by the

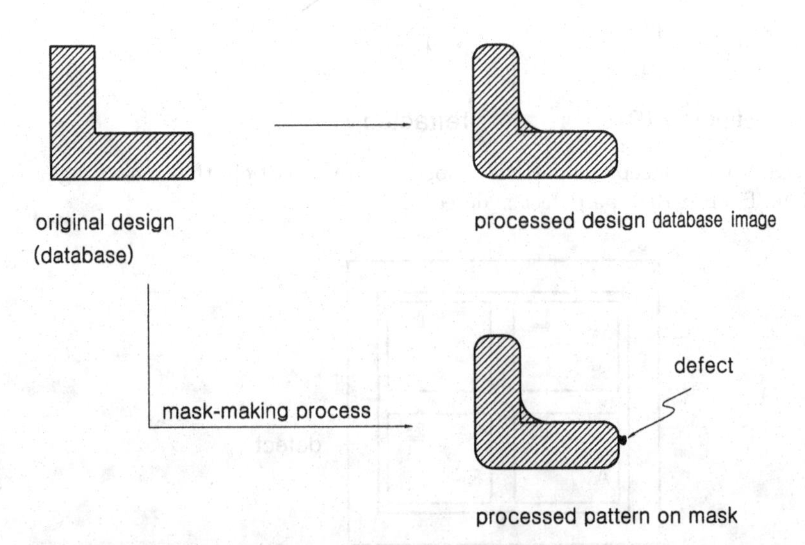

Fig. 8.20. The concept of die-to-database inspection compares the processed database image with the processed pattern on mask.

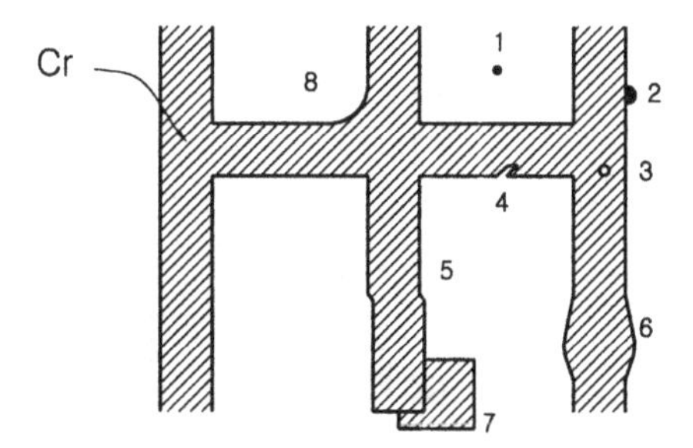

Fig. 8.21. A few examples of defects on mask: (1) pin dot; (2) extrusion; (3) pinhole; (4) intrusion; (5) butting error; (6) CD offset; (7) extra pattern; and (8) corner rounding.

inspection machine; some cannot be captured. Machine capability is expressed in terms of capture rates; for each type and size of defects, the machine has a certain probability of capturing the defects. As the defect size approaches the machine resolution limits, the capturing probability decreases. The actual machine capability is defined as the 100% capturing rate.

8.3.2. *The mask repair*

There are two mainstream mask repair technologies: one with focused ion beams, and the other with laser beams. A focused ion beam (FIB) operates in a similar manner to a scanning electron microscope (SEM), except that the FIB uses a finely focused ion beam in lieu of an electron beam. It employs ion-beam-assisted microchemical vapor deposition for clear-tone defect filling and gas-assisted ion milling for removing opaque defects. When an ion beam hits a substrate surface, the energetic ions impart their energy to the substrate surface atoms, giving off secondary electrons, secondary ions, and sputtered atoms. For a low ion energy level, the collected secondary electrons or ions can give surface image information. With a higher level of ion energy, the incident ions can be used for ion milling an opaque spot, that is, for defect repair, as indicated in Fig. 8.22. An ion beam repair system is composed of several parts. The column holds the ion beam optics, and gallium ions are used in the ion gun. The stage holds the mask to be repaired with an interferometric positioning mechanism.

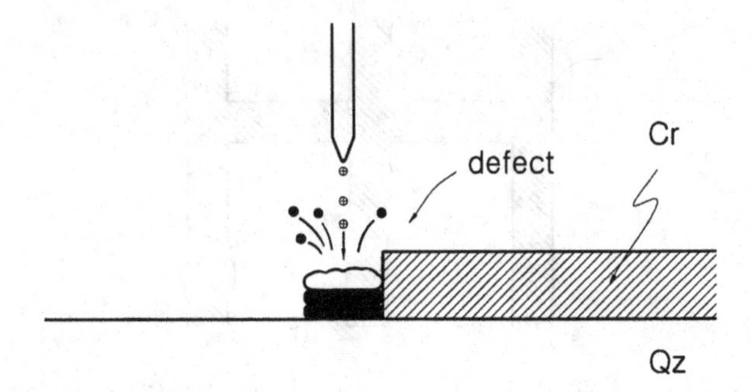

Fig. 8.22. FIB mask repair system uses Ga for milling out the defect.

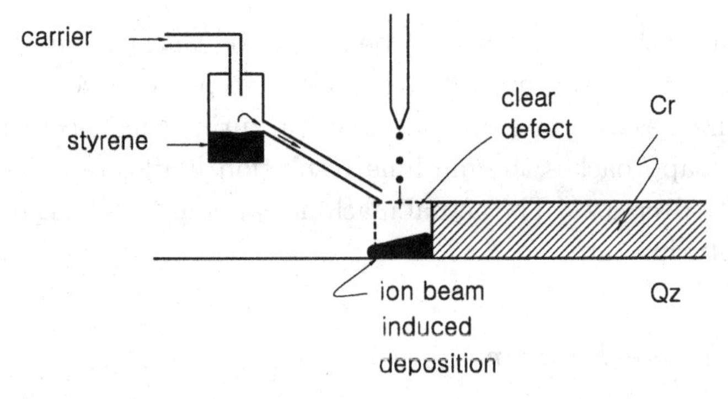

Fig. 8.23. Clear defect (missing Cr) repair uses ion-beam-induced polymer decomposition to form an opaque film.

An electron gun floods electrons onto the substrate to neutralize the ions used for repair. The chemical for the microdeposition is delivered through a nozzle close to the ion gun so that the deposition proceeds right at the clear defect location, as illustrated in Fig. 8.23.

To repair a clear defect, carbon-containing polymers, such as pyrene or styrene, are often used as the reacting gas. The energetic ions result in decomposition of the carbon-containing species and formation of a carbon film, an opaque film:

$$C_8H_8 \quad \text{or} \quad C_{16}H_{10} \xrightarrow{\text{ions}} C(s) + \text{others} \tag{8.2}$$

To repair an opaque defect, ion milling is used. The incident ions knock off the chromium atoms to remove defects. Gas-assisted etching (GAE) can be used to enhance the defect removal rate. Halogen gases are good candidates for Cr removal, while XeF_2 can be used for phase shifter material such as MoSiON:

$$Cr + Br_2 \xrightarrow{\text{ions}} CrBr_{6(g)} \tag{8.3}$$

$$MoSiON + XeF_2 \xrightarrow{\text{ions}} MoF_{2(g)} + SiF_{(g)} + \text{others} \tag{8.4}$$

The drawback of ion beam repair is that the gallium ions often get implanted into the quartz substrate, causing transmittance loss. A proper postrepair procedure, such as a wet etching, is required to retrieve the lost transmittance. During the repair operation, some of the deflected ions can hit the quartz surface, as demonstrated in Fig. 8.24, knocking the Si atoms out of the quartz surface. Such a phenomenon is called river bedding. Severe river bedding also causes transmittance loss. It shows up on wafers as printable defects. In general, the ion beam can be focused onto a beam diameter of about 1/10th of a micron, much smaller than a laser beam.

Laser beam (UV) repair technology, on the other hand, uses photon energy to initiate photolytic CVD reactions, leading to deposition of material for repair of clear-tone defects and laser ablation for repair of opaque defects. The laser energy profile incident on a substrate surface is fairly close to the Gaussian distribution. Figure 8.25 demonstrates a schematic of the laser repair scheme. The laser beam ($Nd:YVO_3$ of 355 nm or Nd:YLF of 349 nm) rasters on the substrate surface at the defect coordinates transferred from the inspection machine. The defect coordinate information drives the stage to the defect location through a stage control mechanism. Imaging is taken

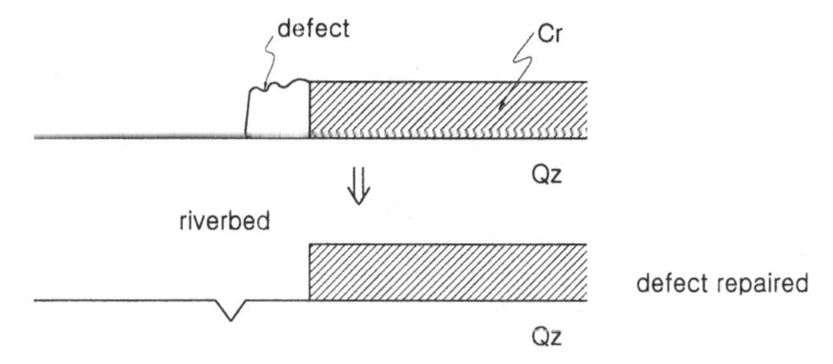

Fig. 8.24. Riverbed results from defect ions during repair.

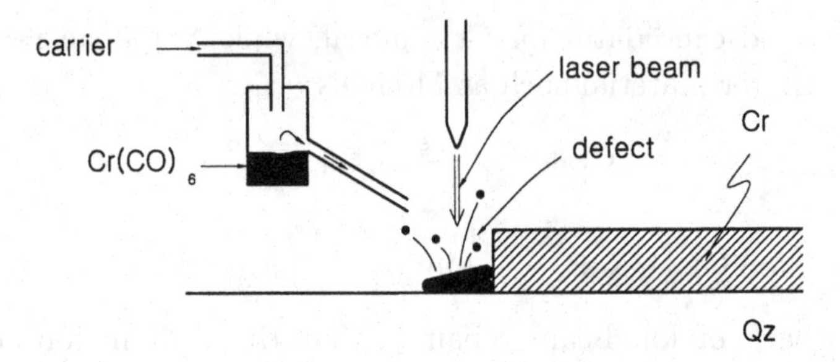

Fig. 8.25. Laser repair system removes defects via laser ablation; clear defects via laser-induced deposition of Cr.

through an optical microscope. The deposition chemical, chromium hexacarbonyl, is fed through a nozzle to the laser rastering position, where the photolytic decomposition occurs:

$$Cr(CO)_6 \xrightarrow{hv} Cr(s) + \text{others} \tag{8.5}$$

The reaction product, chromium, is deposited in the clear-tone defect area. It is very difficult to control the deposited film profile and the spread area for microscale chemical reactions. As long as the deposition covers the area of the defects, it is considered an acceptable repair, providing that the resulting dimension is within less than 10% of the desired value.

Repair of opaque defects employs laser ablation; the photon energy is locally absorbed in the bulk of the Cr in the defect area. The bulk temperature increases with the laser flux and dwelling time (nanosecond range). The instantaneous local temperature can rise above 1000°. The heat is also conducted vertically toward the substrate bulks (both Cr and quartz) and spread laterally. As the bulk temperature rises to the evaporation temperature, the Cr in the defect area is ablated.

The drawback of the laser repair system is that a large amount of instantaneous heat is generated, which could cause damage to the repaired edges, such as rolled up edges, or quartz damage and loss of transmittance. There have been efforts to shorten the laser pulse time from the nanosecond to the femtosecond range to shorten the effects of the generated heat. With laser beam repair technology, the laser beam can be focused to as fine as 1 μm. This is an order of magnitude larger than an ion beam diameter. As a result, the ion beam can be used for technologies from 0.25 μm down to 0.13 μm and possibly beyond, while laser beam technology is used for 0.35 μm and up. Edge misalignments (the alignment between original and the repaired line edge) are often seen during repair. For the 90-nm mask repair tool, the edge alignment accuracy is about 15-nm (3σ). Consequently, an iterative procedure is often required to achieve good edge alignment accuracy for advanced mask repair.

Is repair quality acceptable? The ultimate answer, of course, lies in the wafer printing results. However, when a mask goes to a wafer fab without confidence in the repair quality, it can be a very risky undertaking as a large number of wafers can be at risk. The conventional checking procedure is based on the operator's or engineer's educated judgment. Recent development has resulted in a subjective checking procedure. This involves looking at the repaired area with an aerial imaging system. The system includes major components of an optical column, allowing for adjusting NA, sigma, and illumination. The major difference between a scanner and an aerial image tool is that the latter looks at only a single spot, instead of the whole field. An aerial imaging system shows the intensity profile and a contour plot. To ensure good quality of the repair process, the aerial images are often calibrated against actual wafer prints to account for resist performance. The wafer results set the upper and lower limits of the aerial imaging results in terms of the CD tolerance percentage, for example, $\pm 6\%$, which are then used to judge the quality of all repaired locations on a mask.

8.3.3. *Mask cleaning*

Once the mask reaches the cleaning stage, it should be free of hard defects. Mask cleaning is a necessary step before pellicle mounting. It cleans out the soft defects and residues of the repair process. Soft defects adhere to the mask surface through various mechanisms such as Van der Waals forces, electrostatic forces, physical adsorption, or chemisorption. Van der Waals forces result from dipole–dipole interaction or the electronic polarization of the surface atoms and molecules when two solid surfaces come into close contact. Van der Waals forces increase with the size and shape of the interacting molecules. Longer molecules tend to give larger forces than more spherical or symmetric molecules due to the larger extent of polarization. Flakes of resist material or chipped-off repair carbon films adhering on a mask surface often demonstrate Van der Waals forces.

Electrostatic forces are the other commonly seen forces that make particles stick to a mask surface. Their existence is attributed to the coulombic effects that occur between solid surfaces. The excess static charges on a solid surface polarize the solid surfaces of incoming solid particles, which are possibly carried by laminar fluid flow or by random motion, turning their surfaces into charges of opposite polarity, hence attracting particles onto the solid surface. Tiny particles seen on a mask surface are often visible examples of electrostatic forces. As a solid surface is exposed to ambient gas, some gas molecules could land on the solid surface and become adsorbed. Adsorption decreases with increasing substrate temperature but increases with molecule size. Precipitation, such as ammonium sulfate powder, which is seen on a dried mask surface, often results from the fact that ambient ammonia and sulfur dioxide are physically adsorbed on the mask surface, and they later react to form ammonium sulfate powders. Chemisorption is a result of adsorption, with chemical reactions occurring between the surface and the adsorbents. The binding force of chemisorption is relatively strong as compared to the other forces.

The above mentioned soft defects are supposed to be cleaned with mask cleaning procedures. Typically, a wet mask cleaning bench consists of a series of tanks containing ammonia and sulfuric acid and a de-ionized water rinse. Ammonia and hydrogen peroxide solution (typically NH_4OH:H_2O:H_2O_2 of 1:5:1) is also called RCA Standard Clean-1(SC-1). SC-1 can remove organic and metallic particles and flakes. Hydrogen peroxide is a strong oxidant that oxidizes the particles, and then ammonium hydroxide dissolves the oxidized products by solvation. Sulfuric acid (H_2SO_4:H_2O_2 of 4:1 at 80°C–100°C) oxidizes organic (hydrocarbon) residues, such as resist flakes or pellicle glue residues, at about 100°C. There are several cycles of quick-down rinses (QDR) between the ammonia and sulfuric acid solutions to avoid the formation of ammonium sulfate salt. Also, extensive rinse cycles are needed before the cleaning cycle is done. Failure to use enough water rinse can give rise to the formation of ammonium sulfate, white powders, on the quartz surface.

Wet cleaning recipes must be optimized in terms of defect removal efficiency and CD loss of the mask patterns, or transmittance and phase angle loss of phase shifting materials. Increases in concentrations of SC-1 and ammonium solution as well as increases in wet bench temperatures tend to increase cleaning efficiency, but they result in some negative effects such as binary mask CD shrinkage. Ammonia attacks the phase shifter material (MoSiON), causing the transmittance, phase angle, and CD changes. These variations must be characterized and routinely monitored for each wet bench to avoid mask scrap due to repetitive cleaning cycles. Generally, when a wet bench line is designed for binary mask cleaning, it may not be optimized for phase shifter masks (PSM). Furthermore, to avoid cross contamination in manufacturing, the two types of masks are cleaned in separate wet bench lines.

Following the QDR, the final step in mask cleaning is the IPA cleaning step, in which IPA is sprayed onto the mask surface and forms a thin film falling along the mask surface. Without the IPA film, the DI water dries as a film breaking down into small islands due to surface tension. Particles are easily trapped in the islands, leading to the formation of stains on the mask surface. The IPA film basically displaces the islands and drags them along the film during drying. As the IPA film dries out, the particles are removed as well.

8.3.4. *Pellicle mounting*

After the first mask inspection, repair, and cleaning are done, a pellicle is mounted on the mask. The pellicle is a thin polymer film about 0.8–1.5 μm thick, glued to an anodized aluminum frame, which is about 4–6 mm high. The pellicle frame is then mounted on the mask surface, as illustrated in Fig. 8.26. The primary requirement for a pellicle is its high light transmittance (>99%) and transmission uniformity (>99.9%). To meet these requirements, the composition and

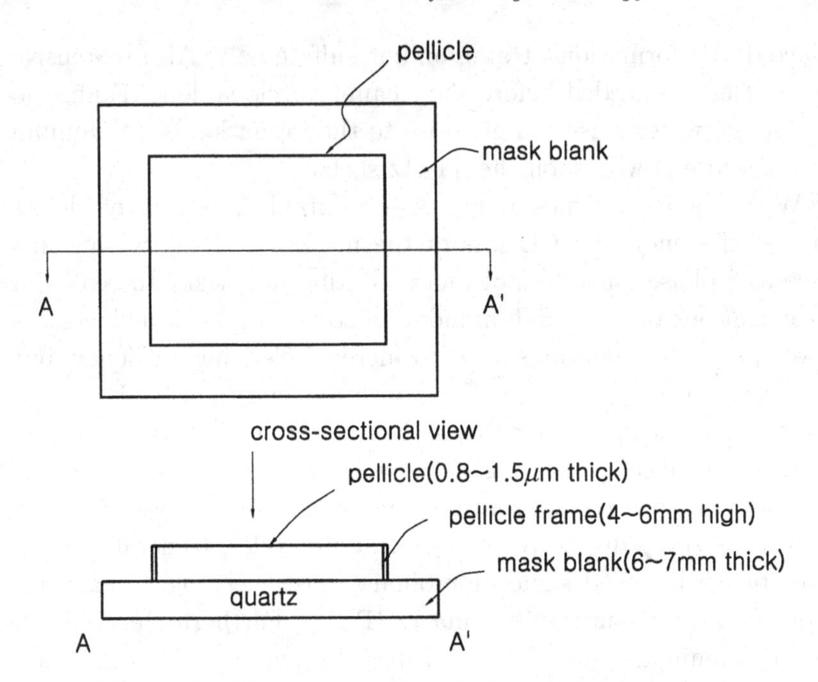

Fig. 8.26. The structure of a pellicle mounted on a mask.

thickness of the polymer must be changed as operating wavelengths shorten.

A pellicle serves different purposes. It prevents particles from landing on the mask surface. The particles would land on the pellicle surface instead. As a result, the particle is out of focus range, becoming nonprintable. A 6″ mask of 6.35 mm thickness has a pellicle mounted on one side. A 5″ mask is much thinner, 2.2 mm thick, and it requires pellicles mounted on both sides. Because the mask is thinner than a 6″ mask, particles on either side would be printable on wafers without the pellicles. With the pellicles mounted, the masks do not need to be recleaned frequently. Most of the particles on the pellicle surface can be carefully blown away with an air gun.

A good pellicle material should possess the following characteristics:

(a) high transmittance
(b) high illumination endurance

(c) excellent light transmittance uniformity
(d) readily available and low cost.

Owing to transmittance requirements at different wavelengths, pellicle composition must vary with the operating wavelengths to have maximum transmittance. Pellicles are composed of cyclic polymers for 365 nm exposure wavelength and cyclic fluoropolymers for 248/193 nm exposure wavelength. The pellicle composition that is optimized for a long wavelength exposure may not be appropriate for a shorter wavelength exposure as the shorter wavelength may burn the pellicle. Pellicle transmittance for a fixed wavelength tends to decrease after a large number of exposures. Sometimes a so-called ghost image can be observed; that is, the circuit patterns get vaguely printed on the pellicle surface. The appearance of the ghost image indicates transmittance decay.

8.4. Resolution Enhancement Technology

Shrinking device geometry has been the never-ending game for the semiconductor industry. Photolithography is the workhorse that enables the game to move forward by pushing the resolution limits. Obvious approaches for pushing resolution limits are to use either a shorter wavelength or a larger NA, which can be realized in the equation discussed earlier:

$$\text{Min. resolvable feature sizes} = k_1 \frac{\lambda}{\text{NA}}. \tag{8.6}$$

NA of a lens corresponds to its physical size and indicates its diffracted light collecting capability, as shown in Fig. 8.27. For a resolved image, the NA corresponds to the minimal lens size in air that can collect the zero- and first-order diffracted light. Any pitches that are smaller than the minimum resolution capability would push the first-order diffracted light a farther apart, beyond the numerical aperture of the lens, and hence the image blurs. On the other hand, if a pitch is larger than the minimum resolution capability, higher-order diffracted light can be collected by the lens, which results in a better image. Improving the resolution limit by increasing the NA

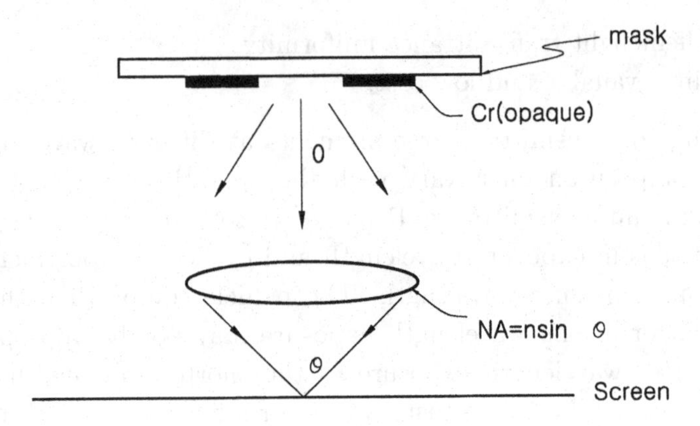

Fig. 8.27. NA of a lens equals $n \sin \theta$. It indicates the diffracted light collecting capability.

would sacrifice the depth of focus (DOF). Furthermore, maintaining superior lens quality, such as low aberration or distortion, can be very costly and difficult for the large NA. On the other hand, using a shorter wavelength would demand that several things be changed altogether, such as the resist, mask blank, and pellicle materials. Oftentimes, the market demand for the end product outruns the evolution of the NA and wavelength. As a result, with the same wavelength and numerical aperture, one often needs to push the k_1-factor of Eq. (8.6) to achieve smaller printed feature sizes with good performance and reasonable process latitudes.

One effective way to push the k_1-factor is to use resolution enhancement techniques (RETs). Some RETs can be realized by changing the mask patterns and structures, others, by changing the illumination approaches. Commonly used resolution enhancement techniques include phase shifting masks (PSM), off-axis illumination (OAI), subresolution assisting features (SRAF), and optical proximity correction (OPC).

8.4.1. *Phase shifting mask technology*

When a beam of light goes through a material with an extinction coefficient of k and refractive index of n, its intensity is attenuated according to Lamber's law:

$$I = I_0 \exp\left(\frac{-4\pi kd}{\lambda}\right), \tag{8.7}$$

where k is the extinction coefficient of the material, d is the thickness of the material, and λ is the wavelength of the incident light. A material is called transparent if the k is zero; it is called absorbing if the k is not zero. For a chrome-on-glass mask (binary mask), the chrome is so strongly absorbing that no light can pass through, while the quartz is so transparent that all light can pass through. Both the refractive index and the extinction coefficient are functions of the material and the operating wavelengths.

Phase shifting technology takes advantage of the fact that when light passes through a material with a refractive index n, its speed slows down by a factor of $1/n$. This creates an optical path difference between the light that goes through the material and the light that does not, as shown in Fig. 8.28.

The optical path difference is,

$$\text{OPD} = (n-1)d, \tag{8.8}$$

and the corresponding phase difference is:

$$\text{OPD} = \frac{2\pi(n-1)d}{\lambda}, \tag{8.9}$$

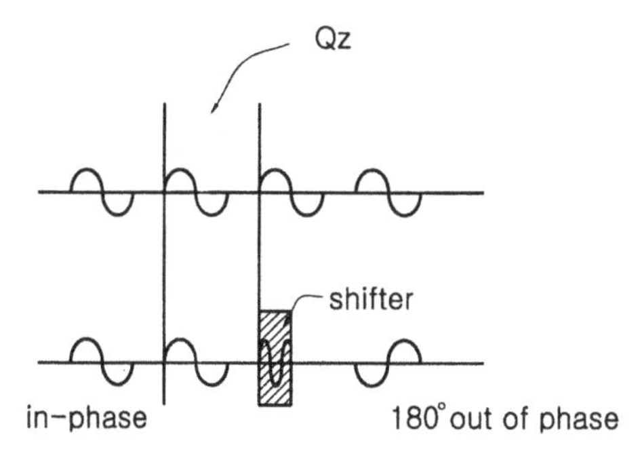

Fig. 8.28. A phase shifter with a designated n, k, and thickness can be used to shift the phase of a light wave.

where n and d are the shifter's refractive index and thickness, respectively. Now, if one desires to have a phase difference of 180°, then the required shifter thickness, d, is

$$d = \frac{\lambda}{2(n-1)}. \tag{8.10}$$

It can be observed that the higher the refractive index or the lower the operating wavelength, the thinner the shifter is required to be achieve a desired value of phase shift. With a weak phase shifter material, such as a rim type or an attenuated PSM, the transmittance is relatively low, ranging from 5% to 15%. The purpose is to pull the attenuated transmitted light rays to the opposite phase and hence improve the contrast of the aerial images, as illustrated in Fig. 8.29.

Attenuated PSM is widely used for hole patterning. Figure 8.30 demonstrates the 0.4-μm line patterning performance improvement in terms of aerial image slope. With attenuated PSM, the slope

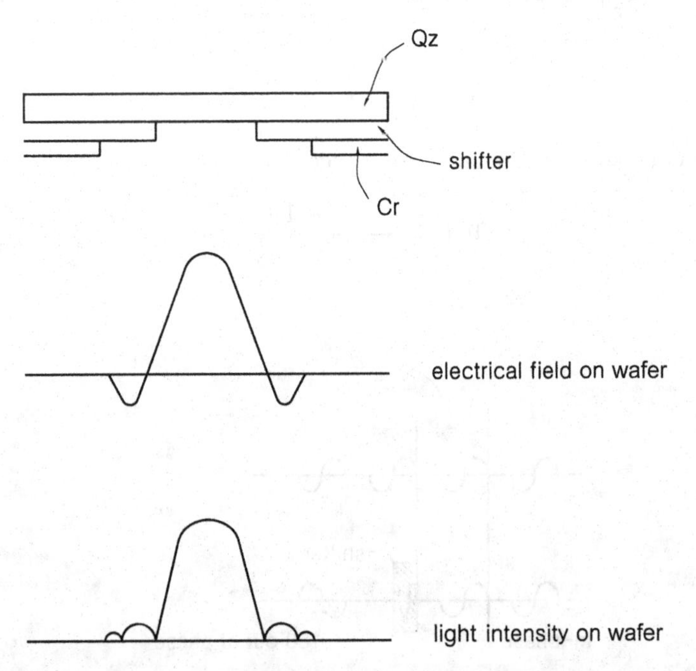

Fig. 8.29. The rim-type PSM using 5–15% of phase shifting material to enhance the edge contrast.

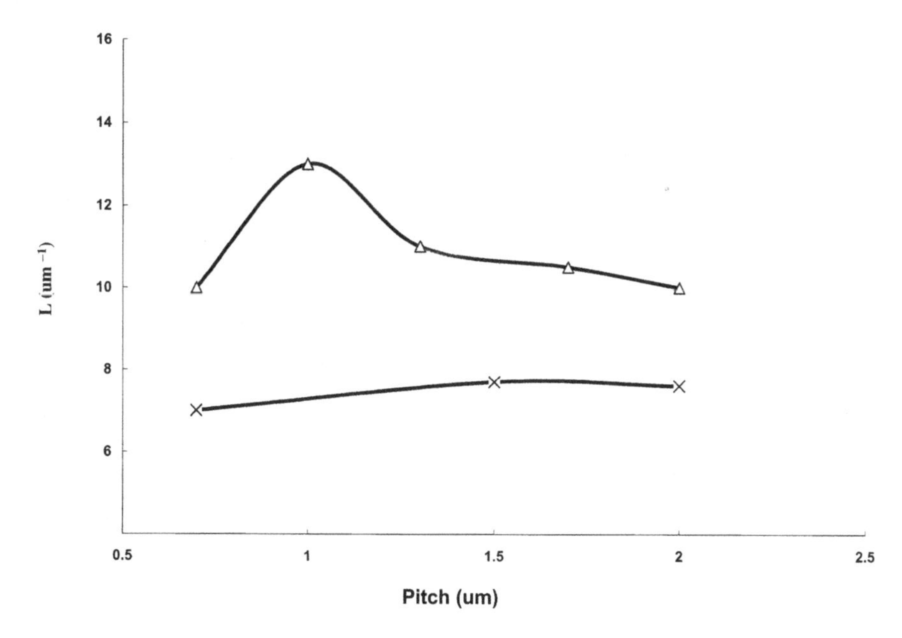

Fig. 8.30. A rim-type PSM for 0.4-μm line (Δ) results in better aerial image (L) sidewall slope than conventional masks (X).

is greatly improved compared with the conventional binary mask. As a result, the DOF can also be significantly improved. Using Eqs. (8.7) and (8.10), one can choose a proper material (n, k) with a proper thickness to meet both transmittance and phase shifting requirements.

The mask manufacturing processes for these types of masks are more complicated than those for binary masks. Figure 8.31 illustrates a typical contact hole PSM mask-making process flow. It starts out with a resist (PR) coated blank, PR/Cr/MOSiON/Qz, that is, a Cr layer on a shifter on a quartz blank. After e-beam writing, the chrome layer is etched through with plasma etching using chlorine and oxygen. Then, the resist is removed, and a second plasma etching is carried out on the shifter using fluorine chemistry, with the chrome as the etching mask. Before shifter etching, the chrome pattern dimensions are measured to check if the target CD is reached. If not, the shifter etching recipe is selected so as to have different

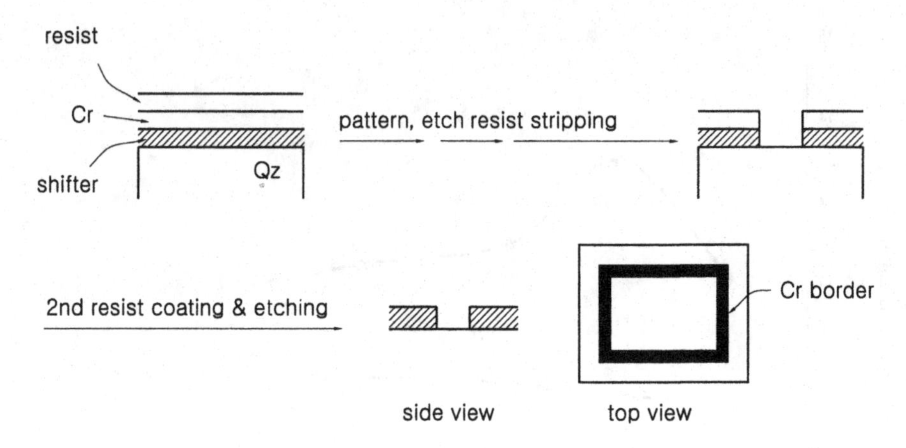

Fig. 8.31. The process flow for making attenuated PSM.

undercuts to achieve the final target CD on the etched shifter pattern. As a further complication, the shifter etching also more or less attacks quartz. The shifter etching recipe must be tuned such that it cannot have too much quartz loss because the refractive index of quartz is not the same as air; in other words, it also causes phase shifting. The defined Cr layer on the shifter then undergoes a second exposure and etching to remove the chrome layer in the pattern area, leaving chrome only in the frame area, where overlay and alignment marks are placed. The process flow for the rim-type PSM for holes is very similar to that of attenuated PSM, except that the second exposure, instead of a blanket exposure, leaves Cr around the shifter edges. The most difficult part of making a PSM is the mask repair, especially for intrusions at a line edge or a hole edge. Because this type of defect repair is done by depositing an opaque carbon film, its transmittance is nearly 0% for DUV light, and it evidently does not render any phase shifting. As a result, if the missing edge is larger than a certain percentage, the repair will not succeed. Normally, for a hole pattern, a defect is considered nonrepairable if two sides need repair and each exceeds 30% of the edge length.

An alternating phase-shifting mask is a strong phase shifting technology in that it shows more pronounced improvements in resolution and photoprocess latitude as compared to attenuated

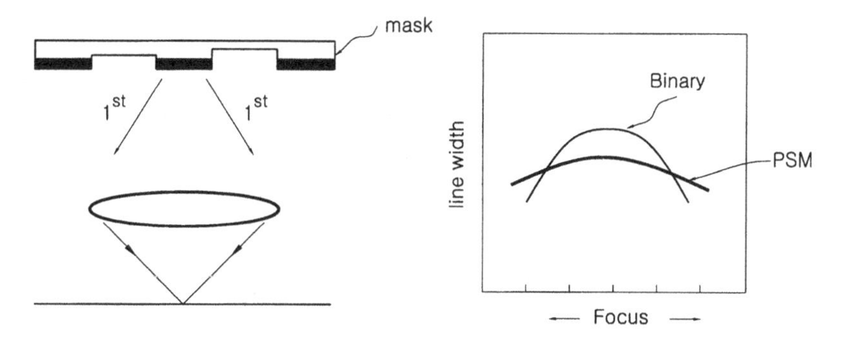

Fig. 8.32. (a) Alternating PSM image formed via two first-order diffracted beams, (b) resulting in great DOF improvement.

PSM. Figure 8.32 shows the neighboring spaces where the light can go through and take opposite phases alternately. This causes the zero-order diffraction to be absent, leading to the interference of two first-order beams for image formation. These two beams are symmetric with respect to the optical axis; hence there is very significant improvement in focus latitude. The reason is that for the conventional mask, three beams are used to reconstruct the image. As the imaging plan moves along the optical axis, the optical path differences (OPD) vary, and the image quality degrades. However, for two-beam imaging with alternating PSM, the OPD between the two symmetrical first-order diffracted beams stays the same. Obviously, the alternating PSM renders much better resolution and focus latitude. Furthermore, it is shown to give rise to smaller mask error factors (MEF) as compared to conventional masks. However, application of alternating PSM technology is relatively limited in comparison to attenuated PSM because its application is limited to periodic patterns.

Alternating PSM mask making is a lot more complicated than that of attenuated PSM, as shown in Fig. 8.33. The phase shift is created by etching through the quartz to a depth having a phase shift. First, the chrome pattern is created, as in the creation of conventional binary masks. Next, the areas that are assigned as zero-degree are exposed, and the pattern is etched into the quartz with plasma etching, using fluorine-based chemistry. After that, the 180° area is exposed and etched. The depth difference of the two areas must be

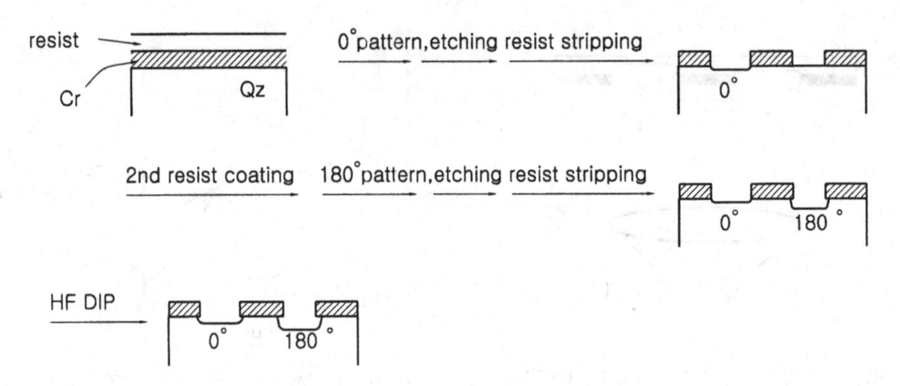

Fig. 8.33. A typical process flow for making an alternating PSM.

accurately controlled to have the desired phase shifting (difference). Next, a light HF wet etching is employed to round off the corners and eliminate some quartz defects. The most difficult part of alternating PSM manufacture is inspection and repair. This is because the quartz defects, such as quartz pits or humps, are still transparent but have different phase shiftings. Unlike in traditional inspection, in this method, special inspection techniques are needed to differentiate the phase angles of the defect area from normal ones. Quartz bumps pose the most challenging issue for repair. The repair technique must remove the quartz bump and yet leave no damage on the quartz that could lead to local transmission loss. Alternating PSM has not been as popular as its attenuated counterpart, mainly because design, software, and mask making are not as straightforward. Despite all these drawbacks, it is used for production of products with very large wafer volumes per mask set. For such circumstances, the long mask-making cycle time is relatively more acceptable as compared to the mask lifetime, and wafer print verification can be used in lieu of a mask inspection tool.

8.4.2. *Off-axis illumination*

Off-axis illumination (OAI) is a relatively cost-effective RET as it does not require mask-making process changes or exposure tool upgrades. The principle is straightforward. A fixed optical system has a specific NA, capable of resolving a minimum pitch, p. For the case of a pitch smaller than p, the first-order diffracted light will fall out of the lens' collecting range, resulting in no image formation. If one makes the light illuminate the mask surface obliquely, with an angle of ω with respect to the optical axis, the zero-order and one of the first-order diffractions (with different intensities) will now be collected by the lens, as indicated in Fig. 8.34. With these two diffracted rays, an image can be constructed. Thus the resolution is improved. However, one can expect that the exposure intensity will be reduced and that the two beams will have unequal intensity. The asymmetry between the zero-order and first-order diffracted beams can be solved by having a second light source from the opposite direction such that the diffractions of the two lights are added and the asymmetry is canceled out; that is, the zero-order diffraction of one light source coincides with the first-order diffraction of the other. Now we have two symmetrical diffracted rays, rendering improved DOF. Figure 8.35 shows the commonly used apertures for off-axis illuminations of annular, dipole, and quadrupole types. OAI optimization is pitch- and pattern-orientation-dependent. An OAI optimized for one pitch may not be suitable for the other.

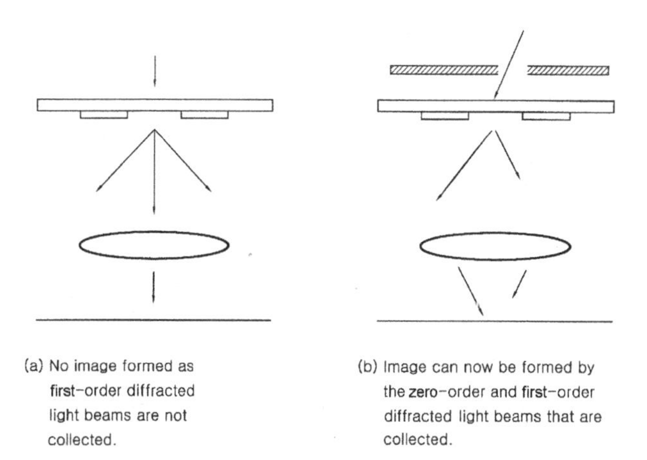

Fig. 8.34. Comparing the image formation with (a) regular illumination and (b) off-axis illumination.

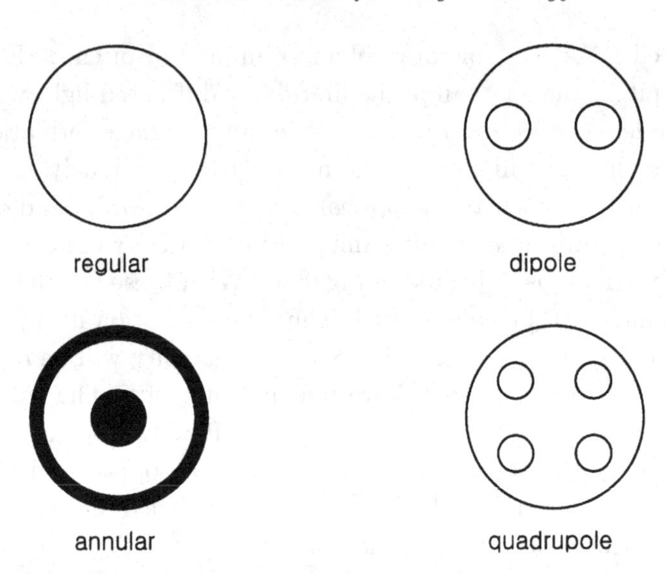

Fig. 8.35. Types of off-axis illumination.

8.4.3. *Optical proximity correction*

One common phenomenon in wafer printing is that the printed wafer dimensions tend to deviate from the designed ones as pattern proximity changes. The other phenomenon is the two-dimensional corners' rounding as the features approach the exposure tools' resolution limits. Examples are shown in Figs. 8.36 and 8.37. The impacts of these proximity effects on circuit performance are tremendous; they sometimes lead to circuit failure. One-dimensional CD variation of the gate layer directly affects the transistor current, which relates to circuit timing. If the CD is below the lower limit, it causes leaky transistors; no signals can be held. CD variations on interconnected layers, such as metal or silicided lines, result in resistance variation or RC delay variations. Smaller CDs on interconnected lines can also result in electromigration or even circuit burn out during operation. Line end shortening, if it happens on a polysilicon gate over an active area, such as that shown in Fig. 8.38, can result in total transistor failure as the source is shorted to the drain area. Optical proximity correction (OPC) is often used to solve the problem. There are two main approaches to implement OPCs: the rule-based

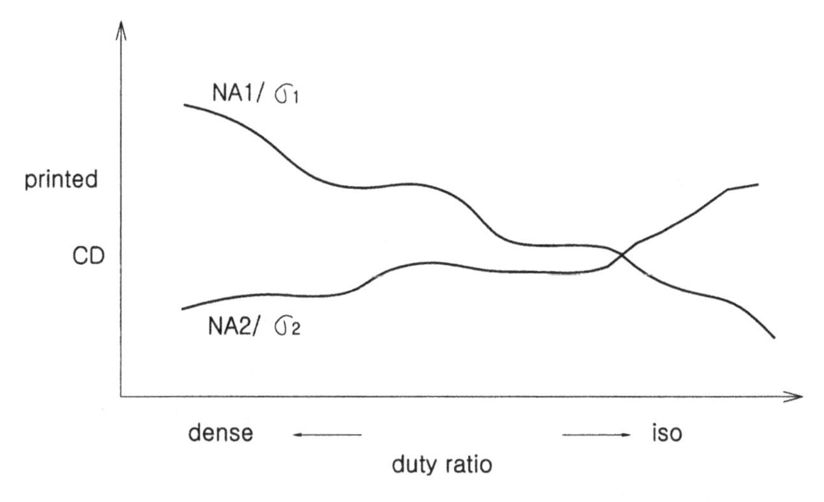

Fig. 8.36. One-dimensional optical proximity effect: printed CD varies with surrounding pattern density.

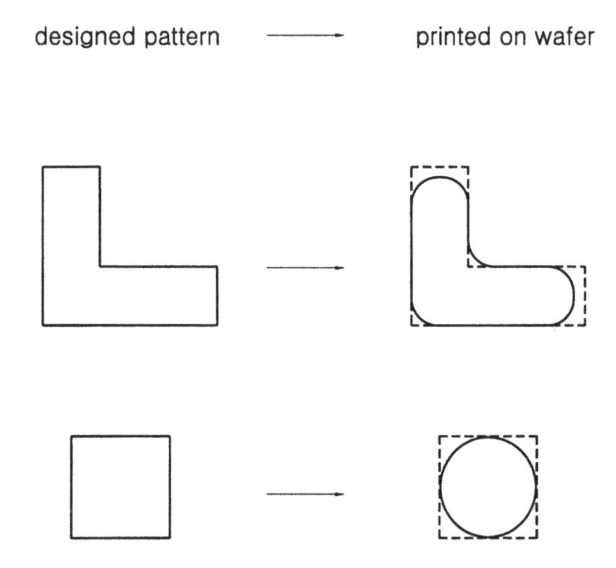

Fig. 8.37. Patterns' corner rounding caused by exposure tools' resolution limit.

approach and the model-based approach. For rule-based OPC, the correction values that are added to the features are basically taken from a lookup table derived from experimental results. Model-based OPC is based on mathematical modeling of the mask patterns. Both

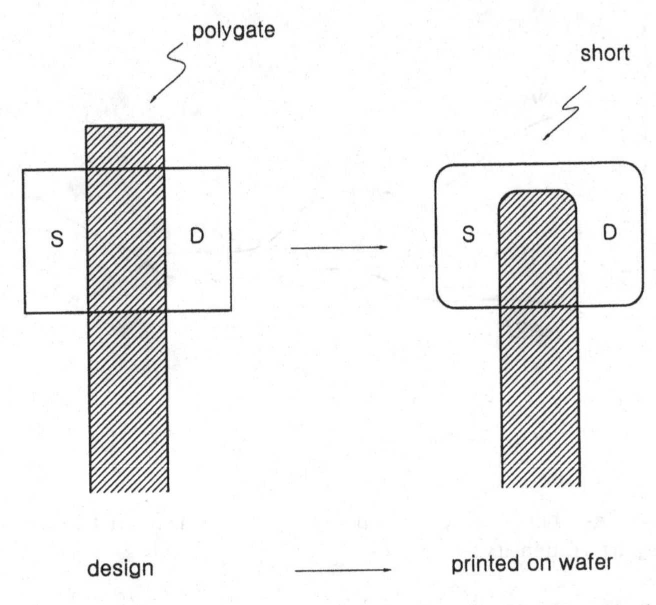

Fig. 8.38. Line end shortening effect causes gate end cap to be severely rounded off and shortened as it is printed on wafers.

methods require iterative approaches to reach the optimized condition where the residual differences between the printed patterns and the designed patterns are as small as desired.

To come up with the rules for rule-based OPC, a large number of one-dimensional and two-dimensional test patterns must be designed to simulate different layout scenarios. The mask pattern is then printed on wafers, measured, and compared with the design values. If the comparison shows inconsistencies, corrections can be added to the mask patterns until matched results (the discrepancy is within tolerance) are found.

On the other hand, instead of using a lookup table, model-based OPC uses correction values that are derived from mathematical modeling of the mask patterns. The model setup procedure starts out with a large number of designed patterns that simulate all the possible patterns and proximities. The patterns are then simulated with the model. The corrected pattern is made into a mask and printed on a wafer. The wafer results are compared with the original design. Depending on how much the residual errors are to be tolerated, the correction values can always be refined. The finer the corrections are, the smaller the residual errors will be. However, nothing comes free; fine corrections result in increases in mask-making cost and cycle time.

8.4.4. *Subresolution assist features*

Another important category of OPC is the addition of subresolution assist features (SRAF). Line width biasing, which is often used in rule-based OPC, does ensure, to some extent, the line width uniformity across the whole design pattern. However, the overlapping process window (latitude of energy defocus window) of the different lines in different proximities can be unacceptably small. The addition of SRAF into the large empty space (around isolated lines), as illustrated in Fig. 8.39, improves the photolithography process window. For a given space, the number of assist features needed, and the distance between the main pattern and the SRAF, must be optimized in terms of wafer print results.

Because the photolithography technology evolution lags behind product design needs, RETs are extensively applied until the next generation of photolithography technology becomes available. Figure 8.40 shows resolution enhancement techniques used for various technology generations. Above 0.35-μm technology, hardly any RET techniques are used. After 0.25 μm, some manual OPCs and

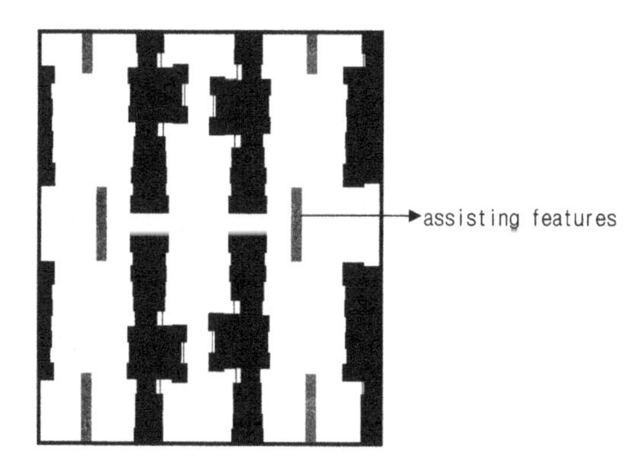

Fig. 8.39. The assisting features for improving photoprocess.

	0.35 um	0.25 um	0.18 um	0.13 um	0.09 um
PSM	X	X	O	O	O
Serif	X	O	O	O	O
Rule base	X	O	O	X	X
SRAF	X	X	O	O	O
Model base	X	X	X	O	O

Fig. 8.40. Types of RET used in various technologies. O, used; X, not used.

hole-attenuated PSMs are used, but in general, the RETs are relatively simple. Below 0.18 μm, the industry starts to see the use of the SRAF and rule-based OPC applications. After 0.18 μm, extensive use of RETs becomes a must. Oftentimes, more than one RET is needed for each technology.

Figure 8.41 illustrates the percentage of masking layers with RETs in a set of masks; one can see that the percentages go

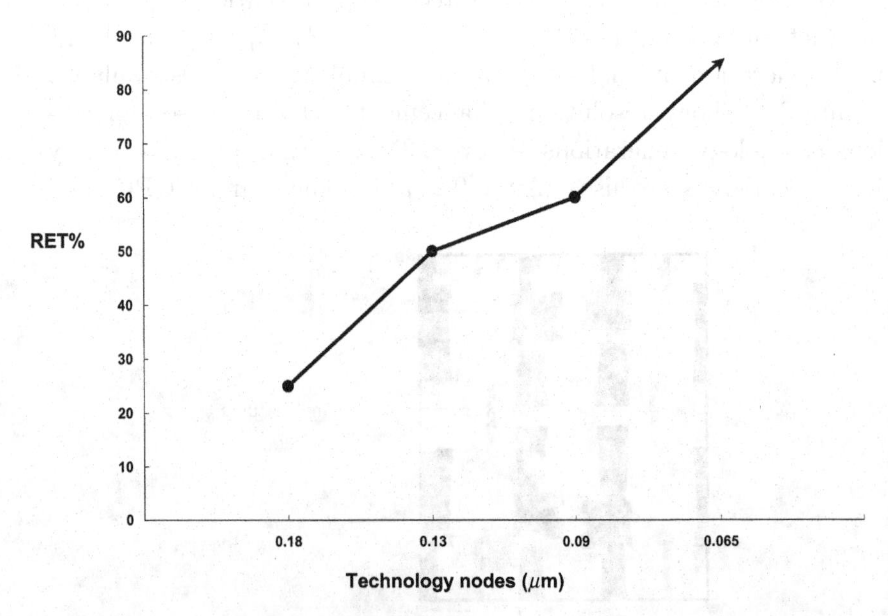

Fig. 8.41. In a set of masks, the percentage of mask layers with RET increases dramatically as technology shrinks.

straight up. This brings up the mask cost and lengthens mask manufacturing cycle time as well. The use of RETs significantly complicates mask manufacturing. Apart from the PSM making issues discussed earlier, OPC also poses some very challenging issues in mask making. The fine OPC jigs and jogs require the minimum writing grid of the electron beam to be used, leading to long writing time. Up to 1.5 times longer writing time is needed for some cases in 0.13-μm technology, as compared to that without OPC. The fine jigs and jogs also require a special mask inspection algorithm and are very difficult to repair. Furthermore, the metrology of such a pattern cannot be measured with an optical metrology tool due to its inappropriate resolution and proximity limitation. It must be measured with a CD SEM tool with a restriction that the CD measurement location must be on a long, smooth line — having no jigs and jogs within a few microns.

Chapter 9

DOPING TECHNOLOGY

A doping process is inevitable in making *pn* junctions, which are the fundamental building blocks for semiconductor devices. In Section 9.1, the readers are exposed to the fundamentals of doping processes, dopant diffusion, and ion implantation. The process modeling greatly helps to elaborate the phenomena and the impact of each process parameter. The detailed mechanism and modeling of the diffusion and implantation are introduced in Sections 9.2 and 9.3, respectively. The substrate damages resulting from implantation must be removed with annealing to retain proper device performance. The defect formation mechanism and annealing are both discussed in detail in Section 9.4. Finally, the applications of the doping processes in making a device are discussed in Section 9.5. By the end of this chapter, the readers should have a very good understanding of how a device structure is formed as related to diffusion and ion implantation.

9.1. Introduction

Doping is a process of adding various atoms into a silicon substrate to alter its resistivity and type. Adding three-valence atoms, such as boron or gallium, results in a *p*-type semiconductor material; adding five-valence atoms, such as phosphorous or arsenic, results in an *n*-type semiconductor material. Semiconductor device structure formation is all about doping different silicon areas. The critical aspects of a doping process include the dopant type, amount, depth, and junction profile. The most commonly used doping methods in device manufacturing are diffusion and ion implantation.

Diffusion is a process that introduces the precursor dopants to the substrate surface; this is followed by a drive-in diffusion at high temperature to allow the dopant to spread isotropically into the substrate bulk. Diffusion is often conducted in a furnace system, as illustrated in Fig. 9.1. The system is mainly composed of a quartz tube surrounded by resistive heating elements. Wafers are arranged vertically on wafer boats. The dopant source can be gas, liquid, or solid. For the solid source, as shown in Fig. 9.2, the dopants are made into a wafer form and arranged alternately with wafers on quartz boats. For the liquid source, the carrier gas is bubbled through the liquid bulk to bring out the dopant gas into the diffusion furnace. For the

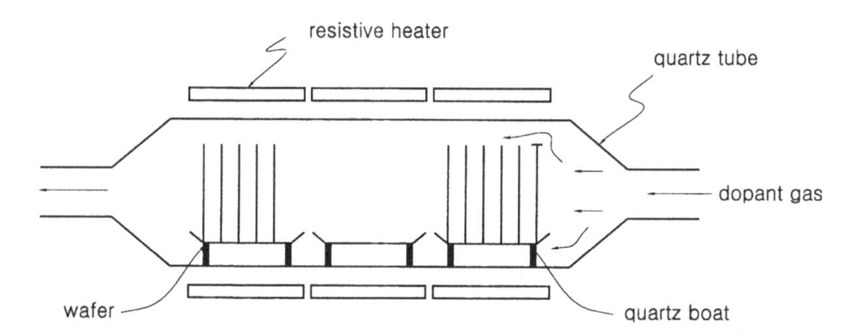

Fig. 9.1. Configuration of a dopant diffusion furnace.

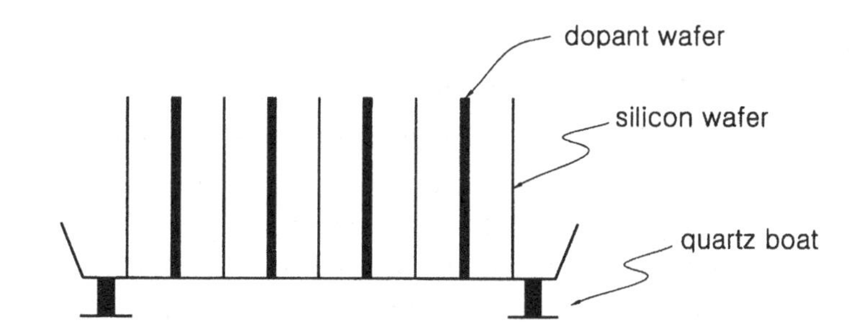

Fig. 9.2. Solid dopant source is often made into wafer forms and arranged alternately with silicon wafers that are to be doped.

gas sources, dopant gases are premixed with the carrier gas. Gas and liquid dopant sources are most often used in advanced semiconductor industry processes. In general, a diffusion furnace batch can handle up to a couple of hundred of wafers at a time. The dopant gas is introduced from one end, flowing through the tube and exiting toward the other end. The dopant gas flows convectively through the annular cross section formed by the wafer edges and the quartz tube wall. In the meantime, dopant transport in between wafers is driven solely by diffusion. A diffusion process takes from one to several hours, depending on the amount of dopant required and the depth of the junction desired. Boron and phosphorous are the two major dopants commonly used for diffusion.

Typical reactions for dopant diffusion are as follows.
For boron, the reactions take place as

$$B_2H_6 + 3O_2 \rightarrow B_2O_3 + 3H_2O\,.$$

This is followed by a silicon surface reaction:

$$2B_2O_3 + 3Si \rightarrow 4B + 3SiO_2\,.$$

For phosphorous,

$$2PH_3 + 4O_2 \rightarrow P_2O_5 + 3H_2O\,.$$

This is followed by a silicon surface reaction:

$$P_2O_5 + 5Si \rightarrow 4P + 5SiO_2\,.$$

The boron and phosphorous atoms then diffuse downward isotropically. When solid sources are of interest, B_2O_3 and P_2O_5 are directly made into wafer forms. P is a widely used liquid-type dopant in $POCl_3$, which is often used in the industry for polysilicon gate doping purposes. It first reacts with oxygen to form P_2O_5, which then reacts with silicon to form phosphorous atoms for diffusion.

A dopant diffusion process is divided into two steps: predeposition and drive-in diffusion. During the predeposition step, the desired amount of dopant is introduced to the substrate surface. The amount is controlled by the dopant's solid solubility, which is a function of temperature. The predeposition step is followed by a drive-in diffusion, in which the dopant supply is cut off and the dopant that were introduced during the predeposition step are allowed to diffuse downward, as illustrated Fig. 9.3. The diffusion front spreads isotropically to a desired depth. The result of a diffusion process is expressed in terms of two parameters: the amount of dopants introduced and the resulting concentration profile. Together, they dictate the sheet resistances and junction depths.

Dopant diffusion in solids proceeds through a variety of mechanisms or paths. A substitutional diffusion mechanism occurs when the dopant atoms move from one vacancy to the next, as shown in Fig. 9.4. A crystal at a temperature above absolute zero contains a certain number of vacancies, and the crystal lattice is in a fairly

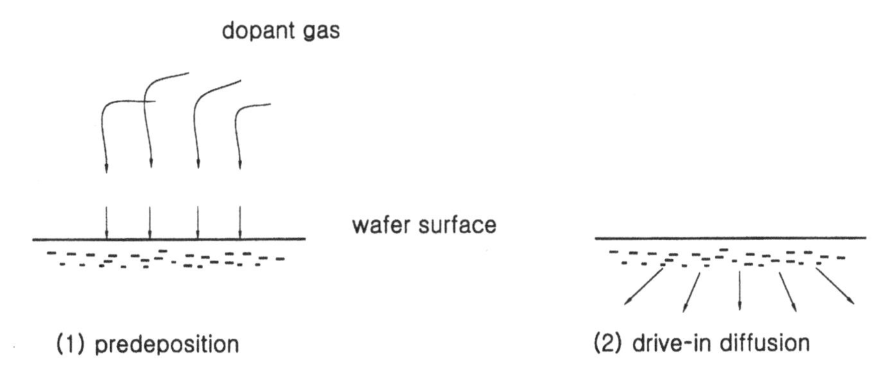

Fig. 9.3. The two-step dopant diffusion process: a predeposition followed by a drive-in diffusion step.

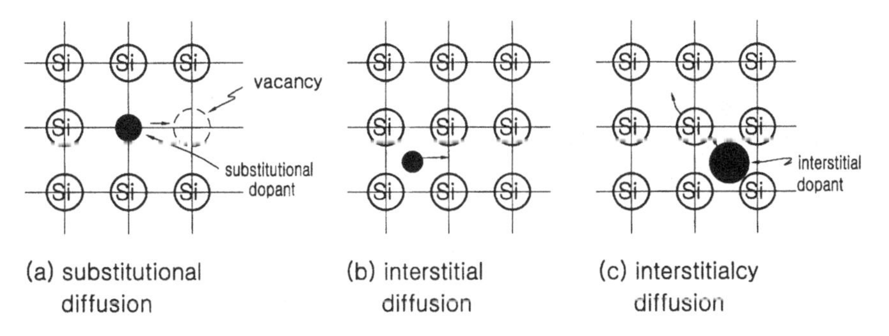

Fig. 9.4. The three typical solid state diffusion pathways.

violent oscillation state. There is a probability that the substitutional atom can hop from its site to the neighboring vacancy. Therefore, as the vacancy moves to the left, the atom moves to the right. In the bulk of the crystal, vacancies and atoms can move throughout the bulk. The interstitial diffusion mechanism occurs when there are more atoms than available lattice sites. The extra dopant atoms move from one point to the next, primarily through the empty space among crystal lattice sites. To make this happen, the space between two lattice sites must be wide enough to allow the dopant to go through. Therefore, the smaller the dopant is, the easier it will be for the dopant to move through. If the interstitial dopant atoms are not small, they are likely to move forward by pushing the lattice atoms off their lattice sites. This is called the interstitialcy diffusion mechanism.

Ion implant is another indispensable method of doping the silicon substrate, in addition to the diffusion process. Ion implantation is a process in which dopant precursors are ionized and accelerated toward the substrate. With the accelerated energy, the dopant ions penetrate into the substrate bulk on impinging on its surface. Figure 9.5 illustrates this process. Dopants are ionized into ions with

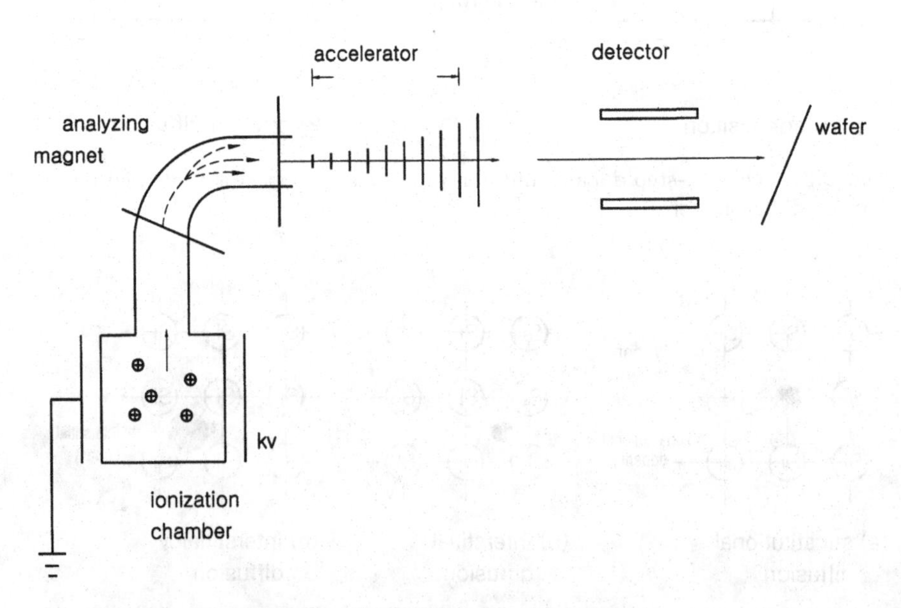

Fig. 9.5. Schematic of an ion implanter.

various charges and are then extracted out of the ionization chamber. As soon as the ions are extracted out of the chamber, they pass through an analyzing magnet. The analyzing magnet separates the desired species from the undesired ones. The desired charged dopants are then accelerated toward the wafer surface. The fast-moving charged dopants penetrate into the wafer surface to a depth that is a function of the incoming ion and its energy. An electron flow is continuously supplied to the wafer being implanted for charge compensation. Ion implantation can accurately control the number of ions that are implanted into the substrate as the charges can be counted, essentially, one by one. The energetic particles cause damage in the silicon substrate; furthermore, the dopants must be placed to substitutional sites to be electrically active. Therefore ion implantation must be followed by an annealing process for the purpose of damage removal and dopant activation. Theoretically, all dopants can be introduced into the silicon substrate with ion implantation.

As compared to diffusion, the ion implantation is more precise in terms of dopant dosage and depth control. Ion implantation is basically an anisotropic doping method. One can place the dopants into a desired depth with minimal lateral spread due to lattice scattering. These two characteristics are very critical in advanced device engineering for gate lengths of shorter than 0.13 μm. Figure 9.6 demonstrates an advanced device structure with several special features. Shallow source/drain junctions are used for short gate lengths; lightly doped drain (LDD) is used for boosting hot carrier immunity, and anti-punch-through (APT) is used in doping substrate to enlarge the

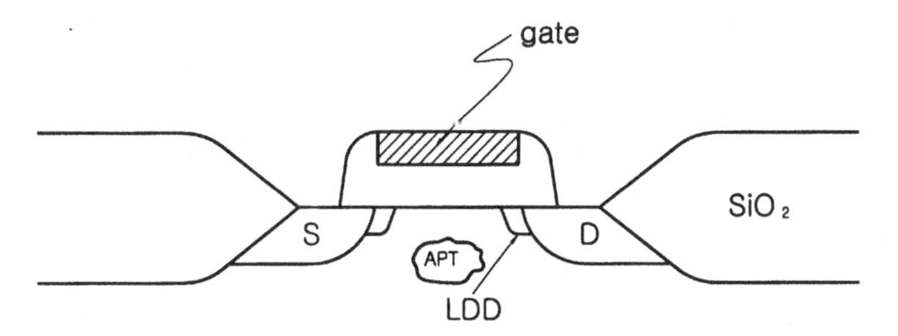

Fig. 9.6. A device structure with LDD and APT implants.

device window. LDD and APT are both difficult with diffusion due to diffusion's isotropic nature. This is because in the case of diffusion, by the time the dopants are driven to a desired depth, the source and drain are possibly shorted already. Ion implantation is indispensable in these applications. Compared to diffusion technology, ion implantation systems are a lot more sophisticated, and therefore they are more expensive in terms of system, maintenance, and operating costs.

9.2. Dopant Diffusion

A dopant diffusion process is initiated by the precursor gas reacting with oxygen and silicon to form atomic boron or phosphorous, which then diffuses into the silicon substrate, as a result of thermal energy and concentration gradients. During predepositon, desired amounts of dopants are introduced at the silicon surface, as indicated in Section 9.1. This is accomplished by exposing the surface to a dopant precursor ambient at a constant temperature. Along the course of predeposition, the surface concentration is kept at the solid solubility value, C_s. The solid solubility of various dopants depends on the dopant type and process temperatures. Mathematically, the diffusion process is described as

$$\frac{\partial C}{\partial t} = \nabla \cdot (D \nabla C), \tag{9.1}$$

where C is the dopant concentration, which is a function of position and time during predeposition. D is the diffusivity, which is a function of temperature, and it can be expressed in Arrhenius form, $D = D_0 \exp(E/kT)$. The left-hand side of the equation represents the dopant accumulation rate in a unit volume; the right-hand side represents the diffusion fluxes. In a solid such as silicon, dopant diffusion can vary with dopant concentration, stress, and defect density. The diffusivity is a characteristic of the dopant type and process temperature, as illustrated in Fig. 9.7. This figure shows that diffusivity increases when the size of the dopant atoms decrease. If one

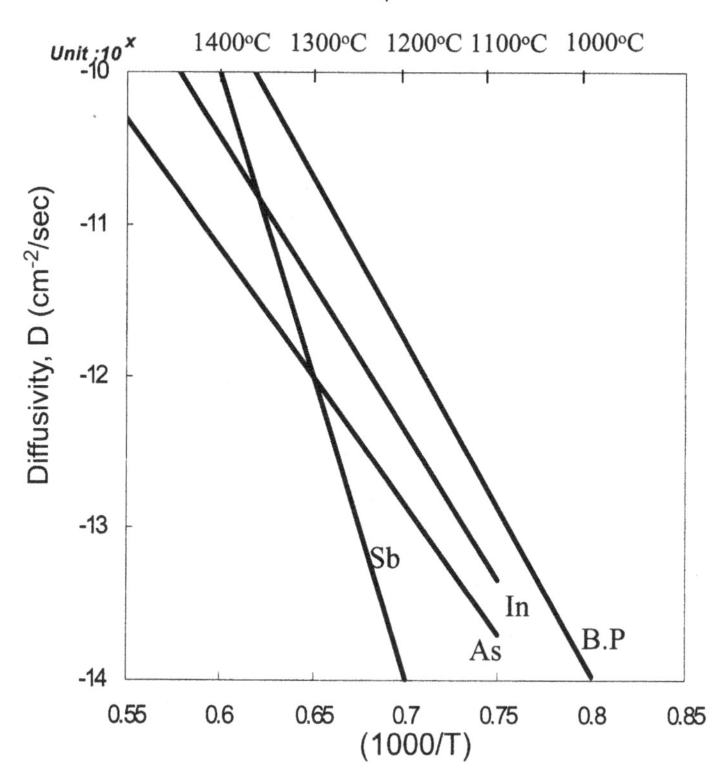

Fig. 9.7. Diffusivity of various dopants.

looks at one-dimensional diffusion and a spatial and concentration-independent diffusivity, Eq. (9.1) can be simplified to

$$\frac{\partial C}{\partial t} = D\frac{\partial^2 C}{\partial x^2}\,. \tag{9.2}$$

For predeposition, the boundary conditions are at the silicon surface, $x = 0$ and $C = C_s$. The other boundary condition is approximated as $C = 0$ at $x = \infty$, implying that the region in which the diffusion takes place is very small as compared to the silicon substrate thickness. On the other hand, the initial condition is $C(x, 0) = 0$. This equation can be solved with the Laplace transform method; the solution, namely, the dopant concentration profile, is expressed as

$$C(x,t) = C_s \text{erfc}\left[\frac{x}{2\sqrt{Dt}}\right]. \tag{9.3}$$

One can define the dosage as Q, and

$$Q = \int_0^\infty C(x,t)dx = 2C_s\sqrt{\frac{Dt}{\pi}}. \tag{9.4}$$

It is apparent that the amount of dopant, and the dosage, introduced onto the silicon surface is a function of temperature and time. Temperature determines the surface concentration, the dopant solubility in silicon substrate, and the diffusivity. Meanwhile, longer predeposition times result in more incorporated dopants. Figure 9.8 illustrates the dependencies. Graphically, the dosage equals the shaded area enclosed by the solution curve.

The drive-in diffusion, which drives the dopants deeper into the silicon substrate to a desired depth, follows predeposition. Drive-in diffusion pushes most of the dopant atoms to substitutional sites to become electrically active. During drive-in diffusion, the dopant supply is cut off; essentially, the dosage, Q, is spread into the substrate. The mathematical model is the same as Eq. (9.2) for predeposition, except for the initial and boundary conditions. At the surface, a constant surface concentration changes to an ever-decreasing concentration. If one further assumes that the predeposited profile is an

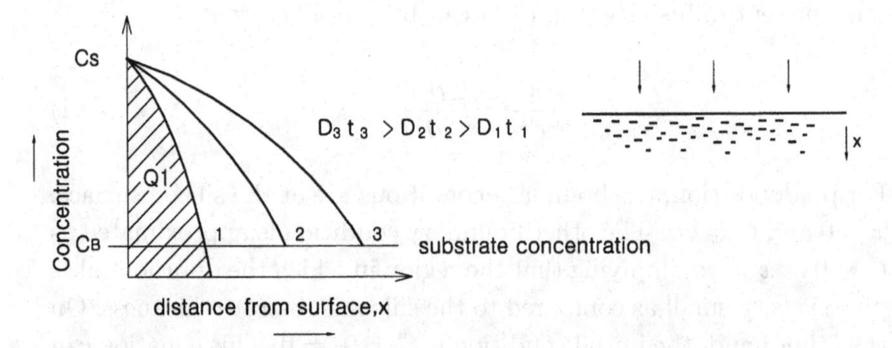

Fig. 9.8. The concentration profiles of various predeposition processes.

impulse function at $t = 0$ and $x = 0$,

$$Q = \int_0^\infty C(x)dx\,. \tag{9.5}$$

This assumption is normally true, if one compares the dopant profiles after the predeposition and after drive-in diffusion. The resulting solution of the drive-in diffusion is a Gaussian distribution:

$$C(x,t) = \frac{Q}{\sqrt{\pi Dt}\exp[x^2/4Dt]}\,. \tag{9.6}$$

The solution is illustrated in Fig. 9.9. The longer the drive-in diffusion time is, the further the dopants will spread into the substrate. However, one should note that the area under the solution curve must remain constant since Q is a constant. How far the dopants can diffuse depends on the value of Dt. The larger this value is, the further the dopants will diffuse. Along the course of device making, there are several thermal cycles that drive the existing dopants to spread out further. For example, well implant takes place at the very beginning of device making. Up to the time that the contact is formed, it further experiences several thermal cycles such as well drive-in, field oxidation, polysilicon gate doping, presource and drain oxidation, source and drain diffusion, BPSG flow, and contact anneal. It is the sum of the Dt, $\sum_1^i D_i t_i$ that determines the final profile of the well dopant

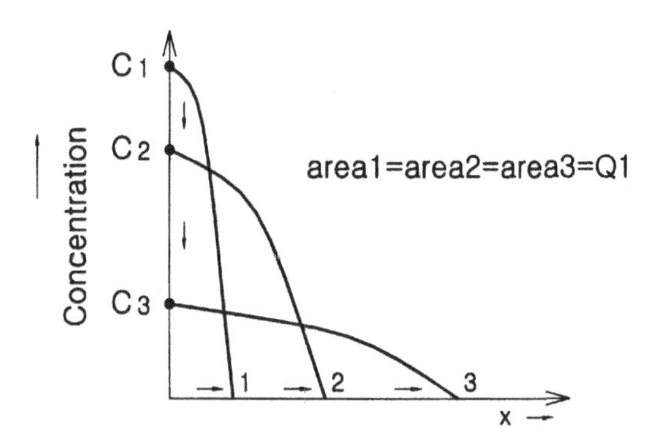

Fig. 9.9. A drive-in diffusion conducted with three conditions D_1T_1, D_2T_2, and D_3T_3 after the same predeposition.

at contact etching. One often defines the thermal budget as

$$\text{Thermal budget} = \sum_{1}^{i} D_i t_i \,. \tag{9.7}$$

The allowable thermal budget decreases with advancing technology as the gate length decreases.

The device junction depth is the depth at which the concentrations of both types of dopants are equal, when dopants diffuse into a doped substrate of the opposite type, as illustrated in Fig. 9.10. Mathematically, by knowing the substrate concentration, C_B, the junction depth can be found by equating $C_B = C(x, t)$ and finding out what the x value is. For the case of predeposition:

$$C_B = C_s \text{erfc}\left[\frac{x}{2\sqrt{Dt}}\right], \tag{9.8}$$

namely,

$$\text{Junction depth } x_j = \text{erfc}^{-1}\left[\frac{C_B}{C_s}\right] \cdot 2\sqrt{Dt}\,. \tag{9.9}$$

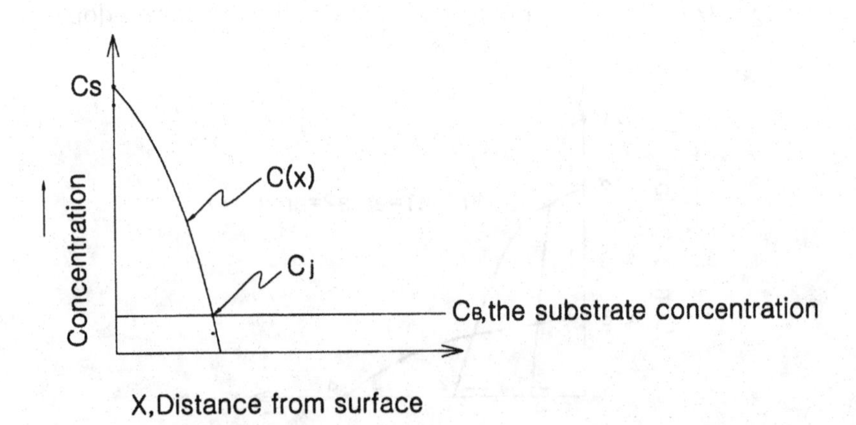

Fig. 9.10. The junction depth is defined as the depth at which the diffusing dopant concentration equals the substrate concentration.

Similarly, for the drive-in diffusion,

$$x_j = 2\sqrt{Dt}\left[\ln \frac{Q/\sqrt{\pi Dt}}{C_B}\right]^{1/2}. \tag{9.10}$$

The significance of the junction depth is that it determines the resistance of the doped area, and it correlates to the penetration distance of the lateral diffusion. The lateral diffusion can be simulated by solving Eq. (9.1) multidimensionally and correlating the horizontal to the vertical dimension, as shown in Fig. 9.11. It is apparent that for a fixed surface concentration in predeposition or a fixed dosage in drive-in diffusion, the higher the substrate concentrations are, the shallower the junction depths will be. The deeper the vertical junctions are, the further the lateral diffusion will be. In particular, a deeper junction is more prone to result in a higher leakage rate between the source and drain for a given transistor gate length, as illustrated in Fig. 9.12. Device engineering is all about tuning the thermal budgets and doping processes, including diffusion and ion implantation; the attempt is to achieve high saturation current with low leakage currents between source and drain. One

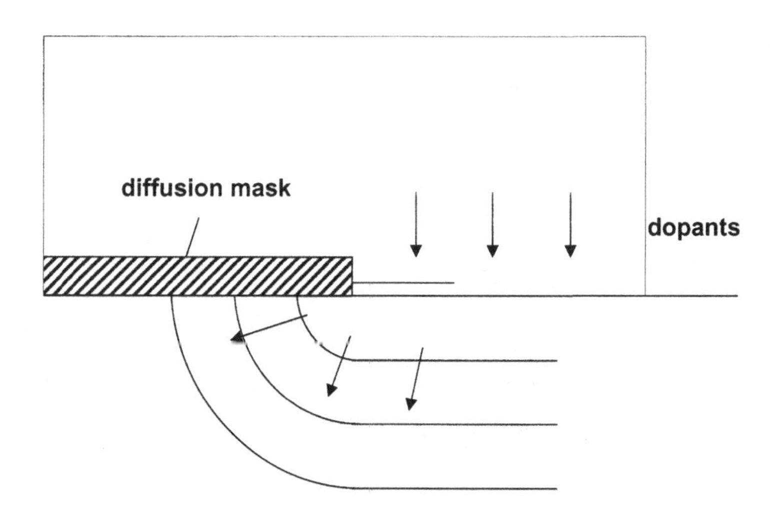

Fig. 9.11. Isotropic dopant diffusion. Dopants penetrate laterally into the substrate under the diffusion mask.

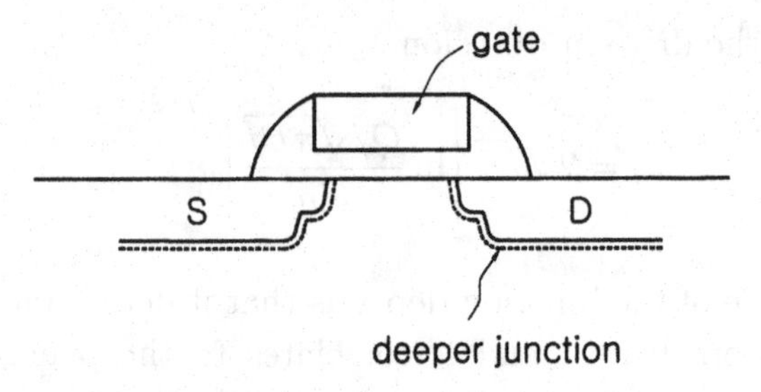

Fig. 9.12. For a fixed gate length, the deeper junction is more prone to induce source (S)-to-drain (D) electrical shortages.

item worth noting is that once the device engineering is complete, the thermal cycle has to be kept as low as possible during back-end processes, or the device characteristics or performance may not meet expectations.

A dopant atom has different diffusivities in different diffusion media. For a specific dopant, one can properly choose a material that renders a low diffusivity to be used as a diffusion mask. This means that after a certain period of diffusion time under the specified conditions, the dopant atoms cannot penetrate through the material. However, they can significantly spread into the underlying silicon through the mask open windows. This material is called a diffusion mask, as shown in Fig. 9.13. With diffusion masks, a substrate can always be selectively doped. Figure 9.14 indicates a process flow that selectively dopes and forms the well structure on a silicon substrate.

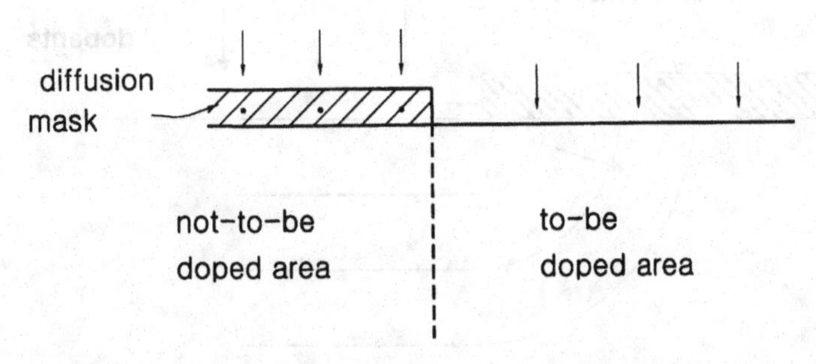

Fig. 9.13. A diffusion mask that defines the to-be-doped area and the not-to-be-doped areas.

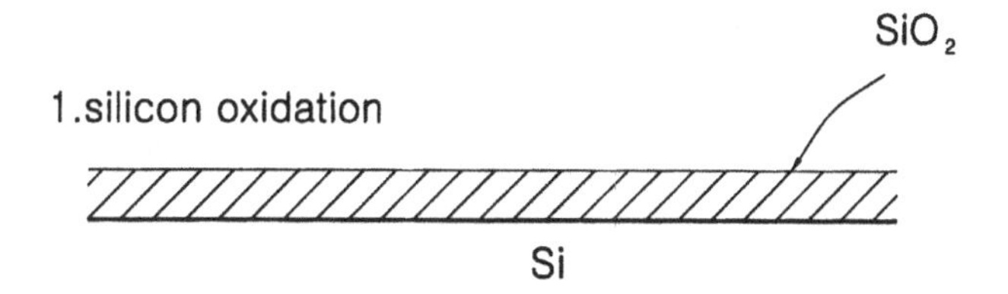

2.photolithography: to define the diffusion window

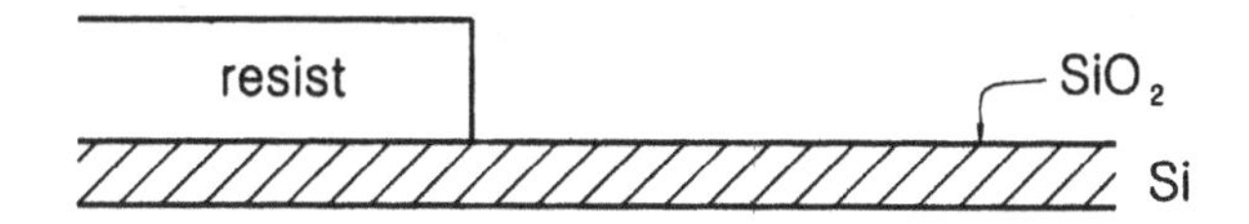

3.Etching: to open up the window,forming diffusion mask pattern

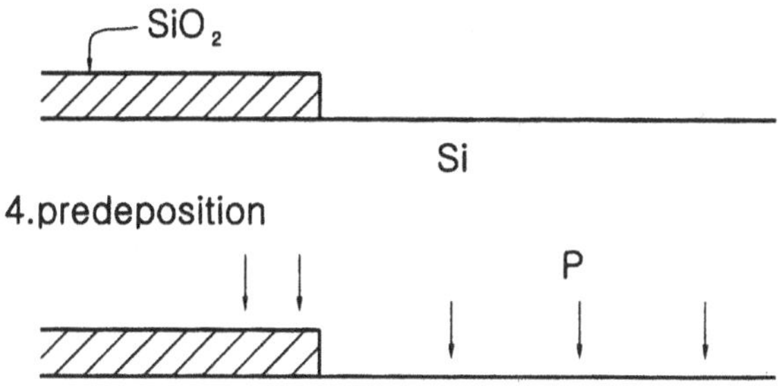

5.Drive-in diffusion: to form the n-well regions

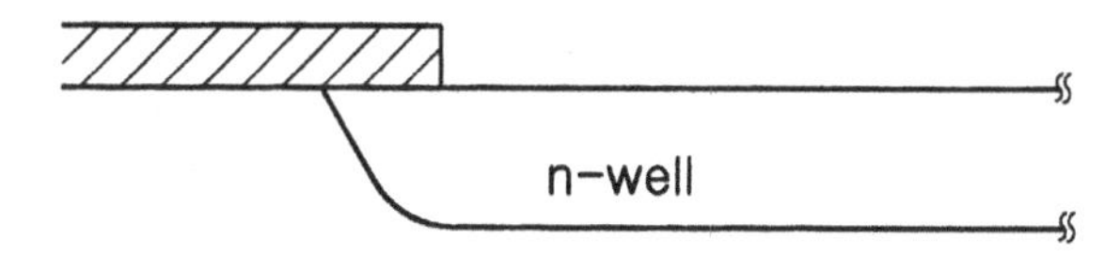

Fig. 9.14. Forming n-well pattern with a two-step diffusion approach.

The extent of lateral diffusion actually limits the applications of diffusion in advanced device engineering.

9.3. Ion Implantation

Owing to accuracy and its anisotropic nature, ion implantation is extensively used in major device manufacturing procedures such as source–drain implant, LDD implant, APT implant, and so on. Ion implants are accurate in terms of both dosage and energy control. With good dosage accuracy, the number of implanted atoms can actually be counted individually; with the high energy accuracy, as set by the accelerator, the penetration depth of each ion can be predicted. In the ionization chamber, various ionized species are formed. For example, for the boron difluoride, possible ionized species can be detected such as $B^+, BF^+, BF_2^+, BF_3^+, B^{+2}, BF^{+2}, \ldots$, and so on. For phosphine, the likely ionized species are $P^+, PH^+, PH_2^+, PH_3^+, P^{+2}, \ldots$, and so on. These species are then extracted out of the ionizing chamber altogether. During the ion transport, the ions with different charges are screened with a magnet analyzer. As a charged particle is accelerated by an electrical field, it receives energy from the field:

$$\text{energy} = qE\,. \tag{9.11}$$

The magnet analyzer takes advantage of the fact that the trajectory of a moving charged particle can be bent by a magnetic field such as

$$\vec{F} = q(\vec{v} \times \vec{B}) = \frac{mv^2}{r}\,. \tag{9.12}$$

This bending force causes the particle to move with a circular path of radius (r) and with a centrifugal force, mv^2/r, as shown in Fig. 9.15. Furthermore, the particle's kinetic energy equals the energy that it

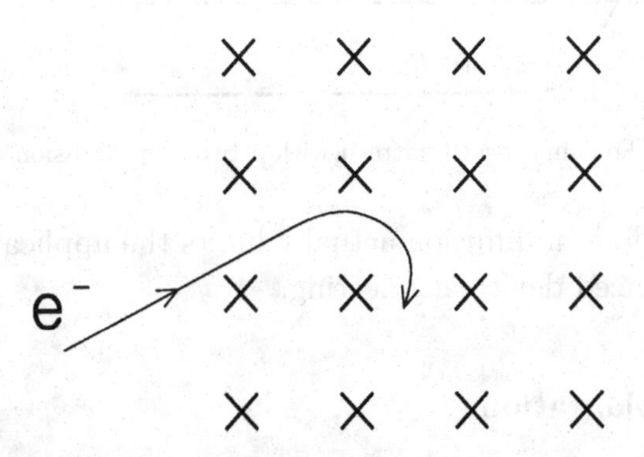

Fig. 9.15. A moving electron is bent by a magnetic field that points inward into the paper.

received from the electrical field:

$$\frac{mv^2}{2} = qE\,. \tag{9.13}$$

Combining Eqs. (9.11), (9.12), and (9.13), the radius of the moving charged particle can be calculated:

$$r = \frac{1}{B}\left[\frac{2mE}{q}\right]^{1/2}. \tag{9.14}$$

In other words, with a fixed magnetic and electrical field, the bending radius of the moving charged particle is a function of the particle mass to charge ratio, m/q. With this, only the right species can pass the tiny metal slit toward the accelerating region, as illustrated in Fig. 9.16; therefore one can precisely select the desired species for the implantation. After the analyzer magnet, the charged particles are further accelerated to a desired energy level toward the wafer surface. The ion traveling path has to be in high vacuum to minimize the possibility of colliding with other residual molecules from chambers or pipelines. Such collisions can lead to further ionization or detraction of its traveling path.

Important parameters for an ion implant process are implant dosage and energy. Along the course of implantation, the dosage is

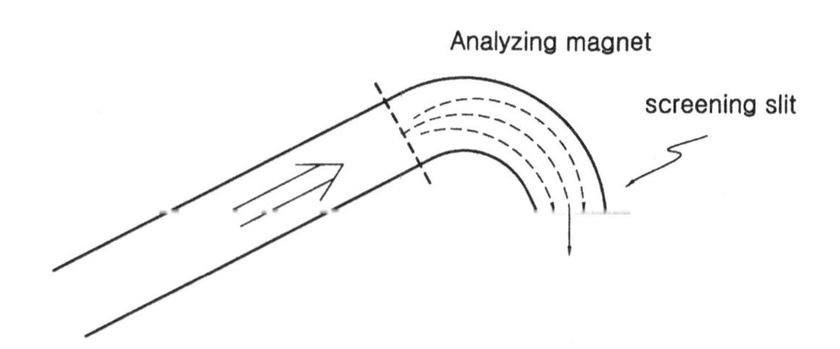

Fig. 9.16. A beam of charged particles passes through an analyzing magnet split into a number of subbeams. Only the particle with the desired m/q ratio can pass through the screening slit.

measured with a Faraday cup and calculated as

$$Q = \frac{\int_0^t I dt}{qmA},$$

where Q is the implant dosage, I is the current density, t is the elapse time, m is the charge number of the ions, q is the charge of an electron, and A is the area to be implanted. The ultimate goal of an implant process is to have good uniformity, say $<1\%$, on dosage and energy across wafer and wafer to wafer. Apart from the mechanical and electrical control mechanisms of the hardware, contaminations have been found to degrade the uniformity. There are two types of contaminations: autocontamination and heterocontamination. The former indicates the existence of undesired species with the same mass/charge, m/q, ratio as the desired one. The latter refers to when the desired species arrive at the wafer surface with an energy that differs from the preset energy. These two types of contaminations cause the final sheet resistance to differ from the target values.

On impinging on the wafer surface, the charged particles, with the accelerated energy, are able to penetrate into the substrate surface. Along its penetrating course, the charged particle sees the substrate as an array of nuclei surrounded by electron clouds. As the particle penetrates into the wafer substrate, it slows down by losing its energy via two mechanisms: the nuclear stopping and electronic stopping,

$$\frac{dE}{ds} = \left(\frac{dE}{ds}\right)_n + \left(\frac{dE}{ds}\right)_e, \tag{9.15}$$

where E is the ion energy and s is the ion traveling distance. The left-hand side of this equation represents the total energy loss of a charged particle along its traveling distance. The first term on the right-hand side of the equation represents the energy loss due to nuclear stopping, while the second term represents the energy loss due to electronic stopping. Once the charged particle hits a nucleus, its energy is transferred to the target nucleus. This causes the target nucleus to move off its original lattice site, and it gives rise to physical damage of the substrate. These damages take different forms; they can be point defects, line defects, or an amorphous layer. In addition, on collision, the charged particle's traveling trajectory gets deflected. The collision between the projectile ion and the target atom or ion can be simplified as a rigid body collision, but with attraction forces between them. Now, assuming a simple screening function (when the attraction potential is inversely proportional to the square of distance), the nuclear stopping power can be approximated as a constant:

$$S_n(E) = \left(\frac{dE}{ds}\right)_n = 2.8 \times 10^{-15} \left[\frac{z_1 z_2}{Z^{1/3}}\right] \left[\frac{M_1}{M_1 + M_2}\right], \qquad (9.16)$$

where subscript 1 stands for the incoming ion and subscript 2 stands for the target nucleus; z is the atomic number, and M represents the mass of the ions. Furthermore, $Z^{1/3} = z_1 + z_2$. In other words, for an incoming ion and the target nucleus, the nuclear stopping power is constant, disregarding the incoming ion energy. On the contrary, the electron stopping power is proportional to the square root of the incoming ion energy if one considers that the projectile ion is penetrating through an electron cloud:

$$S_e(E) = \left[\frac{dE}{ds}\right]_e = kE^{1/2}, \qquad (9.17)$$

where E is the incoming ion energy and k is a constant value determined by the incoming ion–target ion pair. Figure 9.17 illustrates the relative magnitudes of electron and nuclear stopping powers. If one examines an incoming ion that is penetrating into the substrate, the ion is initially losing its energy via the electron stopping mechanism. As it slows down, the nuclear stopping mechanism gradually takes over. There is a critical energy above which the electron stopping mechanism dominates; below this energy, the nuclear stopping power dominates. The critical energy increases with increasing mass of ions. Heavier ions, such as Sb and As, are predominantly slowed down by the nuclei stopping power, while light ions, such as B, and P, are stopped by electronic stopping power. Now, to estimate how far the ion can travel before coming to a

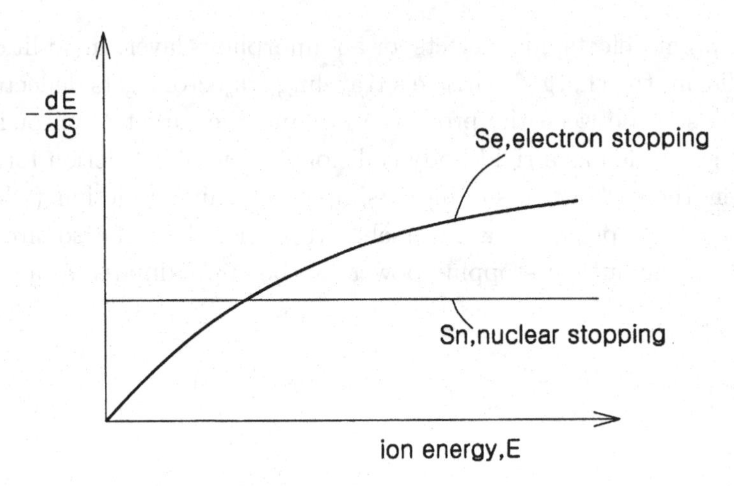

Fig. 9.17. The relative magnitude change of S_e and S_n with ion energies.

complete stop, one can integrate Eq. (9.15):

$$R = \int_0^R ds = \frac{1}{N} \int_0^R \frac{dE}{S_n(E) + S_e(E)}, \tag{9.18}$$

where N is the target atom density (atoms/cm^{-3}).

The ion's traveling distance can be estimated for two extremes:

$$\begin{aligned} &\text{for } E_i \gg E_c, \quad R \cong C_1 E^{1/2}; \\ &\text{for } E_i \ll E_c, \quad R = C_2 E. \end{aligned} \tag{9.19}$$

The above treatment applies to each individual implanting ion; however, when implanting a large number of ions into the substrate, one will not see a single value of the traveling distance, R. Instead, there will be a distribution of distances due to the random scattering nature of the ion–target nuclei and ion–electron collisions. The implanted ion distribution (the concentration profile) in the substrate can be approximated as

$$N(x) = \frac{Q}{\sqrt{2\pi}\Delta R_p} \exp\left[-\frac{1}{2}\left(\frac{x - R_p}{\Delta R_p}\right)^2\right]. \tag{9.20}$$

R_p is the projected range of the traveling distance, as illustrated in Fig. 9.18.

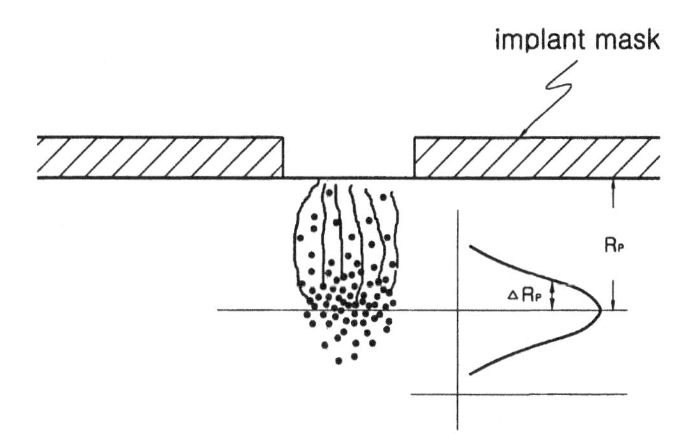

Fig. 9.18. The implanted ion trajectories and the projected range, R_p, with its standard deviation, ΔR_p.

Table 9.1 provides the projected ranges and standard deviations as functions of implant energy for commonly used dopants. One can essentially use this table to estimate what energy to use for different desired implanted junction depths. It can be observed that with the same ion and dosage, the higher the energy, the larger the projected range and deviation will be; namely, the ions penetrate deeper and spread wider. As a result, a lower peak concentration will result. On the contrary, with the same implant energy, heavier ions provide smaller projected ranges and standard deviations than lighter ions. People often implement a shallow boron implant, such as a threshold

Table 9.1. Projected ranges and their standard deviations of dopant implants in Si.

Energy (Kev)	B	P	As
10	0.033/0.017	0.014/0.007	0.0097/0.0036
30	0.098/0.037	0.037/0.017	0.022/0.008
60	0.190/0.055	0.073/0.03	0.037/0.014
90	0.273/0.067	0.111/0.042	0.053/0.019
120	0.349/0.074	0.150/0.053	0.069/0.024
150	0.425/0.083	0.189/0.063	0.085/0.03
180	0.487/0.089	0.228/0.072	0.100/0.034

adjustment implant, with BF_2 instead of boron. Furthermore, for the same reason, advanced device engineering uses heavier dopants, such as antimony, to replace arsenic and use indium to replace boron. Figure 9.19 shows that with Sb at 13 KeV and As at 10 KeV, Sb (antimony) has a shallower projection range. Furthermore, Sb has a far lesser extent of out-diffusion after postannealing as it is heavier than As. These two features make the heavier ions, such as Sb and In, good candidate dopants for forming shallow junctions for advanced devices.

Channeling is a phenomenon that occurs when the incoming ion direction happens to coincide with a major crystal orientation; as a result, the ions can travel much further than the predicted projected ranges due to lack of nuclear stopping power. Figure 9.20 explains channeling in a two-dimensional silicon crystal lattice. The opening formed by the lattice with respect to the moving ion is the largest as the ion direction moves along the crystal axis. Possible approaches to avoid channeling are to predamage the silicon surface prior to the ion implant or to create a misorientation of the crystal axis with

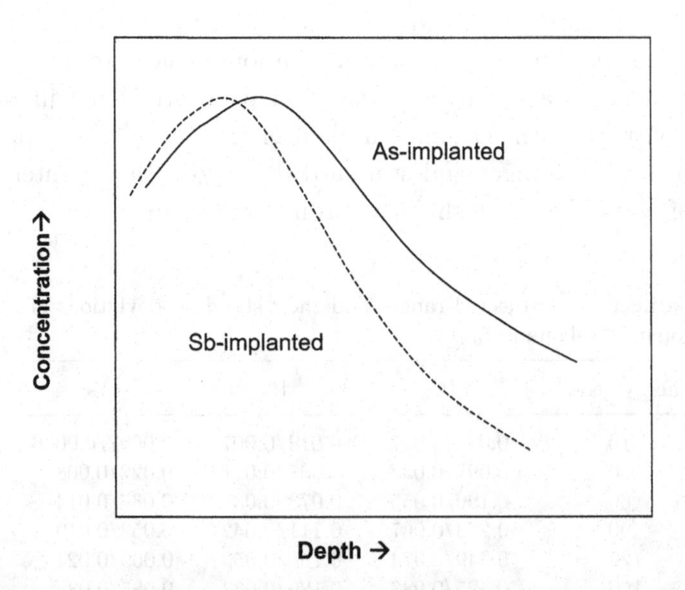

Fig. 9.19. The As-implanted profiles for As and Sb show that Sb has shallower peak than As.

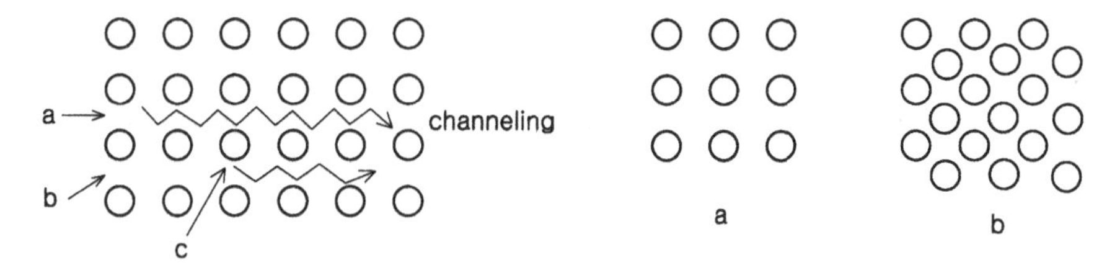

Fig. 9.20. Ion channeling occurs in direction a but is less possible in direction b. For cases c, ions can be scattered into the channeling direction.

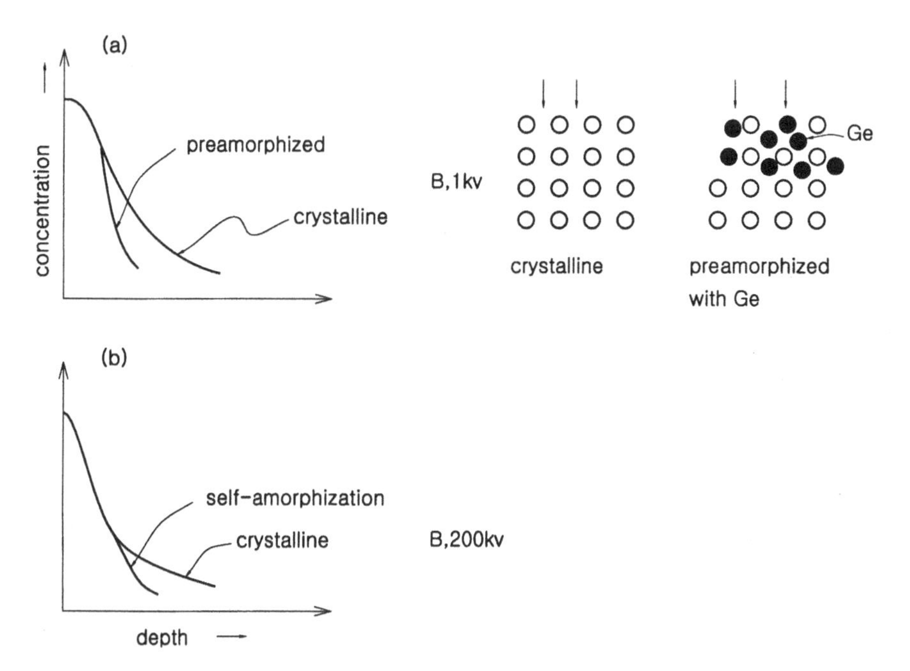

Fig. 9.21. Boron implant of 5×10^{14} at (a) 1 kv and (b) 200 kv. The Ge preamorphized sample gives less channeling.

respect to the incoming ion direction by 7°–10°. Basically, predamaging the silicon surface will amorphize the surface to minimize the channels for the incoming ions. Figure 9.21 illustrates that channeling is diminished when the silicon is predamaged. By misorienting the wafer crystal axis to the incoming ion direction, one can significantly eliminate the channeling as the opening is reduced. However, some probability still exists that the incoming ions can be scattered into the channeling direction. Another approach is to use a heavy molecule containing the desired ion for implant, such as BF_2 instead of B. Heavy molecules are more prone to create an amorphous layer for the subsequent implant ions, and thereby they reduce the ion channeling.

Most of the ion implant processes for making a device are selective in nature; in particular, an ion implant mask is often required. The mask acts as the physical shield, which must be so thick and tough that it can trap the incoming ions within its own bulk. The thickness of the mask must be bigger than the projected range of the incoming ions. For example, when implanting the phosphorous atoms to the PMOS area, such as the HALO implant, the implant in the NMOS area must be totally blocked, or the implant will cause the NMOS device window to degrade. This is illustrated in Fig. 9.22, which shows the NMOS device window made smaller due to the undesired Halo implant (P). The most commonly used implant masks are photo resist, silicon oxide, or silicon nitride layers. Table 9.2 lists the projected ranges and standard deviations for the dopants in an oxide mask.

The implant mask is a sacrificial layer, meaning that it does not exist in the final device structure. Its thickness must be optimized such that it does not exert excessive stress or cause any stripping problems. Generally, one tends to use a thickness larger than theoretical estimates, that is, $> (R_p + 3\Delta R_p)$. However, it is worth noting that when one uses a large-angle implant, the implant mask thickness

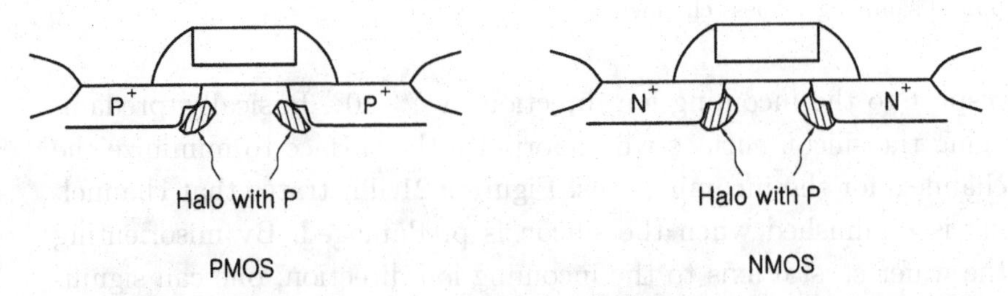

Fig. 9.22. The PMOS Halo implant uses phosphorous. If it is implanted into the NMOS region, it will degrade the NMOS device window.

Table 9.2. Projected ranges and their standard deviations of dopant implants in oxide mask.

Energy (KeV)	B	P	As
10	0.030/0.014	0.011/0.005	0.008/0.0026
30	0.095/0.034	0.029/0.012	0.017/0.0057
60	0.192/0.054	0.059/0.021	0.030/0.010
90	0.282/0.067	0.090/0.031	0.043/0.014
120	0.365/0.077	0.122/0.039	0.056/0.018
150	0.443/0.085	0.154/0.046	0.069/0.021
180	0.517/0.091	0.186/0.053	0.082/0.025

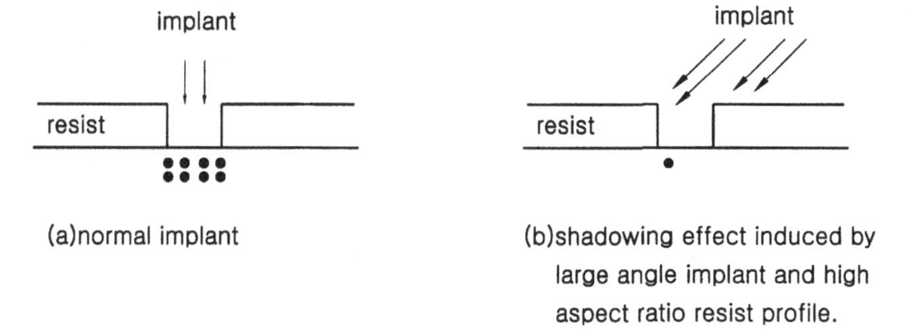

Fig. 9.23. Thick resist on small geometries (high aspect ratio) can cause a shadowing effect, meaning that some areas may not receive any implant.

cannot be too large as it may cause a shadowing effect, as illustrated in Fig. 9.23.

9.4. Implant Damages and Annealing

During ion implantation, the ions are slowed down mainly by electron stopping power at the silicon surface, and then by nuclear stopping power as ions slow down. For the nuclear stopping mechanism, if the energy transfer from the penetrating ion is large enough, it can dislodge the lattice silicon atoms, which creates a damaged site. If the amount of energy is larger than twice the dislodging energy, the dislodged silicon atom can act as a projectile particle, which will induce a subsequent dislodging event. If the amount of transferred energy is many times larger than the dislodging energy, a cascading dislodging event can be initiated. In other words, the defects increase with the increased implant energy. Furthermore, the electronic excitations that the incoming ions created along their traveling courses tend to enhance the point defect diffusion and give rise to the point defect migration and aggregation.

As far as the ion mass is concerned, traveling light ions get scattered off their original directions and tend to travel longer distances compared to heavy ions. Hence, light ions cause damage under the silicon surface, and the damage tends to be longer and scattered, as shown in Fig. 9.24. On the contrary, heavy ions are less prone to be scattered off their original traveling directions, and if the projectile ions are energetic enough, they cause cascading collisions and move a large number of atoms off the lattice sites. Therefore heavy ions such as As or Sb can cause damage starting from the substrate surface. It is plausible that higher implant dosages could cause more damage and hence a higher tendency to result in an amorphous layer. As an implanting ion beam scans across the wafer surface, the total

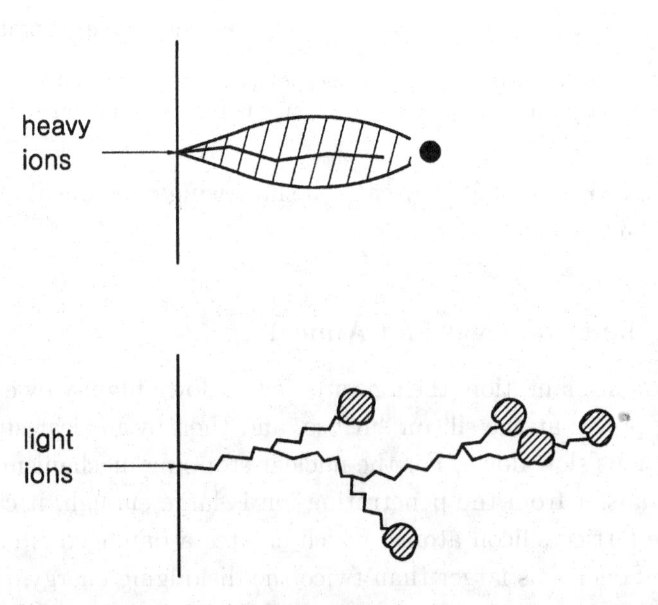

Fig. 9.24. Heavy ions cause damage from the surface downward; light ions cause damage deep underneath the surface.

deposited ions result in a large amount of energy in wafer bulk; as a result, local heating can be detected. This local heating helps the damaged crystal structure to restore itself to its preimplant structure, somewhat. In general, heavy ions, high implant dosages, and low substrate temperatures are more prone to result in the formation of amorphous layers. At 200°C, it takes about 5×10^{14} of P or $10^{14}\,\mathrm{cm}^{-2}$ of As to create an amorphous layer, but B does not result in an amorphous layer due to its light weight. The damage or amorphous layer formations degrade the carrier mobility and reduce the carrier's lifetime. Lower carrier mobility or a lower carrier lifetime translates to a higher resistivity or a lower conduction current. Nonetheless, the formation of a damaged layer can reduce the channeling effect and can enhance dopant diffusion during postannealing, as shown in Fig. 9.25. Furthermore, the diffusion enhancement is less

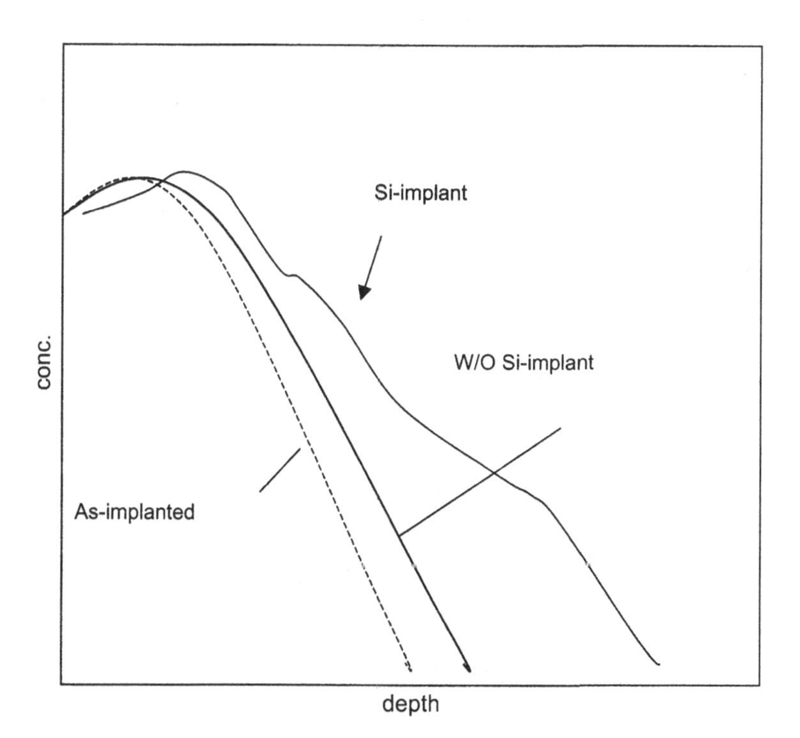

Fig. 9.25. Boron diffusion is enhanced by the predamages caused by Si - implant.

obvious at a higher-temperature anneal, favoring the use of high-temperature postannealing.

A postimplant annealing is required to remove the damage and activate the dopants, and at the same time, dopant redistribution takes place isotropically. To avoid the loss of the implanted dopant atoms, annealing is often conducted under oxidizing ambient, which forms a cap oxide layer, a diffusion mask. Theoretically, a prolonged annealing can perfectly restore the crystal structure to its preimplant structure. However, in real life, an anneal step is often far from able to completely restore the perfect crystal structure due to thermal budget constraints. When a residual damage trail threads through a depletion region of a junction, it leads to junction leakages. It is often better to tune a junction so that the damages are enclosed by the dopant fronts, as indicated in Fig. 9.26. Annealing can be conducted in a conventional furnace with temperatures ranging from 700°C to 1100°C for 1 to several hours, or with a rapid thermal processor with temperatures ranging from 900°C to 1200°C for a few seconds. Annealing must take into account how much thermal budget is allowed in terms of dopant further diffusion (redistribution). Oftentimes, in device manufacturing, several implants can be done and finished with a single annealing step to minimize the dopant

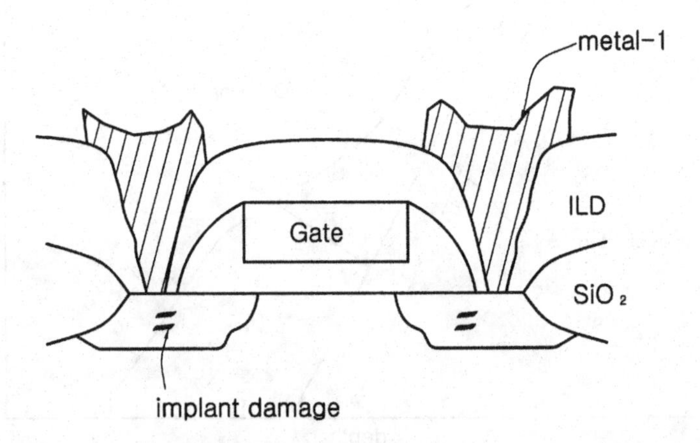

Fig. 9.26. Contact implant damages are supposed to be enclosed by the source drain region to avoid leakage.

redistribution. On implanted wafers, there exists a large amount of implanted ions and silicon atoms that are not at the lattice sites, but, as annealing proceeds, the interstitial silicon and the implanted atoms would race to occupy the vacant lattice sites. The diffusivity of silicon's self-diffusion is one to two orders smaller than that of dopant atoms. When silicon interstitial atoms move to the lattice sites, they restore the crystal structure; when dopant atoms move to the lattice sites, they become activated. Although there is a rate difference, the dopant activation and the crystal structure restoring take place hand in hand. As the annealing temperature increases, point defects may migrate or recombine to form various forms of defects. As annealing is prolonged, the point defects transform and eventually disappear.

Crystal regrowth can proceed at temperatures as low as 400°C–500°C. The regrowth rate is dependent on crystal orientation; for example, it is faster in the ⟨100⟩ orientation compared to the ⟨111⟩ orientation. Figure 9.27 illustrates that the amorphous–crystal interface advances with time. As more crystal structure is restored, the excess amount of dopants is swept across the interface. In the meantime, even at the same temperature, the extent of dopant activation varies with different dopants. For P, full activation can be achieved at 600°C, while for As, only 50% activation is achieved. For most dopants, the full activation is achieved at 900°C–1000°C. Dopant redistribution also takes place, and it is frequently enhanced by defects during postannealing. The process characteristics of rapid

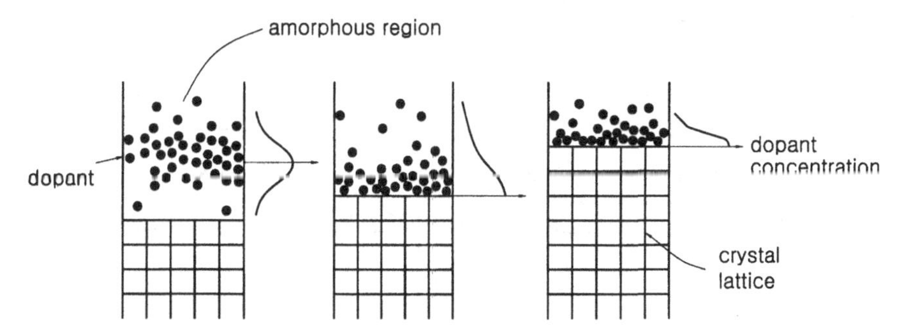

Fig. 9.27. During annealing, excess dopants are swept across the bulk as crystal–amorphous interface advances.

thermal annealing (RTA) are high temperatures with very short processing time. The temperature runs up to 950°C–1100°C for a few seconds. It mainly activates the dopants and anneals the defects without significantly redistributing the dopant atoms. RTA has been widely used to form shallow junctions for advanced device engineering.

9.5. Applications of Doping Technology

In this section, we will review the applications of doping technology that are required for building devices as the technology advances. For accuracy consideration, as a device shrinks, the applications of furnace doping lose ground to ion implantation. This is primarily due to the fact that ion implantation has a better accuracy in controlling both implant dosages and energies. It enables an accurate amount of dopants to be placed at the right depth and location. Most important, unlike diffusion, the ion implant can place the dopant concentration peak in the bulk of the substrate.

For transistor isolation and biasing purposes, a number of transistors are placed in the same "tub." These tubs are called wells. For large devices, such as for 2-μm gate lengths and above, the NMOS process was used. NMOS transistors were placed in the *p*-well, while PMOS transistors were placed in an *n*-type substrate, as indicated in Fig. 9.28. With the large geometry, the *p*-well can be formed with boron diffusion. A long drive in diffusion is needed to push the dopants further down enough for the well structure. The

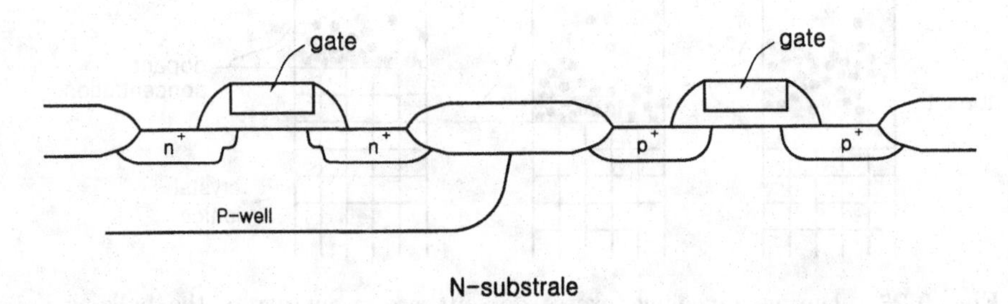

Fig. 9.28. The *p*-well process: NMOS are located in the *p*-well, PMOS, in *n*-substrate.

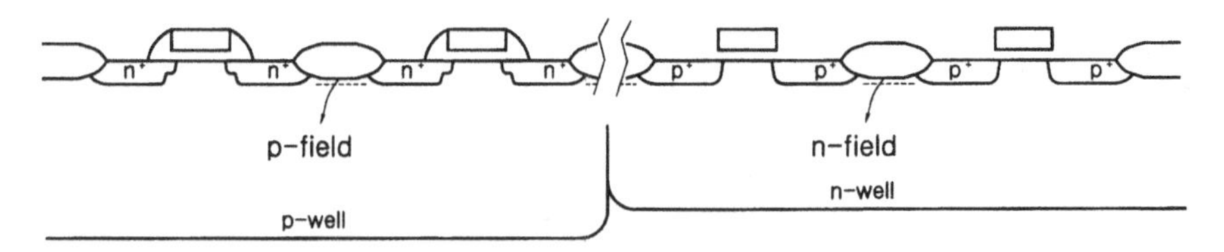

Fig. 9.29. Twin-well process: both NMOS and PMOS are located in wells.

threshold adjustment implant, a boron implant, is employed to adjust the NMOS threshold voltages (V_t) to the desired level. The polysilicon gate is doped with P in a furnace system during the POCL doping process. When the device size shrinks down to below 2 μm, the twin well process is used, for which NMOS and PMOS are placed in p-well and n-well, respectively, to allow for separate transistor optimization for NMOS and PMOS. The source and drain are then doped with P and B.

Beyond 1.5-μm technology, additional device features are added to ensure good device performance, as illustrated in Fig. 9.29. The field isolation implant placed underneath the field oxide is employed in n-well and p-well, respectively, to improve the field parasitic device isolation. The NMOS source and drain are formed with implantation of As instead of P due to its capability to form shallower junctions. An LDD structure is used for NMOS to obtain a graded junction profile and to decrease the electrical field strength in the drain region, and thereby to achieve a better hot-carrier performance. The spacer width defines the graded junction, as indicated in Fig. 9.30. Furthermore, as NMOS transistor length decreases, more precautions must be taken to avoid a source–drain shortage. The junction depths must be decreased, and additional features are required. For example, a 1.2-μm gate length, anti-punch-through implant is indispensable

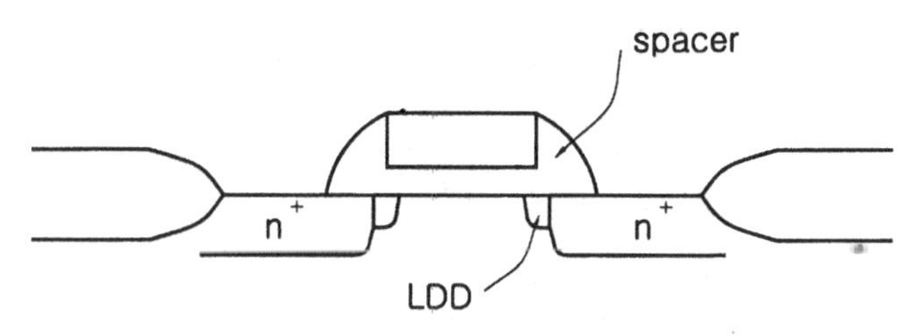

Fig. 9.30. LDD structure is used in NMOS.

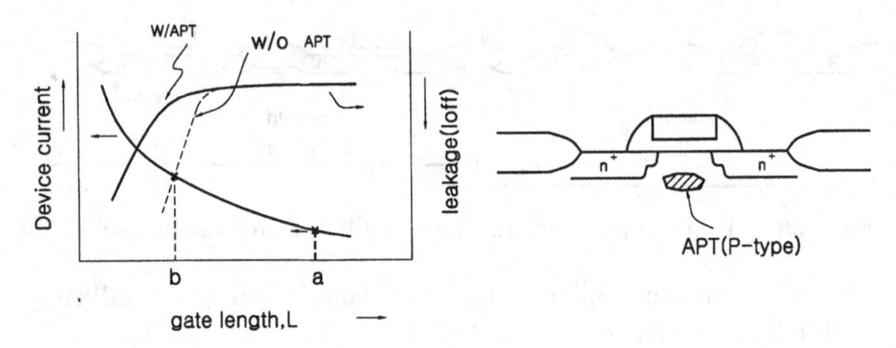

Fig. 9.31. The device window, as defined by a and b, is larger with APT implant than without.

between the source and drain in avoiding a short channel effect. In particular, it improves the short channel device window, as illustrated in Fig. 9.31.

For submicron devices, both NMOS and PMOS need to have LDD structures since the PMOS hot-carrier effect starts to be a concern. PLDD is done with a boron implant. Halo implants are needed to avoid short channel effects. It is done with a large-angle implant, often larger than 7°–10°; 30° is often used. The dopant of the opposite type is placed purposely at the source and drain junction edges to decrease the depletion region width, as illustrated in Fig. 9.32. Types *n*- and *p*-Halo must be done separately for NMOS and PMOS, respectively.

For submicron devices, the size of a contact hole (the electrical contact between the silicon substrate and the first-level metal) shrinks below 1.2 μm, and the contact resistance increases dramatically without a contact implant. The contact implants for the *n*- and *p*-type contacts are implanted with As and BF_2, respectively, after

Fig. 9.32. A Halo implant is implanted with a large-angle implant to place *p*-type dopant at the edge of the *n*-type junction to enlarge the device window.

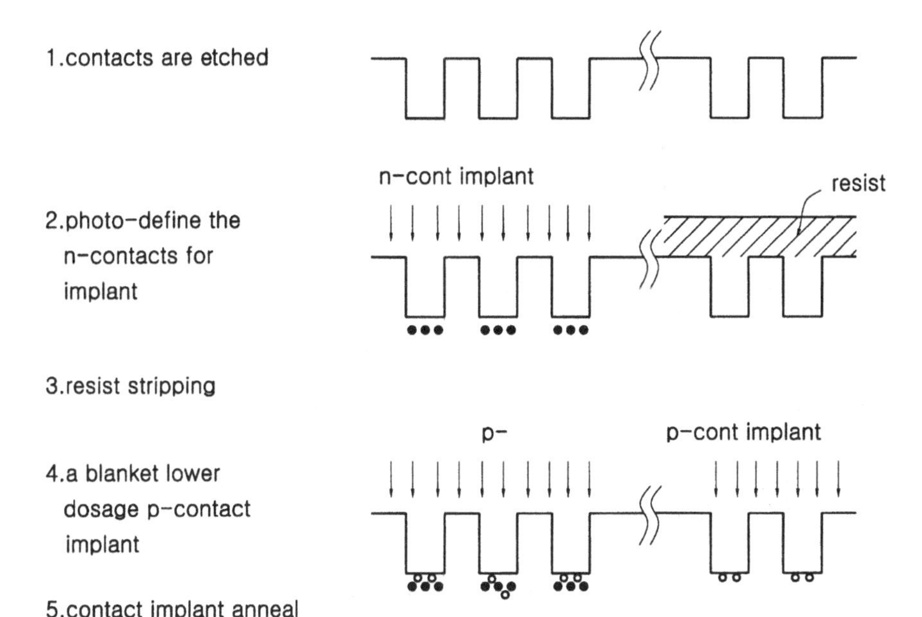

Fig. 9.33. An overcompensating scheme of n-type and p-type contact implants.

the contacts are etched. A resist mask is needed to mask out the n-type during the p-type implant, and vice versa. This is done to prevent n-type implants from being implanted into the p-type contacts. To save a photolithography step, an overcompensating scheme can be used. The p-type implant is done first without a blocking mask, say, with a dosage of 1e15; then the masked n-type contact implant is done at a higher dosage, say, at 3e15, as illustrated in Fig. 9.33. A brief contact implant anneal is done altogether with an RTA step, at about 950°C–1000°C for 30 s, to activate the implanted dopants to lower the contact resistances. For subquarter micron devices, n-doped polysilicon gates are no longer proper due to the threshold voltage matching between n- and p-MOS transistors. The dual gate structure is accomplished with separate implants onto the n-poly and p-poly, followed by the subsequent annealing. With implanted polysilicon gates, the resistances are often too high to be practical. Salicidation on top of the gates is needed to lower polysilicon interconnect resistances.

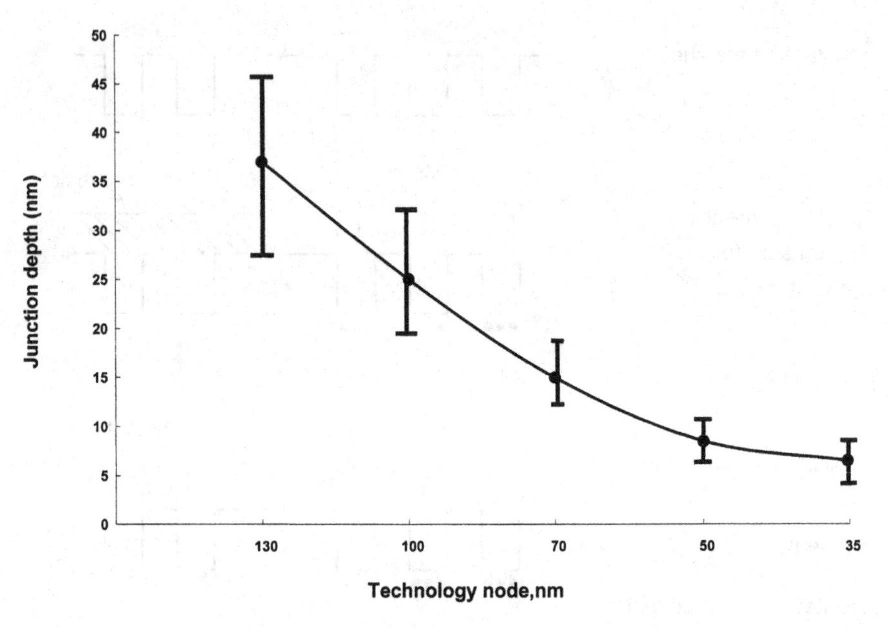

Fig. 9.34. Devices junction depth trend as technology shrinks.

As device geometry continues to shrink, so does the device junction depth, as shown in Fig. 9.34. A few approaches can be taken to achieve shallow junctions, such as a low-energy implant, heavy dopants, preamorphizing, spiking RTA, and so on. A new implant technology, called the differential mode implant, has enabled the implant energy to stay as low as 200 eV, while keeping a reasonable implant current to maintain the same throughput level as the conventional approach, the drift mode. The differential mode essentially extracts the ions from the ionization chamber with a high voltage for a high ion flux (a high current), and then slows the ions down before hitting the wafer surface. In addition to the low-energy implant, preamorphizing also provides short implant ranges. Preamorphizing can be done either with Si or Ge implants prior to the desired dopant implants. On the other hand, heavy ion implants by themselves are also effective in making shallow as-implanted junctions since they tend to create more damage. Examples of this include using BF_2 for B, In for B, or Sb for As. With the same energy level, these heavy molecules all provide shorter implant ranges. Finally, an RTP anneal provides much smaller dopant redistribution and results in shallower junction depths. Spiking the RTP is even more effective in reaching a higher dopant activation and lower extent of redistribution. In comparison to conventional RTP, spiking RTP uses a much higher temperature ramp-up rate, at around 150°C/s. It will reach the highest temperature of 1150°C and soaks for a much shorter time, say, 1 s, and then it ramps down. A shorter soaking time allows for an even higher peak temperature to be used, resulting in more dopant activation and a lower resistance; also, even shallower junction depths can be obtained.

Chapter 10

METALLIZATION AND SILICIDATION

In semiconductor manufacturing, metallization refers to the process of forming metal interconnects that electrically connect various building blocks to comprise a circuitry. In this chapter, we will illustrate various interconnect materials and explain their processes. Section 10.1 introduces the evaporation and sputtering processes, which are used in forming aluminum interconnects. Section 10.2 further elaborates on the sputtering system since it is mostly used for aluminum deposition. The composition of the required metal interconnects are introduced in Section 10.3, which explains the reasons of adding dopants in detail. It is then followed by the explanation of the barrier metal evolution and the role of the antireflective coating (ARC). Section 10.4 deals with a very critical issue — step coverage improvements on the bulk metal and barriers. Beyond submicrometer technologies, metal silicides are widely used to reduce resistances of gate and source/drain areas. In Section 10.5, the processes of various metal silicides and their evolution are introduced.

10.1. Introduction

Physical vapor deposition (PVD) is a process in which the target material is vaporized or sputtered and is then condensed onto a substrate surface. There are no chemical reactions involved in the vicinity of the substrate surface. In PVD, the atoms or molecules in the vapor phase physically adsorb on the substrate surface and thereby form a solid film. In general, a PVD process often leads to amorphous film formation. Typical PVD processes, including vapor condensation, molecule beam epitaxy (MBE), and sputtering, are used for film coating. Among them, evaporation and sputtering are used more often than others for metallization in semiconductor manufacturing. Yet sputtering is the only technique that is widely used in advanced technologies. Table 10.1 summarizes the major differences between CVD and PVD processes. CVD films are formed through preferential chemical reaction pathways; purity control is not as crucial as it is in PVD. For PVD, the purity control requires a high vacuum level in the deposition chamber. Unless special features are added, PVD processes in general result in inferior step coverage when compared to CVD processes.

Evaporation is a process in which atoms or molecules in the liquid gain enough energy to escape from the bulk through the liquid–vapor interface. A process that proceeds in the opposite direction is termed condensation. Consider an enclosed system half filled with liquid, as shown in Fig. 10.1; at any constant temperature, liquid–vapor equilibrium can be reached with time. At equilibrium, the atom or molecule fluxes of the evaporation and condensation are equal. As the temperature rises, more molecules evaporate into the vapor phase, and as a result, the vapor pressure increases. Before an equilibrium state is reached, the molecular flux of evaporation is larger than that of condensation, and the net rate of evaporation flux, F, can be estimated:

$$F = (2\pi mkT)^{-1/2}(p^* - p)\,, \tag{10.1}$$

where p^* is the saturated vapor pressure at T (temperature), and p is the actual vapor pressure.

Table 10.1. Comparison between CVD and PVD process characteristics.

Items	CVD	PVD
1. Chemical reactions	Yes	No
2. Precursors	Chemically adsorb	Physically adsorb
3. Purity control	Preferential reactions	Requires high vacuum
4. Step coverage	Good to excellent	Poor
5. Morphology	Amorphous to crystalline	Mostly amorphous

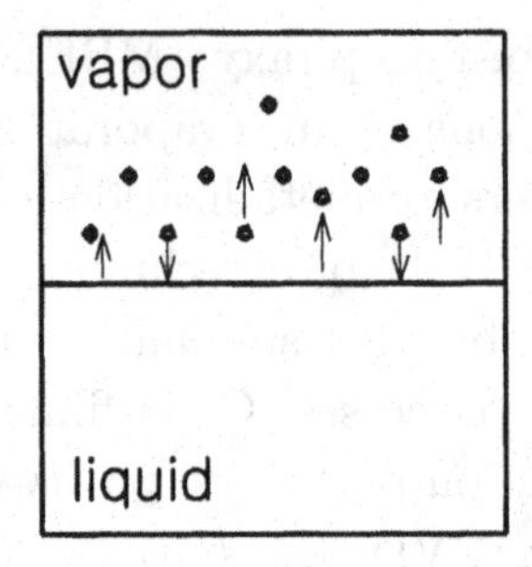

Fig. 10.1. Evaporation and condensation in an enclosed chamber.

For semiconductor manufacturing technology of larger than a few microns, pure aluminum layers are coated on the wafer surface by using the evaporation approach. Various heating sources, such as filament heating or an e-gun, are used to vaporize the aluminum source material. For filament heating, the aluminum source is wrapped around a heated filament such as tungsten; when the filament heats up, the aluminum is vaporized, as demonstrated in Fig. 10.2. Figure 10.3 shows an e-gun heating in which the e-beam is generated and then bent with a magnetic field toward an aluminum load

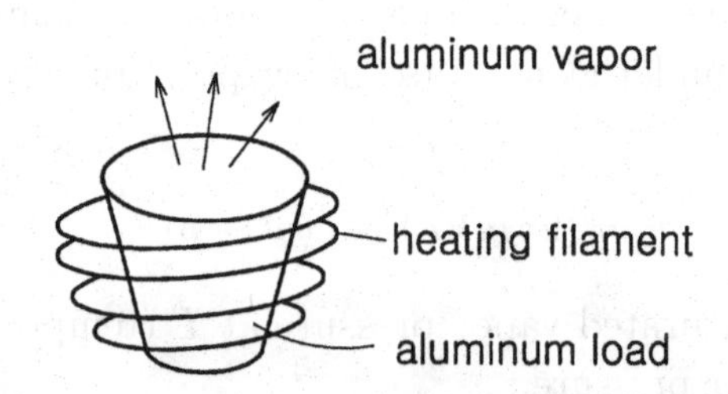

Fig. 10.2. Aluminum evaporated from a heated crucible.

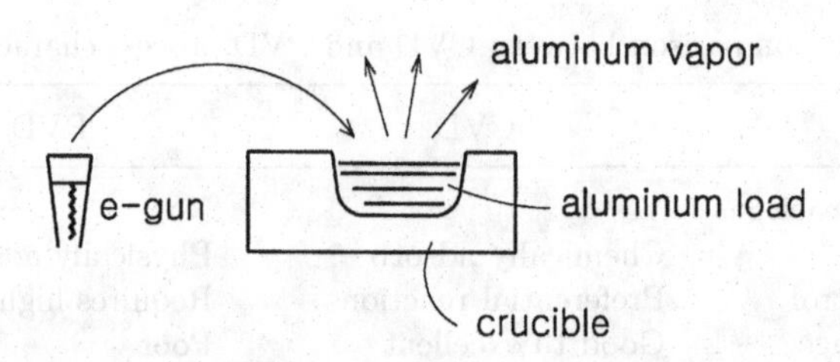

Fig. 10.3. Aluminum evaporation with e-gun heating.

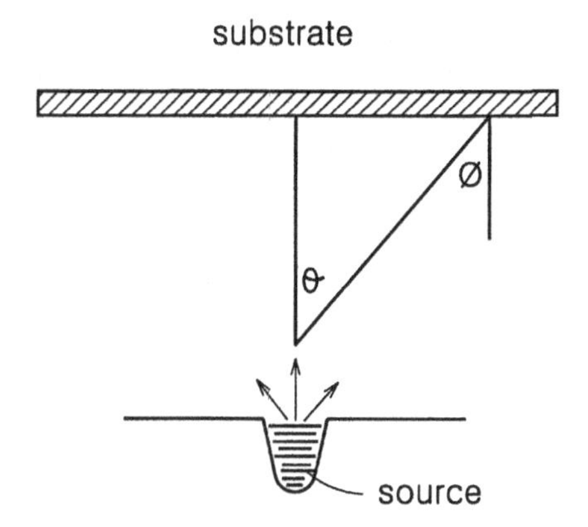

Fig. 10.4. A directed evaporation source with a substrate placed above it.

contained in a copper crucible. To achieve uniform coating across a substrate, as shown in Fig. 10.4, the angular dependence of the evaporation source and the effect of wafer orientation should be eliminated. Figure 10.4 shows a directional effusion source and the angular dependence of the substrate. Assuming that a normal radiation intensity from a source is I, if the radiation intensity on a surface with an angle θ to the normal direction is measured, as indicated in Fig. 10.5, then the intensity becomes $I\cos\theta$. According to Lambert's cosine law, the deposition flux on the wafers located at the edge,

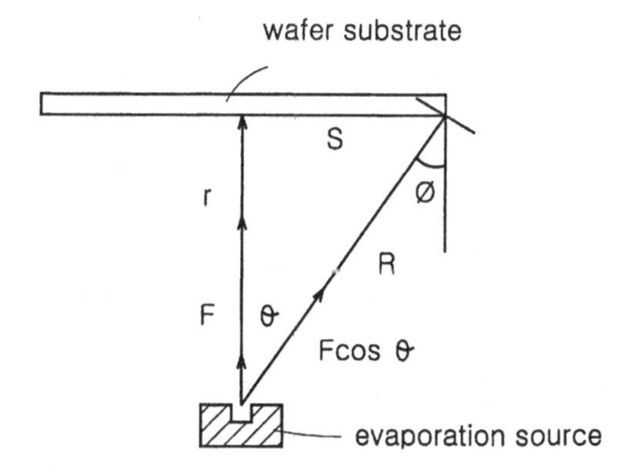

Fig. 10.5. The normal evaporation rate, F, deposits onto the center of the wafer.

V_{edge}, and center, V_{center}, are related by

$$\frac{V_{edge}}{V_{center}} = \frac{r^2}{R^2}\frac{r^2}{r^2+s^2} = \left[\frac{r^2}{r^2+s^2}\right]^2 = \left[\frac{1}{1+\left(\frac{s^2}{r^2}\right)}\right]^2 . \tag{10.2}$$

The edge wafer deposition rate is scaled down by the cosine law. Clearly, for the case of $S \gg r$, meaning the evaporation source is close to the substrate plane and the targeted substrate is far away from the center, it will result in a very nonuniform deposition across wafers (thick around the center locations and thin at the edge). On the other hand, if $S \ll r$, the film thickness uniformity across wafers can be very uniform since cosine law does not prevail. Practically, when $S \ll r$, it is of little use in the production environment. To obtain a uniform film thickness with high deposition rates, the evaporation source should be isotropic to issue evaporation rates equally in all directions. Furthermore, each individual wafer should be oriented toward the evaporation source; in other words, ϕ is zero. Manufacturing experience has demonstrated that with a rotating planetary wafer arrangement, highly uniform film deposition across wafers can be obtained.

Sputtering is a PVD process in which atoms or molecules of the target material are sputtered off and condensed on the substrate surface. As an example, typical equipment is sketched in Fig. 10.6. To avoid chemical reactions, inert gas ions such as Ar ions are often used to sputter off the target material. Ar ions are created in the Ar plasma and are confined between the electrodes. The target is mounted on the cathode, while the wafer is mounted on the anode. In some applications, the target electrode is purposely biased to increase the ion bombardment energy. The Ar ions, accelerated by the plasma sheath impinging onto the target material, impart their momentum onto the target atoms, pushing them off their equilibrium sites to move forward. These moving atoms in turn cause more collisions with other atoms. Eventually, some of the moving atoms escape from the target surface, as shown in Fig. 10.7. A linear collision cascade occurs when each collision involves one stationary and one moving atom and the

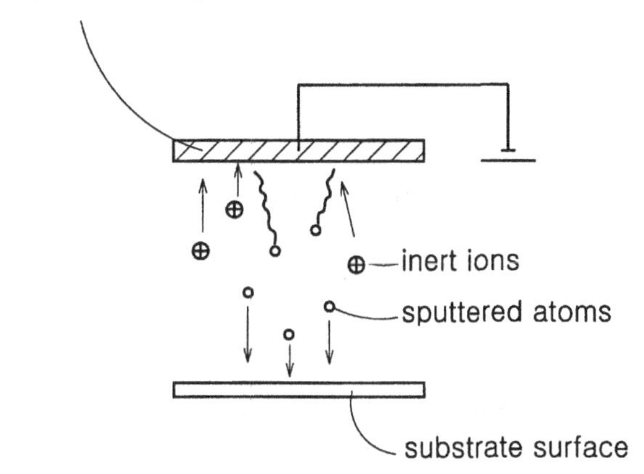

Fig. 10.6. Schematic of a sputtering system.

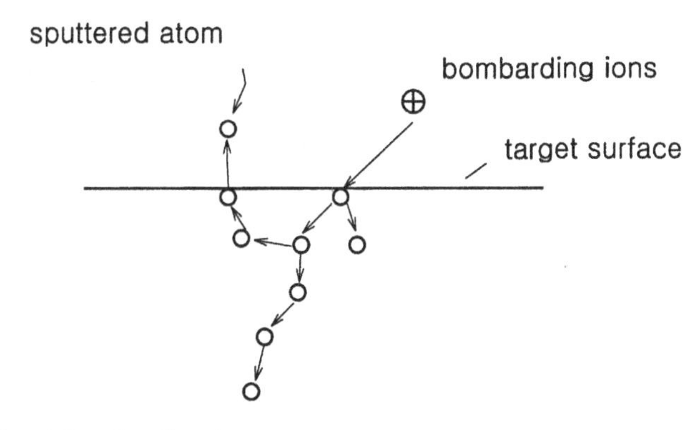

Fig. 10.7. A bombarding inert ion causes a series of collisions. Some atoms are sputtered out of the target surface.

number of recoils is low. On the other hand, in a nonlinear cascade regime, the number of recoils is high; essentially, a large number of atoms are off their equilibrium sites and in motion simultaneously. In this regime, the sputter yield is surprisingly high. The sputtering yield is defined as the number of atoms that are sputtered off the target per incident particle (projectile). The sputtering yield is a function of the target material, the projectile, and the projectile energy level. If the energy level of the impinging particle is too low to transfer enough momentum to the target atoms, no target atoms can be sputtered off. There exists a threshold energy below which the target atoms cannot be sputtered off. Both the threshold energy and the sputtering yield are related to how strong the target atoms are bound together, which corresponds to the material sublimation energy.

Figure 10.8 demonstrates the sputtering yields of Al and W, with Ar ions being the projectile. A few observations can be drawn from this figure. First, with a fixed projectile energy, the sputtering yield for Al is higher than W. Second, when extrapolating to zero sputtering yields, the threshold energy for W is higher than that of Al. These two observations are in accordance with the fact that the sublimation energy of Al is 0.11 eV, which is lower than that of W, 0.49 eV. Third, the sputtering yield increases with increasing projectile energy. Finally, above certain energy levels, the sputtering yield tends to level off and then decrease with increasing projectile energy. This regime indicates the onset of an ion implantation phenomenon. Figure 10.9 shows the sputtering yield of silicon with various projectiles at 45 KeV. It demonstrates that the sputtering yield increases when the projectile atomic number is increased. The heavier the projectile ion, the better the momentum transfer on incident on the target surface will be.

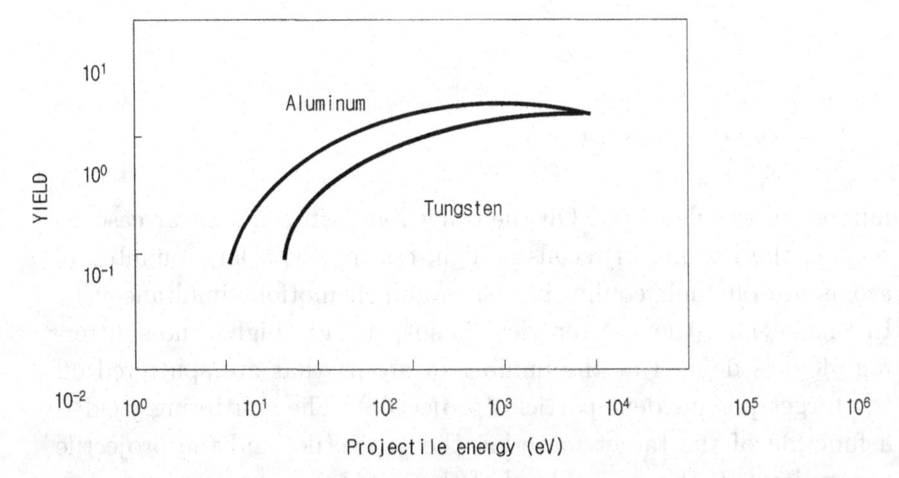

Fig. 10.8. Calculated sputter yield of W and Al at various energies of projectile Ar ions.

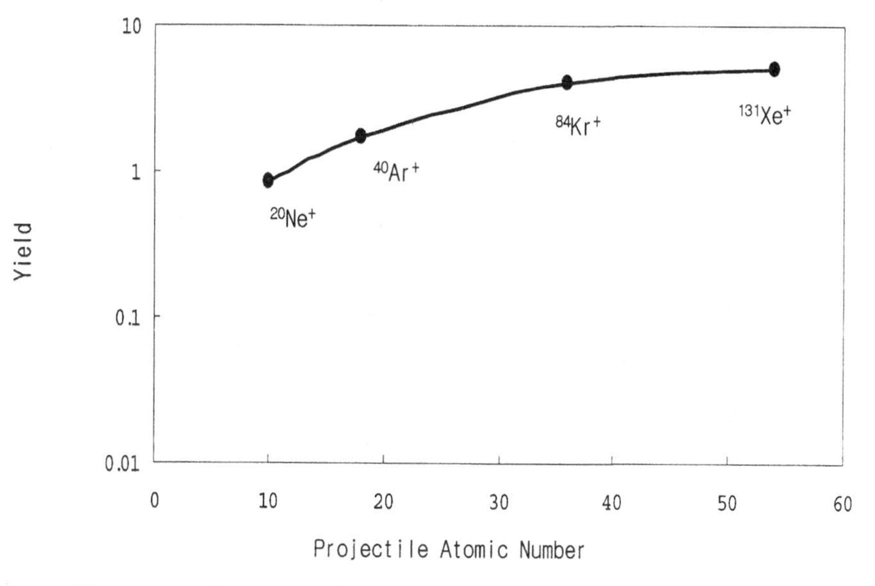

Fig. 10.9. Sputter yield of Si at various projectile ions at 4.5 KeV.

Although the theory of sputtering phenomena is quite complicated, the sputtering yield, Y, can be expressed in a simple form:

$$Y = A(E^{1/2} - E_{\text{th}}^{1/2}), \tag{10.3}$$

where E is the projectile energy; A and E_{th} are constants and dependent on the target projectile combinations.

To find the sputtering rate, R, of a specific sputtering process, one needs to know the ion flux, which can be derived from the ion impinging rate, I_f:

$$R = \frac{YI_i}{q} = YI_f, \tag{10.4}$$

where q is the electrical charge of the ion. Unfortunately, it is often difficult to relate the ion current to the electrical characteristics of a sputtering process. In literature, people have been able to correlate the sputtering rate to more measurable parameters such as power input, P_g:

$$R = C_g(P_g - P_0), \tag{10.5}$$

where P_0 is the turn-on power. C_g is a constant that is dependent on plasma gas species. In other words, for a sputtering process, the sputtering rate increases with increasing system power input.

10.2. Sputtering Systems

Major advantages of a sputtering process are good film quality and low process temperatures, meeting the low thermal budget constraint in the back-end device manufacturing processes. However, the main drawback of sputtering technology is poor film step coverage. Several sputtering techniques have been employed in semiconductor manufacturing, including DC sputtering, RF sputtering, bias sputtering, and magnetron sputtering. To improve the deposited film properties or step coverage, variations of the sputtering techniques, such as reactive sputtering, ionized metal sputtering, and collimated sputtering, have been introduced; these variations will be discussed in Section 10.4.

A DC sputtering system consists of a DC capacitively coupled plasma system, in which the target and substrate are mounted on the cathode and anode, respectively, as shown in Fig. 10.10. The

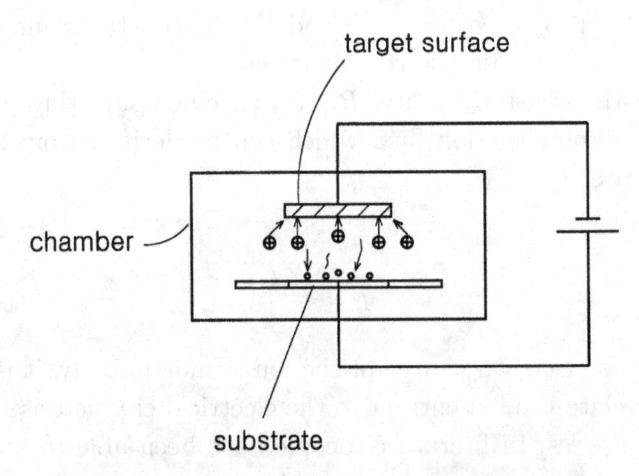

Fig. 10.10. A representative DC sputter system.

target on the cathode is subject to Ar ion bombardment; the target atoms are sputtered out and then condensed onto the substrate surface. To ensure that a large percentage of the sputtered atoms can condense onto the substrate surface, the gap between the target and the substrate is generally made very narrow. With DC sputtering, it is difficult to handle an insulator target surface as the positive ions that are accumulated over the sputtering period cannot be compensated for. RF sputtering is an appropriate alternative since the electrode's polarity alternates; the positive charges accumulated in one polarity can be compensated for in the next polarity. However, if the surface areas of the target and the substrate are equal, they will both be sputtered alternately and equally; this is not desired.

To achieve the desired goal, sputtering, it is a common practice to make the target electrode area smaller than the substrate's area so that the ion flux toward the target is much larger than the one toward the substrate surface, as shown in Fig. 10.10. If the electrode areas are swapped, as Fig. 10.11 illustrates, the sputtering etching on the substrate can be obtained. Further improvement can be made by adding a DC bias on the substrate to achieve a slightly more energetic and directional ion sputtering etching. This technique is often employed before aluminum sputtering to sputter etch or preclean the bottoms of contacts, where native silicon oxide is often formed on the

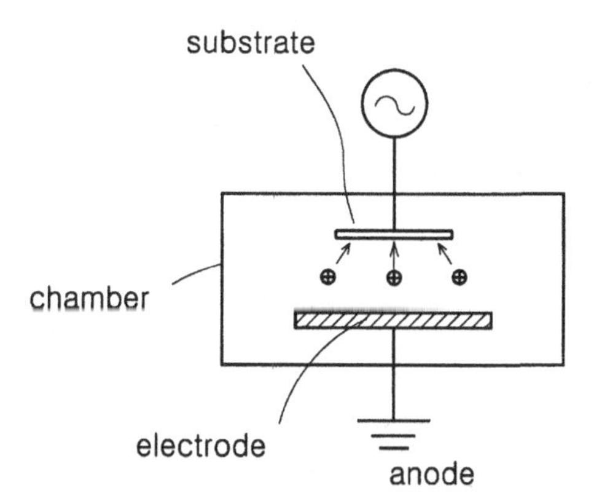

Fig. 10.11. Sputtering etching on substrate can be obtained if the substrate is mounted on the cathode and the target on the anode.

silicon surface. The existence of the native oxide layer, even as thin as 10–15 Å, can be detrimental to the contact resistances.

In conventional sputtering technology, the sputtering rates are often low. To increase the sputtering rates, the ion flux toward the target surface has to be increased by raising the power input to the system. However, a better way to achieve a higher sputtering rate is to use a magnetron sputtering technology. This is done by imposing a magnetic field around the sputtering target so that the electrons can be trapped around the field lines. As a result, the plasma density, the impinging ion flux, and the sputtering rates are increased. Figure 10.12 shows a planar magnetron sputtering target, coupling with a magnetic field. This figure demonstrates that the electrons are in gyromotion around the magnetic field lines. The magnetron sputtering technique has been successfully used to produce films with low impurity level and high quality at acceptable deposition rates.

The enhancement of adding the magnetic field can be explained. Considering a charged particle traveling in a uniform magnetic field with its velocity component parallel and perpendicular to the magnetic field, as shown in Fig. 10.13, it will experience the following force:

$$\vec{F} = q(\vec{v}_H \times \vec{B})\,. \tag{10.6}$$

At the same time, the vertical velocity, V_V, is unaffected and pushes the electron along the magnetic field. Therefore a helical motion is formed. If an electrical field moves more parallel to the magnetic field, then the helix pitch will be increased. When the electrical field is perpendicular to the magnetic field, the electron is forced to circle

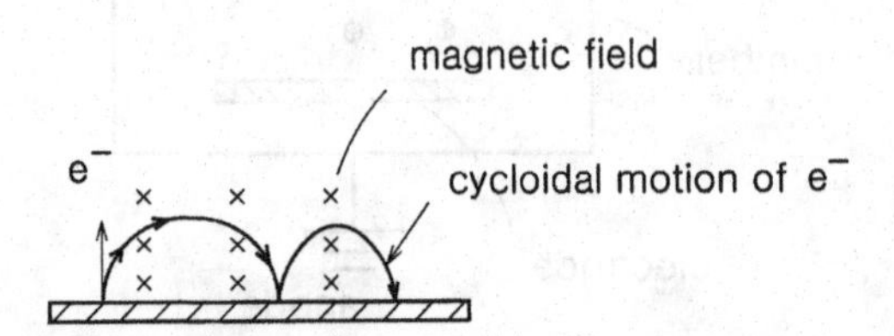

Fig. 10.12. A magnetron sputter system uses a magnetic field to bend the electron motions.

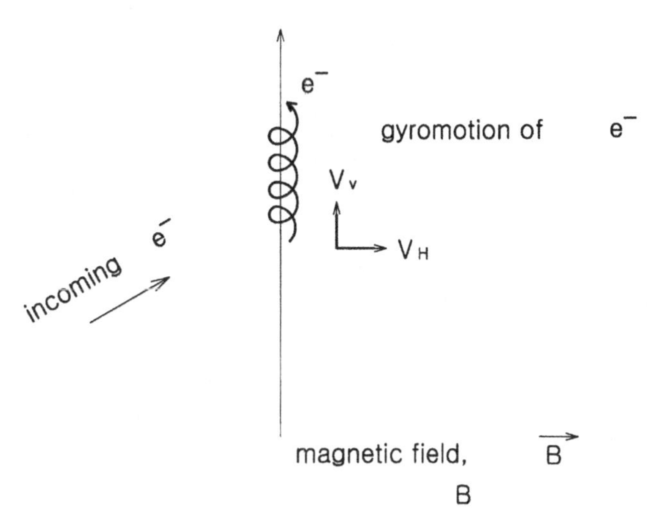

Fig. 10.13. A moving electron undergoes a gyromotion when subject to a magnetic field.

around the field line. These electron gyromotions greatly reduce the probability of electron loss due to colliding with the chamber wall. Therefore the electron density is increased significantly.

10.3. Aluminum Metal System

Among all conducting elements, aluminum is not the best conducting metal. However, it is the most used metal for interconnects in semiconductor manufacturing, primarily because of its combined advantages of low resistivity, abundant availability, and etchability. It is most commonly deposited with the sputtering technology in advanced semiconductor manufacturing. The sputter-deposited film morphology varies with deposition conditions and the underlying layer; oftentimes, they are amorphous or polycrystalline. Aluminum adheres well to silicon oxide or doped silicon oxide surfaces by forming aluminum oxide at the interface. As thin film metal lines of a semiconductor device, aluminum lines can carry a high current density of up to 10^{10} Å/cm^2.

Electromigration is a phenomenon that often occurs in a conducting line. When electrical currents flow in a conducting line, a large number of moving electrons will impart their momentum onto the metal ions and push them along the direction of the moving electron. This will cause a local mass loss, which leads to local material depletion, voids, or breaks in some locations, while causing material accumulation, hillocks, or extrusion in others. Figure 10.14 demonstrates the phenomena of electromigration failures. Once local aluminum mass loss is observed, the local electrical current density is increased due to the shrinkage in cross-sectional area of the conductor. This leads to local joule heating and even more severe electromigration. Electromigration performance, which is measured by the increase of electrical resistances of metal lines, can be described as

$$F_x \propto A \cdot J^{-n} \mathrm{e}^{-E/kT}\,, \tag{10.7}$$

where F_x is the time for x percentage of failures, A is the cross-sectional area, J is the current density, E is the activation energy, k is Plank's constant, and T represents absolute temperatures. Different

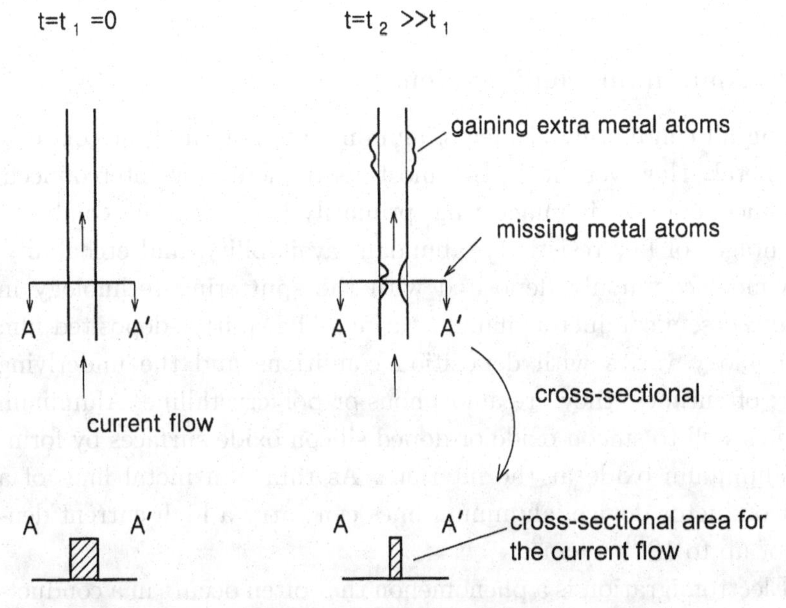

Fig. 10.14. Top view of electromigration of a metal line.

activation energies indicate different mechanisms for material transport. For example, an activation of around 0.5 eV indicates a low temperature grain boundary diffusion as the main material transport mechanism, while an activation energy of around 1.4 eV indicates self-diffusion of Al in crystal bulk. For an aluminum line, the time for the 50% failure rate increases with the mean grain size. In addition, the crystal orientation, which is a function of deposition, also plays a critical role, with $\langle 111 \rangle$ being the preferred orientation. According to the model, the electromigration degrades with shrinking line width. Nonetheless, when the line width continues to shrink until the crystal grain size is larger than the line width, the grain boundary diffusion paths are significantly reduced, as shown in Fig. 10.15. The electromigration performance can be significantly improved. Furthermore, the aluminum electromigration performance can be strongly enhanced by adding small amounts of Cu (between 0.5% and 4%) into aluminum. The role of Cu is primarily to hinder the aluminum diffusion.

Stress migration is a phenomenon in which the film stress causes the material to migrate. As a result of the aluminum stress migration, voids are often observed. In semiconductor manufacturing, it is a common practice to tailor the aluminum film and the capping dielectric layer such that they have as low tensile stress as possible to prevent void formation or stress migration failure.

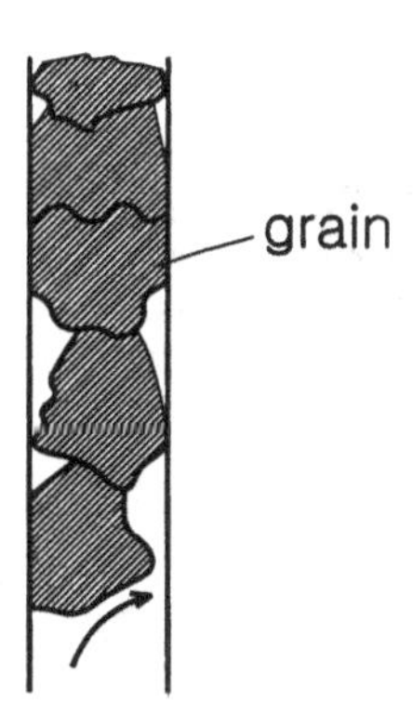

Fig. 10.15. Grain boundary diffusion is significantly hindered if the grain size is comparable to or larger than the line width.

Silicon is often used to dope aluminum to improve its contact integrity with silicon surface. As discussed earlier, one purpose of metallization is to connect individual devices together to constitute a functional circuit. This will require the aluminum metal lines to be in direct contact with the silicon substrates, the source, and the drain areas. It must be noted that silicon has certain solubility in aluminum. During intimate contact, if any of the back-end processes experience temperatures of above 420°C, the silicon will start to diffuse into the aluminum, and it will leave a cavity for aluminum to diffuse into the silicon; this results in spiking. The depth of the spiking increases with increasing temperature and time. The higher the temperature and the longer the time, the deeper the spiking results will be. When the spiking does not penetrate through the junction depth (the heavily doped silicon area enclosing the contact), the silicon cavity often causes high contact resistances. When it does penetrate through the junction depth, as shown in Fig. 10.16, excessive contact leakage can result. In both cases, contact failures will result. To avoid the contact failure due to junction spiking, some silicon (typically about 1%) is added to the aluminum to saturate the silicon in the aluminum. Considering the contact integrity and electromigration, the most commonly used aluminum material is Al doped with 1% Si and 0.5% Cu, instead of pure Al. The aluminum sputtering target must be properly doped with Si and Cu to have the desired film composition.

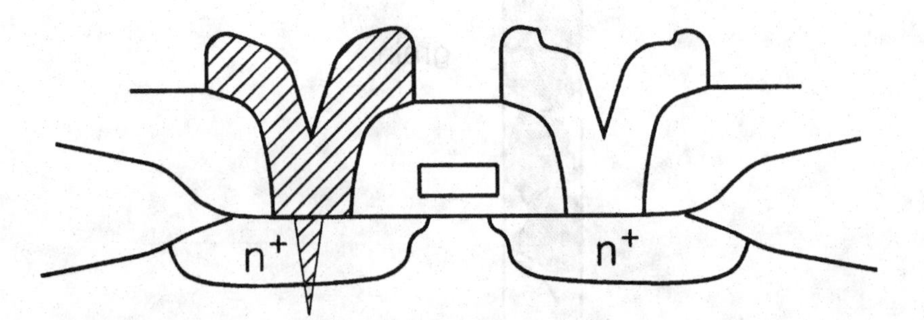

Fig. 10.16. Al spiking penetrates through the contact enclosure, causing a metal-to-substrate shortage.

10.3.1. *Barrier metal systems*

Some back-end process steps, such as PECVD dielectric deposition or alloying steps, with temperatures above or close to 420°C are sometimes inevitable. In addition to silicon doping, to further reduce the possibility of the above-mentioned contact failures due to Si–Al interdiffusion, a barrier metal layer or a barrier metal stack is often required for the back-end processing. In fact, when any two metals are placed against each other, the atom diffusions will lead to intermixing of the two materials. It is certain that the extent of the mixing depends on the time and temperature for which the two materials are brought together. In view of preventing the material intermixing or maintaining the contact integrity, the requirements for a diffusion barrier for the Al–Si system should have some characteristics:

(a) The barrier must adhere well to the underlying oxide, and Si–Al on top of it. A poor adhesion can result in delamination between the metal stacks and leads to poor reliability.
(b) The diffusivities of Al and Si in the barrier should be as low as possible to ensure that a minimal amount of Si and Al diffuse through the barrier within the thermal budget of the back-end processing.
(c) The step coverage of the barrier must be better than that of Al. Manufacturing data have repeatedly shown that conventional contacts can still be functional without Al, as long as there is a thin barrier metal. Therefore it is essential to achieve good barrier metal step coverage. However, the barrier cannot be too thick since its resistance is much higher than that of Al.
(d) The barrier should be thermodynamically stable so that it is not consumed by Si and Al within the period of interest. Once the barrier is consumed, the role of the barrier disappears.
(e) The barrier must be readily etchable and have a low resistivity, like Al.

Barrier layer fails if the metal on one side diffuses to the other side. From a diffusion perspective, two critical parameters in measuring the barrier layers' performance are time and temperature. Theoretically, if a barrier is placed between two metals and processed through a thermal cycle at a very high temperature for a very long time, the barrier will eventually fail. The failures often start out with barrier layer defects or microstructures that render fast interdiffusion paths for both sides of the metals. Alternatively, the failure can be due to the chemical reactions that the barrier experiences with the surrounding metals.

In literature, diffusion barriers can be classified into three major groups: the stuffed barrier, passive barrier, and sacrificial barrier. Stuffed barriers are characterized by the fact that the two materials' interdiffusion is inhibited primarily by the impurity that is stuffed in the barrier layer, as shown in Fig. 10.17. TiW(N) and TiN(N) for Al and Si are known to be typical stuffed barriers. Experimentally, the barriers that are stuffed with nitrogen and oxygen show better barrier properties than those without the stuffing. There are two ways of stuffing barriers. One method is to add a vacuum-break step, meaning that after the barrier is sputter-deposited, the chamber vacuum is broken, and it lets the wafers be exposed to the ambient, followed by the Al layer deposition step. The second method is to introduce nitrogen and oxygen (4:1 ratio) during barrier layer sputtering. The impurities are believed to be stuffed into the grain boundaries at very small percentages (less than 0.1 at%). The major role of stuffed atoms is to block fast diffusion paths such as grain boundaries.

Passive barriers are chemically inert to materials on both sides of them, as shown in Fig. 10.18. Obviously, the diffusivities of both materials in the barrier must be very low. A good example of a passive barrier is TiN as the barrier for Al and Si. Most nitrides, borides, or carbides of transition metals make good passive barriers. Figure 10.18(b) indicates a sacrificial barrier, which is a barrier that

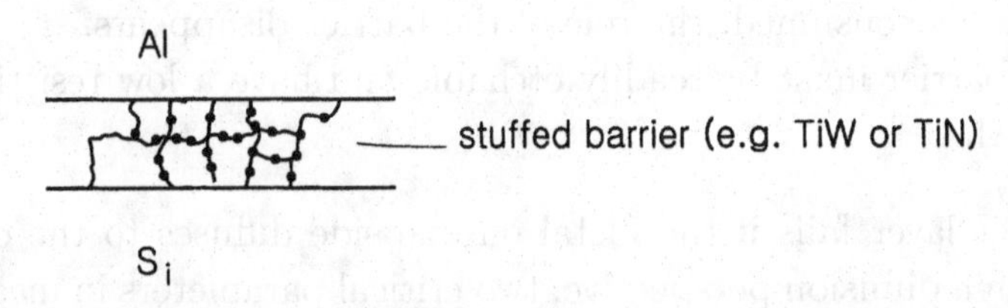

Fig. 10.17. Stuffed barrier: the impurity stuffed in the grain boundaries.

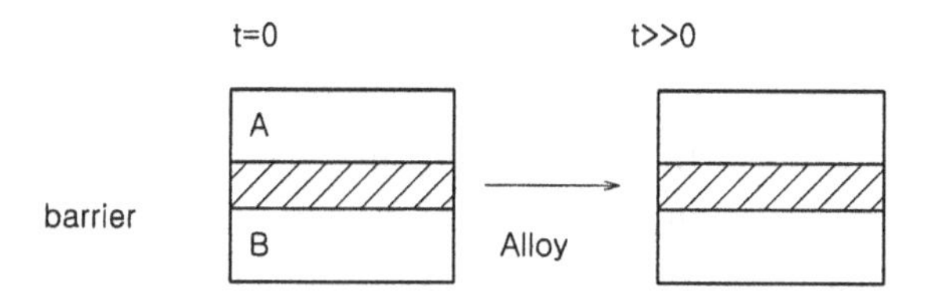

(a) passive barrier: remains intact after t>>0

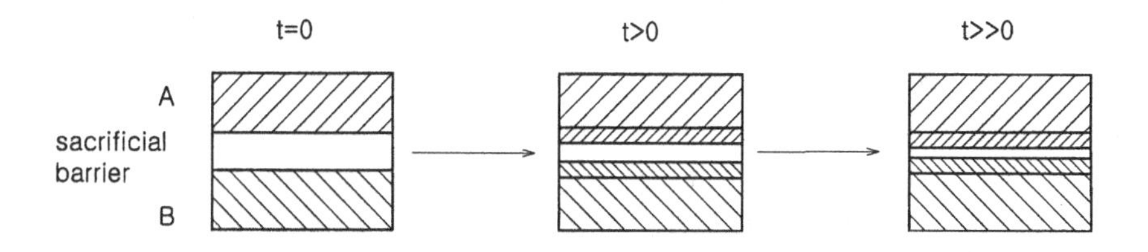

(b) sacrificial barrier: consumed with time

Fig. 10.18. The comparison between a passive and a sacrificial barrier.

reacts with materials on both sides of it. It is consumed over time. However, as long as the barrier is not totally consumed within the period of interest or the lifetime of a device, the barrier function maintains. Ti is an example of a sacrificial barrier that is used for Si and Al. Another example is that the refractory metal silicides in source and drain areas are often used in submicron devices to lower the contact resistance, as will be discussed later. However, as shown in Fig. 10.19, aluminum and silicide form an unstable contact; Al

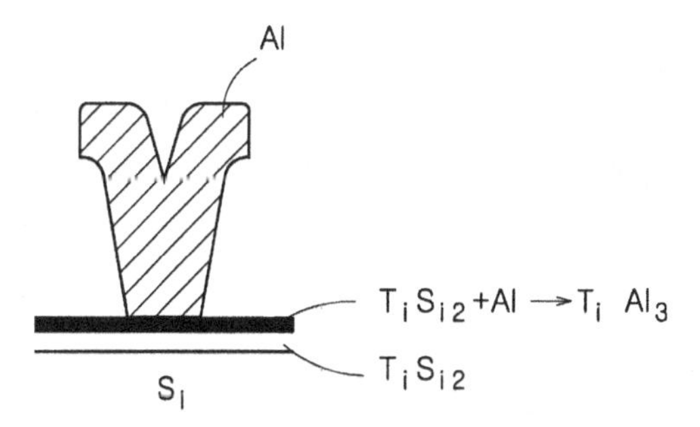

Fig. 10.19. Al contacts silicided surface. $TiSi_2$ is consumed by Al to form $TiAl_3$.

attacks the silicide, forming metal aluminide ($TiAl_3$) starting from the aluminum silicide interface. Once the whole silicide layer is transformed, Al reaches the silicon surface, and the contact integrity will change dramatically. Such a system can be stabilized by using Ti, V, or Cr as the sacrificial barrier layer.

10.3.2. *Antireflective coatings*

As device geometry shrinks into the submicrometer regime, the aluminum surface is too reflective to ensure good-quality photolithography printing. Part of the incident light is reflected off the aluminum surface toward the resist bulk, which is not intended to be exposed. This causes serious notching of the metal lines after developing, which copies onto final metal profiles, as shown in Fig. 10.20. This makes controlling the fine metal line width very difficult. One approach to solving this issue is to deposit a less reflective layer, such as TiN, on top of the aluminum layer. The reflectivity of TiN is much lower than that of aluminum. TiN for this application is often deposited by using a reactive sputtering technique. Nitrogen is introduced into the Ar plasma during the Ti target sputtering. The resulting nitrogen radicals react with the Ti target to form TiN on the target surface, which is then sputtered off.

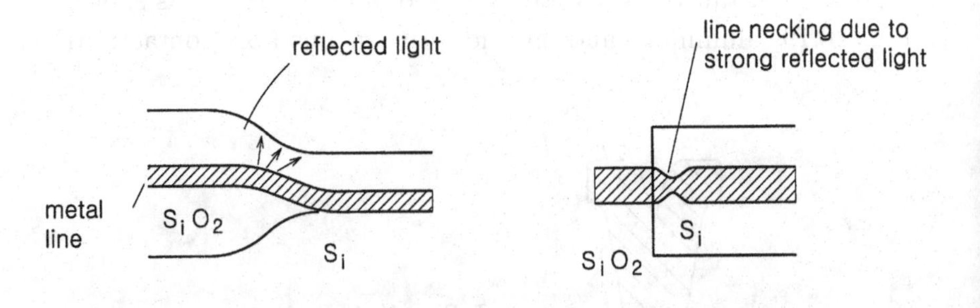

Fig. 10.20. A highly reflective metal surface can lead to strong reflection during exposure, and hence line necking.

Table 10.2. Evolution of aluminum metal structures.

Si/Al
Si/AlSi
Si/AlSiCu
Si/Ti/TiW/AlSiCu
Si/Ti/TiN/AlSiCu
Si/Ti/TiN/AlSiCu/TiN

Advanced aluminum metallization structures often consist of a stack of layers. Table 10.2 shows the evolution of Al metallization and the layer structures. The barrier metals not only help prevent junction spiking, but they also help with the electromigration. The role of Ti is very critical for contact resistances; it is often used in direct contact with the silicon surface. Before the barrier is deposited, if the bare silicon is exposed to ambient, it leads to native oxide formation (about 10 Å or so). If Al, TiN, or TiW directly contact the oxidized silicon surface, the contact resistance will be too high to be acceptable. In such cases, the use of titanium is preferred, and it results in the so-called snow plowing effect with respect to oxygen in the native oxide. It basically dissolves the native oxide into the bulk of titanium, leaving a clean, oxide-free interface for the metal contact. In addition, titanium forms silicide with the silicon surface, rendering a silicided contact. Therefore the contact resistances are significantly decreased. Modern clustered sputtering equipment is designed to deposit these films without the need to break the vacuum. Advanced etching equipment can handle the most complicated stack in a single machine by just switching etching recipes.

10.4. Meeting the Step Coverage Requirements

Poor step coverage is the main drawback of a sputtering process; especially as the device geometry shrinks, the hole aspect ratios increase dramatically, as illustrated in Fig. 10.21. A contact aspect ratio of 5:1 is not uncommon for a 0.35-μm technology. The major concern regarding the poor step coverage of aluminum is not only

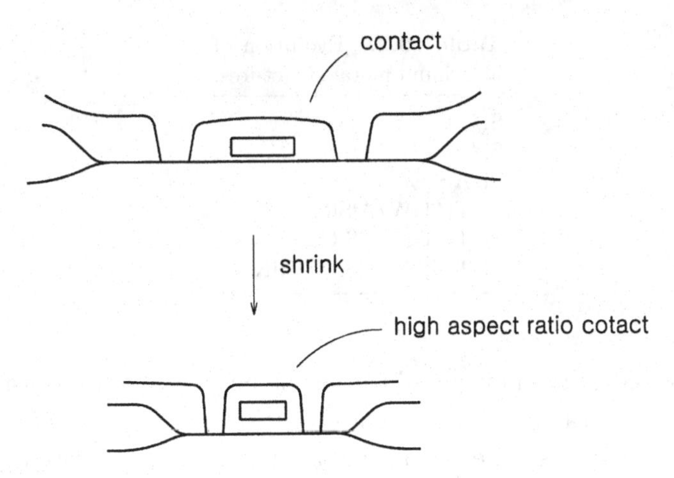

Fig. 10.21. Technology shrinks more aggressively in the lateral direction than in the vertical direction.

related to the electrical connections, but also to the reliability. Several approaches can be employed to cope with the ever-increasing hole aspect ratios. The evolution of the following approaches makes the shrinking feature sizes possible:

(a) *Tailor the contact sidewall profile.* The movement of condensing sputtered atoms onto the substrate surface is essentially in the line of sight. As the aspect ratios increase, it becomes harder and harder for the atoms to reach the bottom. One effective method to improve the probability of the atoms reaching the bottom corners is to tailor the sidewall profile with wet–dry etching or isotropic etching, followed by an anisotropic dry etching process, as shown in Fig. 10.22. This approach will open up the cross-sectional area at the top of the hole, and allow more atoms to reach the bottom corners. Logically, the step coverage increases according to the extent of edge modification. However, it is limited by the design rule (distance) between the contact edge and the neighboring patterns, for example, the distance between the contact edge and the polygate edge. The contact to gate leakage or shortage causes circuit functional failures. Tailoring the contact profiles to achieve better step coverage is widely used for technologies of down to 0.8 μm or for hole sizes of larger than 1.0 μm.

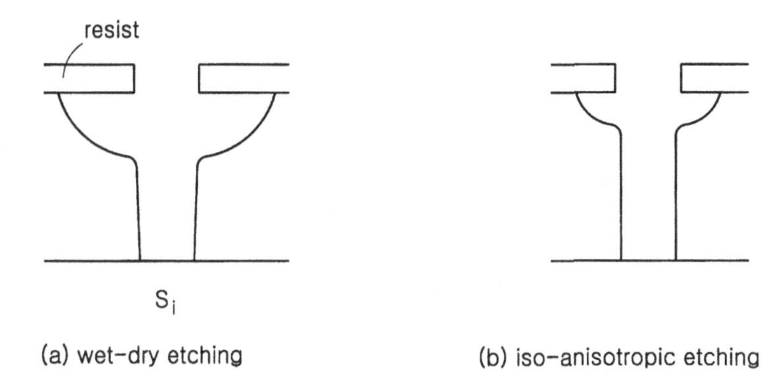

Fig. 10.22. The wet–dry and isoanisotropic contact etching opens up the top of the contacts and greatly improves the subsequent metal step coverage.

(b) *Hot aluminum sputtering.* Hot aluminum sputtering is accomplished by depositing a thin uniform layer of Al seeds at low temperatures; this is then followed by deposition of the bulk of aluminum by sputtering at high temperatures (above 400°C) to increase the adatom surface migration energy. This process achieves significantly improved Al step coverage for a fixed hole geometry; the higher the temperature, the better the step coverage will be. However, the high temperature setting is limited by the possibility of incurring Al layer hillocks or junction spikes at high process temperatures. A strong barrier layer is essential to the success of this approach. There are several other similar approaches such as the laser anneal process, which uses a laser beam to locally melt the metal and force it to flow into the contact hole. One of the drawbacks of this approach is its low throughput.

(c) *Increasing the barrier metal thickness at the bottom corners.* A metallization structure should render low resistance and good electromigration performance. Some argue that as long as the circuit can pass electromigration tests, the poor step coverage is not a concern. The barrier metals, such as TiW and TiN, have much stronger resistance to electromigration than aluminum. This leads to the idea of increasing the barrier metal thicknesses to ensure enough barrier metal at the bottom corners to pass the electromigration tests. A typical metal stack for a 0.8-μm technology is 500 Å Ti/1200 Å TiW/7 kÅ Al. A modified stack structure could be 500 Å/1500 Å/7 kÅ Al. However, as the technology progresses to below an 0.8-μm regime, the ever-increasing hole aspect ratio makes this approach useless.

(d) *Aluminum CVD.* Another alternative approach for overcoming the poor step coverage of Al sputtering is to use MOCVD to deposit blanket Al layers. This process employs the pyrolysis of dimethyl aluminum hydride (DMAH), as shown in Fig. 10.23. DMAH has a vapor pressure of 2 mmHg at room temperature, which enables a viable CVD process at low temperatures. Suitable deposition temperatures range from 400 to 575 K; the DMAH molecules adsorb on a heated substrate and decompose into various fragments, such as Al, H, and CH_3, and so on. The resulting film step coverage is excellent, as demonstrated in Fig. 10.24. An ideal Al film should contain nearly no impurities that are supposed to be out-diffusing away from the deposited film. However, one of the difficulties associated with Al CVD is how to add dopants such as Cu and Si. The 1–4% Cu can be introduced by a postprocess deposition of Al–Cu or Cu, followed by a diffusion step. It can also be introduced with *in situ* doping by using $C_5H_5CuP(C_2H_5)_3$ as the precursor. Al CVD is often carried

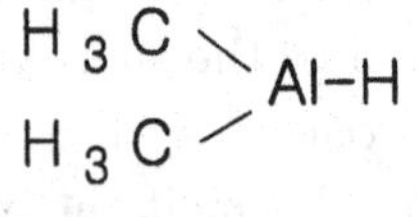

H_3C \
Al–H
H_3C /

Fig. 10.23. The DMAH molecule for Al CVD.

Fig. 10.24. Excellent step coverage is obtained from CVD Al as compared to PVD Al.

out on Ti/TiN surfaces; the surface is pretreated with hydrogen to improve the Al nucleation. Owing to difficulties with introducing stable amount of dopants as well as the inferior film quality from CVD compared to that from sputtering, the Al CVD has rarely been used in volume production.

(e) *Using W-CVD.* The chemistry of W-CVD has been discussed in Section 5.5. Blanket W-CVD is often used to obtain a conformal deposition even over hole patterns with high aspect ratios, as demonstrated in Fig. 10.25. The step coverage of over 85% can be achieved by using this approach; it is typically used beyond 0.8-μm technology. Owing to the relative high resistance of the tungsten film, it is often used only as a plug for filling up the contact or via holes. To achieve this, a thick blanket tungsten film is deposited over a patterned wafer surface that is coated with Ti/TiN or Ti/TiW glue layers. It is very important to have some glue layer even at the bottom corners for the W-deposition. Otherwise, the precursor (tungsten hexafluoride) can attack and penetrate through the silicon substrate and cause contact failures. After the etchback is completed, plugs are formed in the holes. AlSiCu film is then sputter-deposited on the top, followed by metal line patterning. The contact sidewall profile plays a crucial role in the plug process. On one hand, the sidewall angle has to be large enough to have more atoms reaching the bottom corners. On the other hand, large sidewall angles cause large keyholes in the contact holes after etchback. This can trap impurities or chemicals and cause circuit malfunction or reliability failures.

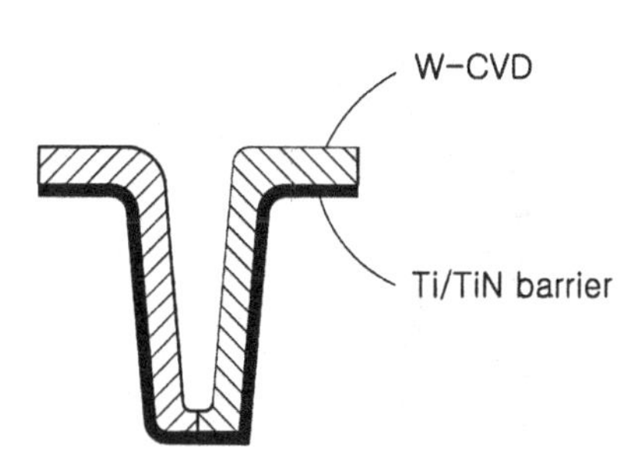

Fig. 10.25. W-CVD provides superior film step coverage as compared with PVD Al.

The etchback step must properly stop on the TiN or oxide surface for the TiW glue layer since excessive overetching results in a recessed plug, as shown in Fig. 10.26. The recess step height makes the subsequent planarization more difficult to accomplish. In the case of stacked via, as shown in Fig. 10.27, the recess can propagate all the way to the top metal layer. On the other hand, there must be proper amounts of overetching to eliminate W-residues over steps or topography, as shown in Fig. 10.28. The best integration approach would be to have a planarized surface before W-deposition to minimize the amount of overetching, or recess (the plug surface being lower than the oxide surface due to overetching), and at the same time, keep the oxide surface residue-free.

(f) *Improving the barrier metal step coverage.* Good step coverage of barrier layers at the bottom corners of contact or via holes are very important for both conventional Al sputtering processes and for advanced Al-CVD or W-CVD processes. For the former, the barrier layers provide good electromigration results even when hardly any Al exists at the bottom corners. For the latter, the successes of Al-CVD or W-CVD hinge on the existence of uniform barrier layers. As contact and via hole aspect ratios increase with shrinking

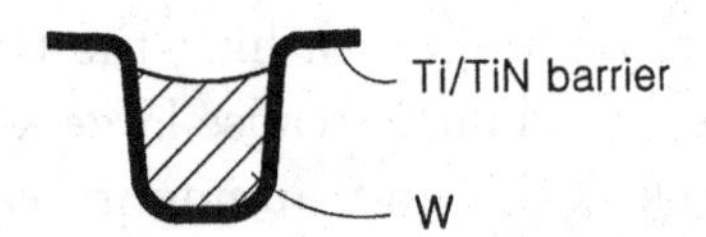

Fig. 10.26. Excessive overetching results in recessed W-plugs.

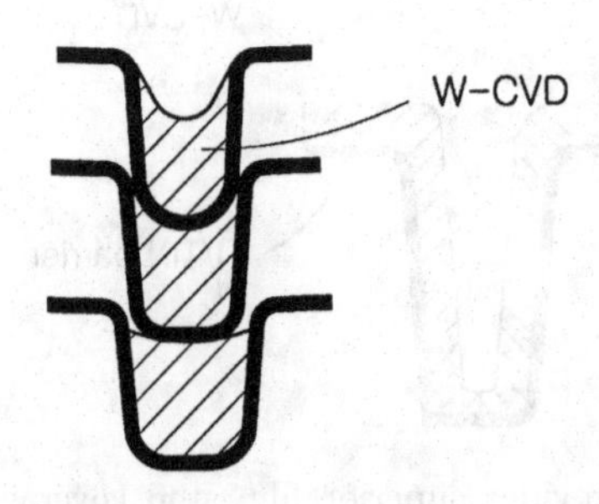

Fig. 10.27. Recessed plugs on a stacked–via structure can create significant step height on the top layer.

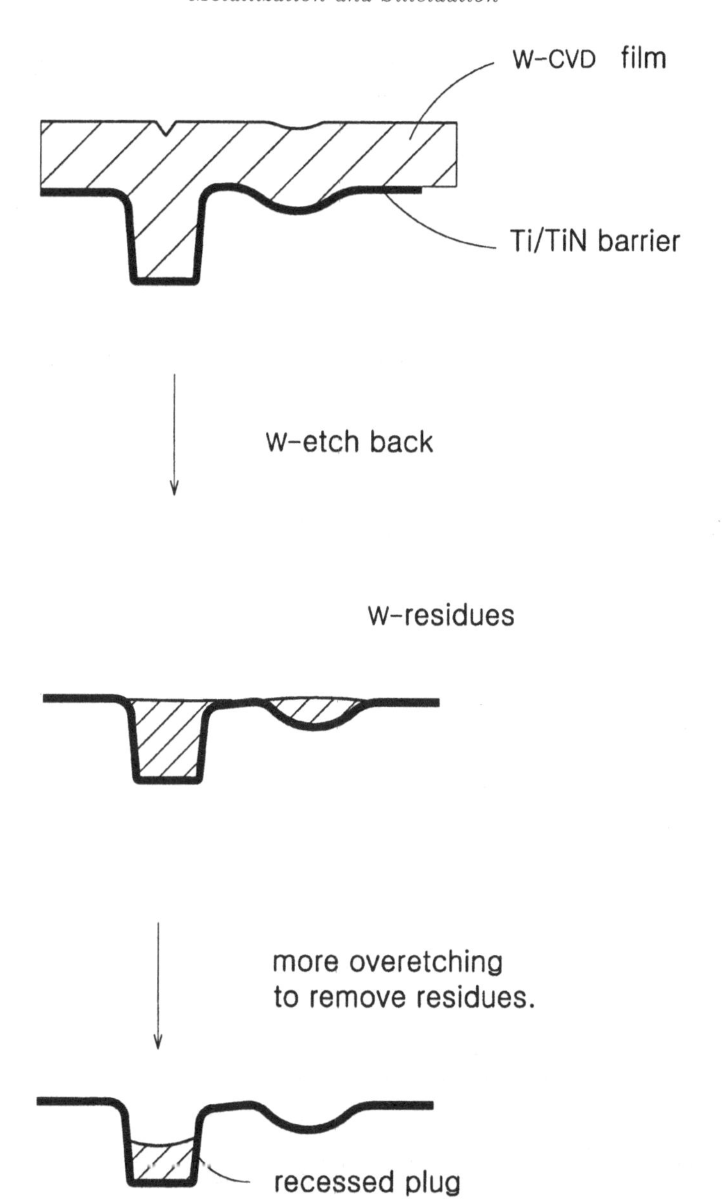

Fig. 10.28. Extra overetching is required to remove W-residues at steps or topograph.

technology, as shown in Fig. 10.29, it is more and more difficult to obtain proper barrier metal thickness at the bottom corners of a hole by using conventional sputtering technology. Some volume production technologies, such as collimated sputtering, IMP Ti, and CVD TiN, have been developed and used in volume production to meet the requirements.

Collimated sputtering occurs when the traveling directions of randomly scattered sputtered atoms are physically filtered. Only those scattered atoms with smaller angles can pass through a collimator and transport in the line of sight toward the contact or via bottom and corners. Figure 10.30 shows a schematic of the

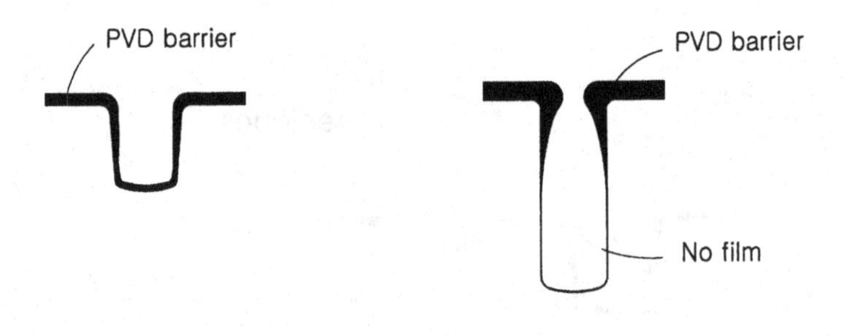

Fig. 10.29. PVD barrier metal sputtering provides high aspect ratio contacts with poor film step coverage.

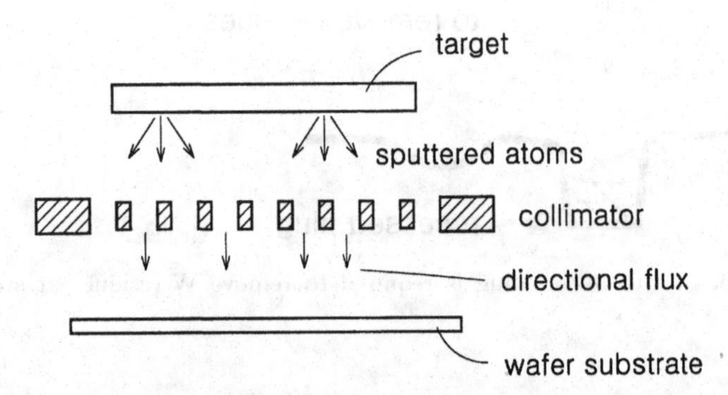

Fig. 10.30. Collimated sputtering uses a perforated plate to spatially filter the depositing atom flux to improve step coverage.

collimated sputtering system. As we know, molecular or atomic scattering increases with increased pressures. More scattering tends to compensate for the extent of collimation. As a result, the lower the pressure is, the better the collimation will be; yet the lower the pressure is, the lower the deposition rate will be. Figure 10.31 compares the deposition rates in a hole with collimators of different aspect ratios. It is shown that although the layer thickness at hole bottom can be increased with a high aspect ratio collimator, the overall deposition rates are reduced.

Ionized metal plasma (IMP) sputtering is another alternative approach used to increase the step coverage of barrier metals. The idea is simple; in addition to the conventional sputtering chamber, an RF coil is added to ionize the metal atoms. The positively charged metal ions are then accelerated toward the substrate, which is negatively biased, as illustrated in Fig. 10.32. This technique has been used for both Ti and TiN barriers.

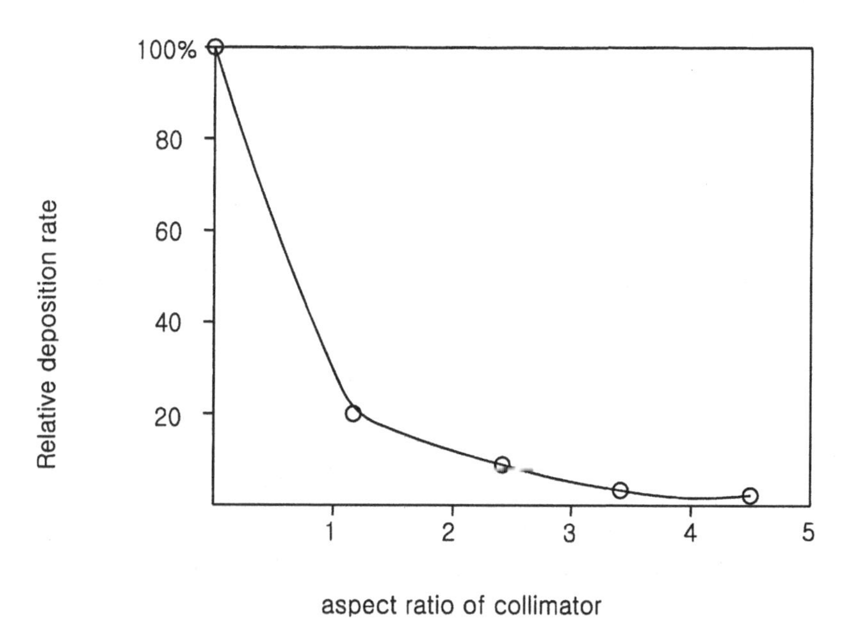

Fig. 10.31. Deposition rate reduces dramatically with increasing aspect ratio of the collimator.

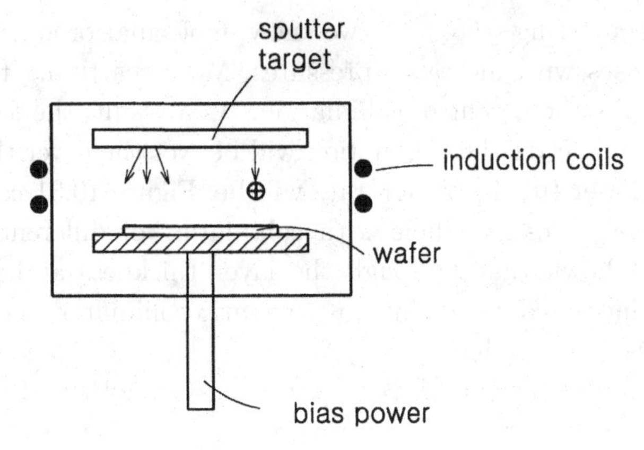

Fig. 10.32. IMP uses induction coils surrounding the chamber to ionize sputtered target atoms.

CVD TiN is relatively mature, compared to CVD Al. It can be deposited via CVD or MOCVD:

$$\text{CVD}: \ \text{TiCl}_4 + \text{NH}_3 \longrightarrow \text{TiN} + \text{others}. \tag{10.8}$$

The reaction takes place at temperatures above 350°C; however, to keep the chlorine contamination in the film low, the reaction temperatures must be kept above 600°C. As a result, this process is not suitable for the second-layer metal and above. The resulting film is slightly Ti-rich, with a resistivity of about 70 $\mu\Omega$ cm.

MOCVD takes place at lower temperatures around 350°C–450°C:

$$\text{TDMAT} + \text{NH}_3 \longrightarrow \text{TiN} + \text{others} \tag{10.9}$$

$$\text{TDEAT} + \text{NH}_3 \longrightarrow \text{TiN} + \text{others}. \tag{10.10}$$

The molecular structures of tetrakis di methyl amido titanium (TDMAT) and tetrakis di ethyl amido titanium (TDEAT) are shown in Fig. 10.33. These organometallic precursors produce films without chlorine, but with carbon contamination. The addition of ammonia helps to decrease carbon contamination. The resulting film is N-rich, with a stoichiometry of around 1.2 and resistivity of about 500 $\mu\Omega$ cm.

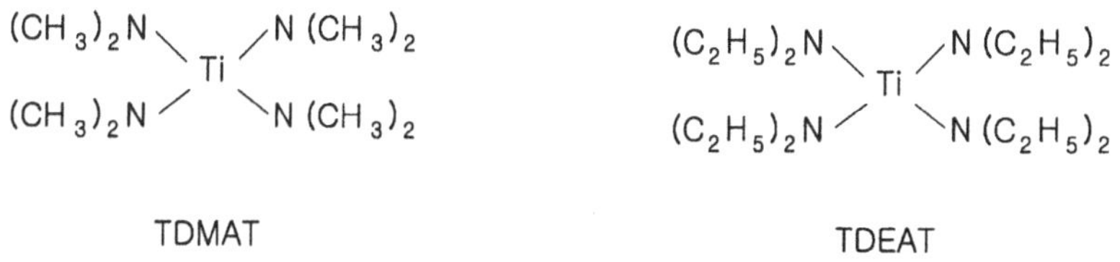

$$\begin{array}{c} (CH_3)_2N \quad\quad N(CH_3)_2 \\ \diagdown \quad \diagup \\ Ti \\ \diagup \quad \diagdown \\ (CH_3)_2N \quad\quad N(CH_3)_2 \end{array} \qquad \begin{array}{c} (C_2H_5)_2N \quad\quad N(C_2H_5)_2 \\ \diagdown \quad \diagup \\ Ti \\ \diagup \quad \diagdown \\ (C_2H_5)_2N \quad\quad N(C_2H_5)_2 \end{array}$$

TDMAT TDEAT

Fig. 10.33. TDMAT and TDEAT molecules for CVD TiN.

10.5. Why Silicides?

As technologies move forward, the gate and interconnects shrink both in vertical and horizontal dimensions. With this shrinking trend, if the gate or interconnect materials remain the same, the resistances will increase dramatically, and the device speed will degrade unacceptably. Refractory metal silicides are often used as alternative materials for the gate level interconnects due to their low resistances, as compared in Table 10.3. The reason that silicides are used instead of aluminum is partly due to their better thermal stability. Nonetheless, the silicide integrity does degrade as the subsequent processing temperatures exceed 700°C or so.

There are two types of silicide applications in manufacturing; one is the polycide gate formation, and the other is the self-aligned silicide formation. Tungsten polycide, as depicted in Fig. 10.34, is often used for gate structure without silicide in the source and drain surface. The self-aligned silicide process, as illustrated in Fig. 10.35, forms the silicide not only on the gate, but also on the source and drain areas. The self-align nature stems from the fact that the metal

Table 10.3. The resistivity of various metal disilicides, Si to metal consumption ratios, and each nm silicide formed per nm of metal consumed.

Silicide	Resistivity ($\mu\Omega$ cm)	nm of Si consumed/ nm of metal	Silicide formed/ metal
Ti-disilicide	13–17	2.51	2.51
Mo-disilicide	22–100	2.59	
Ta-disilicide	8–45	2.21	
W-disilicide	14–17	2.53	
Co-disilicide	14–20	3.64	3.52
Ni-disilicide	40–50	1.83	2.34

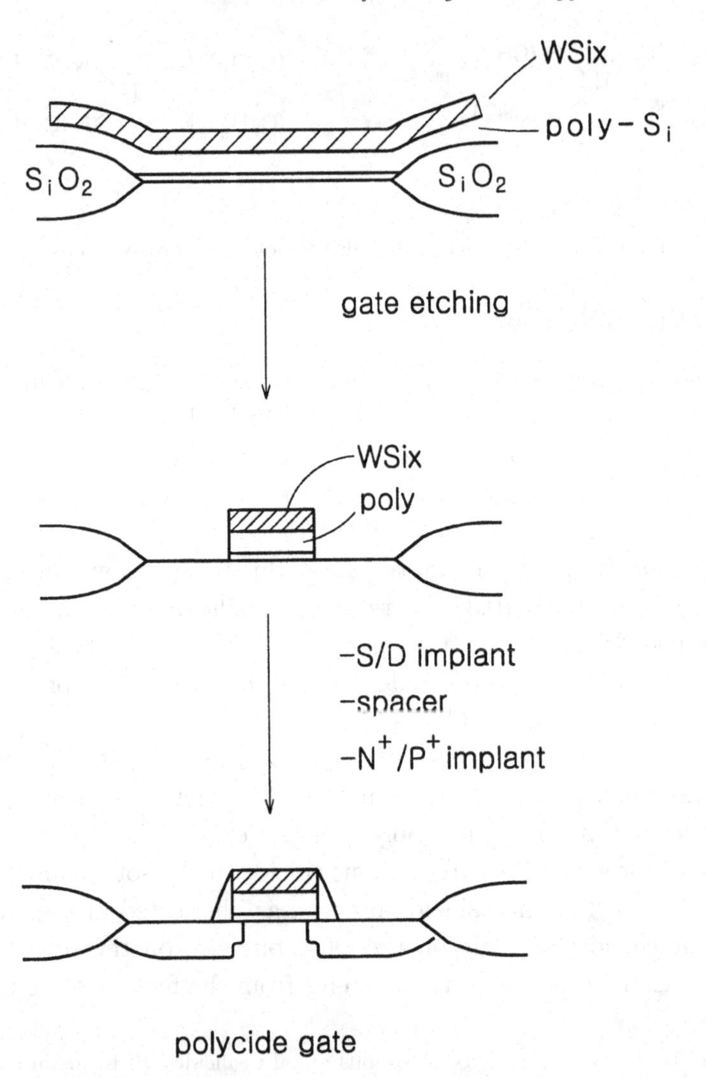

Fig. 10.34. A simplified process flow for polycide gate formation.

reacts with silicon, but not with oxide. Therefore the silicide can be formed selectively on gate, source, and drain areas. Although specific products may have different requirements, in general, tungsten polycide is used for technology nodes starting from 0.8 μm and below. Beyond the 0.5-μm node, Ti-salicide is used; beyond 0.18 μm, Co-salicide is becoming popular, and beyond 0.1 μm, Ni-salicide is used.

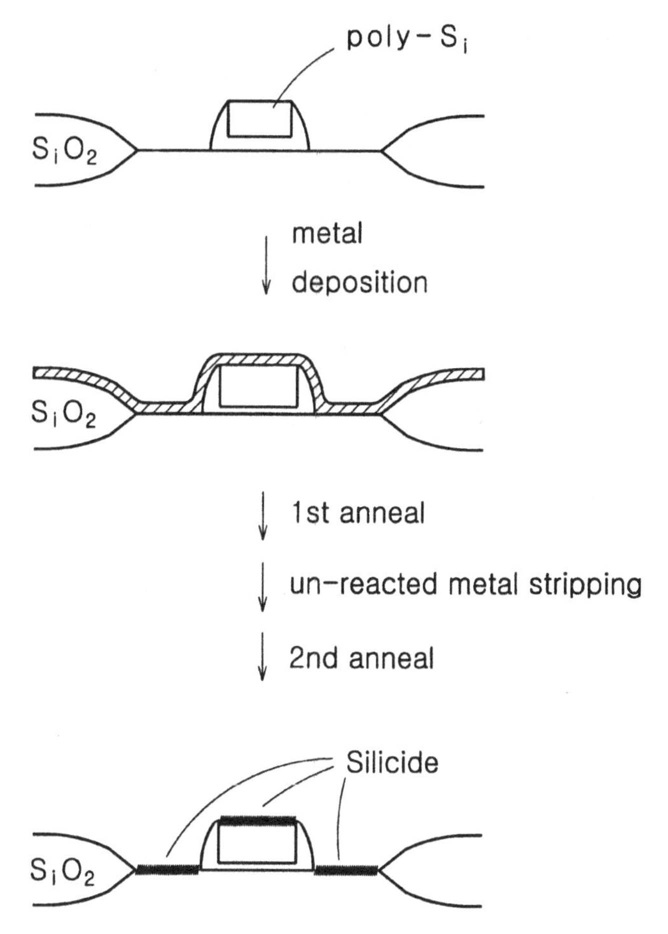

Fig. 10.35. The formation of self-aligned silicide, which forms only on silicon surface.

For applications in manufacturing lines, a good silicide should possess the following characteristics:

(a) *Low resistances.* With the cost and effort that is required to successfully add a layer of silicide in a device structure, the use of silicide should provide a resistance of about one order of magnitude lower than doped polysilicon or silicon.

(b) *Compatible with subsequent processes in terms of thermal stability.* Silicides have better thermal stability than aluminum, but still, the postsilicide processing temperatures should not exceed 700°C without proper protections; otherwise, the silicides may degrade or peel off and cause contamination. On the other hand, silicide integrity may degrade, defeating the original purpose of using silicides. Integration work should take the potential degradation and other integration issues into account in cases such as ILD flow and annealing.

(c) *Good etching characteristics.* These characteristics refer to two issues. One issue is the etching of the silicide on polysilicon gate structure such as tungsten polycide gates. In this case, the etching should be able to achieve a vertical profile for the dual-layer structure, as discussed earlier. On the other hand, for the source and drain areas, the silicide should be able to withstand the oxide etching without any silicide loss; otherwise, resistances will increase.

(d) *Low junction leakages.* Excessive salicide thickness or improper annealing may result in junction leakage. Furthermore, silicide metal atoms form deep trapping centers in the silicon bandgap, which will likely cause leakages. When dealing with silicide, the postsilicide processes must be carefully isolated to prevent the metal atoms or ions from penetrating into the silicon.

(e) *Worthwhile in terms of increase in cost versus performance enhancement.*

Ti-salicide is the most used salicide at present. To form the salicide, a layer of titanium is deposited on the gate structure by using metal sputtering techniques after the spacer is etched. The thickness of the deposited titanium is determined by the desired final sheet resistance of the silicide. The first RTA is conducted at temperatures ranging between 630°C and 700°C. The higher the temperature, the lower the sheet resistance will be, as indicated in Fig. 10.36. However, at higher temperatures, the process is more prone to bridging between the gate, source, and drain, as indicated in Fig. 10.37. The reason for this is twofold. First, at higher temperatures, titanium can react with spacer oxide. Second, when titanium reacts with silicon to form silicide, the silicon is the diffuser. At higher temperatures, silicon diffuses extensively along the spacer and reacts with

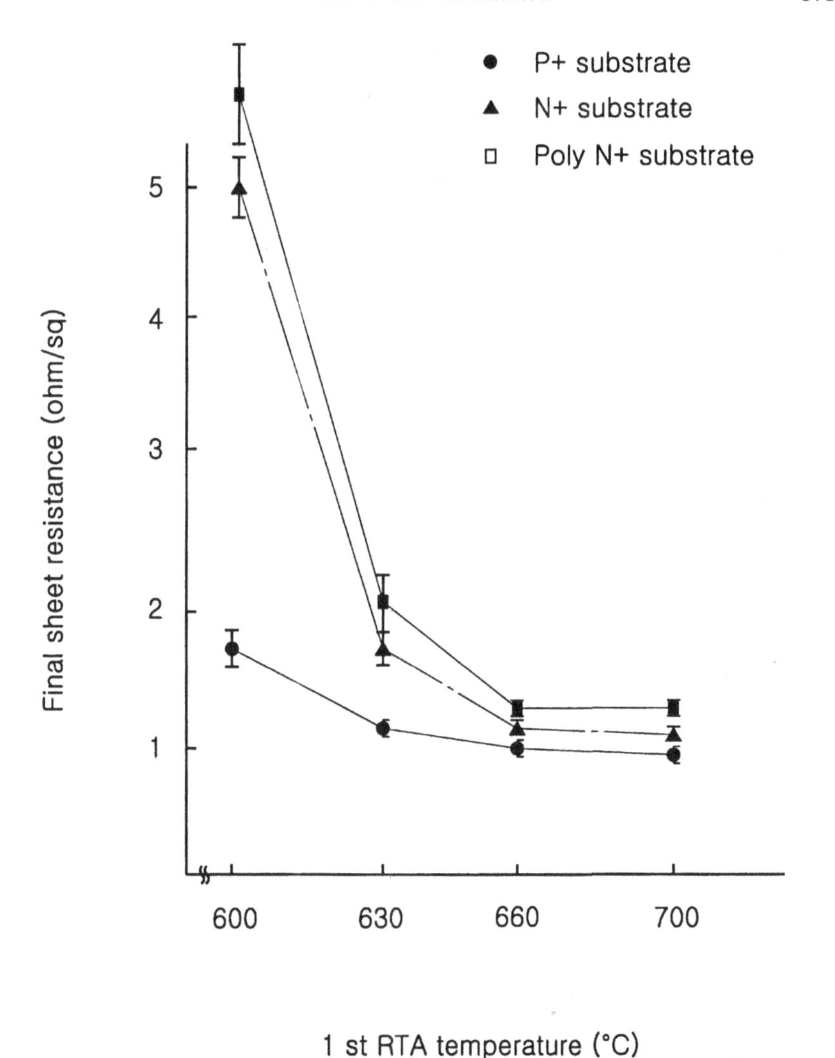

Fig. 10.36. The silicide sheet resistances decrease with anneal temperature (Ti: 1000 Å, 2nd anneal at 850°C, 20 s).

titanium to form silicide on the spacer sidewall. Any residual amount of silicide formed on the spacer sidewall will cause either electrical shortage or reliability concerns. The bridge-free process window, as defined by a rectangle in the temperature–time domain, of the first anneal can be defined by conducting experimental studies on anneal

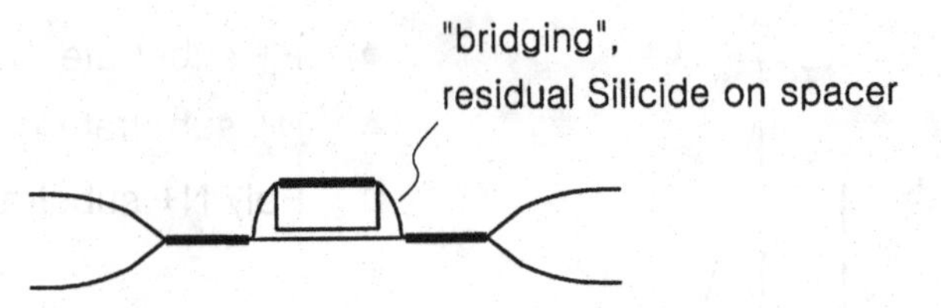

Fig. 10.37. Bridging effect, the residual silicide on spacer, causes gate to be shorted to source/drain areas.

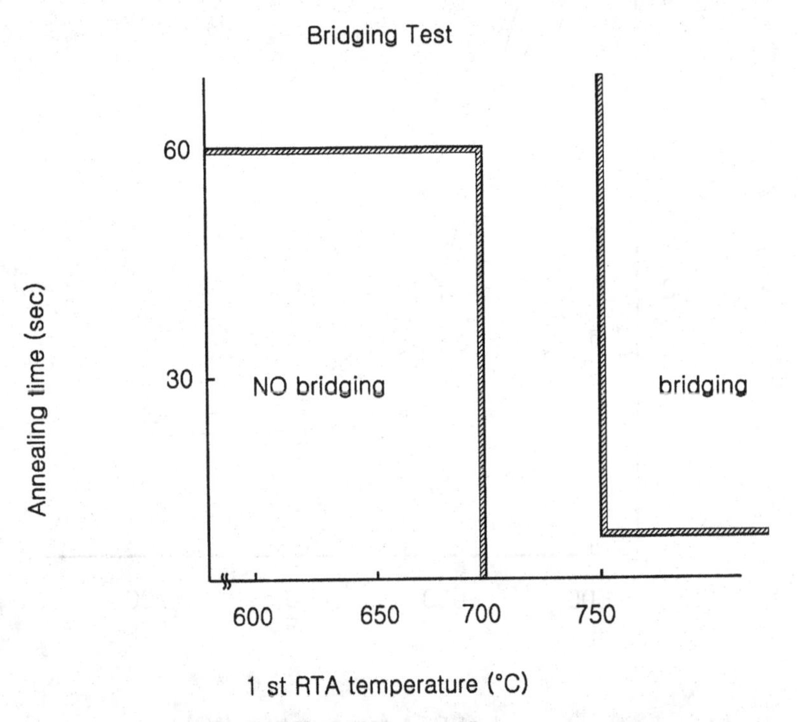

Fig. 10.38. The process window defined by experiments for the bridge-free process.

times and temperatures, as illustrated in Fig. 10.38. Bridging can be minimized by using nitrogen anneal ambient, which results in TiN on top of silicide and TiN on top of oxide. TiN slows down the silicon's out-diffusion, thereby minimizing the bridging. With tungsten silicide deposition, any trace amounts of native oxide on the silicon surface cause the depositing tungsten silicide to peel off. The titanium silicide formation process is more forgiving on the surface native oxide because titanium can snow plow the trace oxide; that is, it pushes the oxygen toward the bulk of silicide and titanium and leaves a clean interface for further silicidation. However, it is certain that excessive oxide retards the titanium silicide formation.

After the first anneal, any unreacted TiN can be selectively etched off using a ammonium hydroxide and hydrogen peroxide solution. This will result in titanium silicide of C49 phase on the gate, source, and drain areas. The C49 phase has a high resistivity (about 60–70 $\mu\Omega$ cm) and can be converted to a C54 phase with a second anneal. The C54 phase has a much lower resistivity (about 13–16 $\mu\Omega$ cm). Figure 10.39 shows the sheet resistances of the $TiSi_x$ as a function of the as-deposited titanium sheet resistance on different substrates. It can be observed that the sheet resistances vary slightly with substrate types as well. The salicide integrity can be degraded by excessive postsilicide thermal cycles such as the ILD or contact anneals. Figure 10.40 shows that the silicide resistance increases with increasing anneal temperatures and times. Basically, at temperatures higher than 900°C, the silicide integrity degrades, and it decomposes into silicon and silicide islands, and thereby the sheet resistance will increase.

The major drawback of using titanium silicide is associated with the technology scaling issue. As the underlying line width shrinks beyond 0.8 μm, the titanium silicide limitations will become apparent; sheet resistances will increase dramatically with the shrinking geometry, as indicated in Fig. 10.41. Furthermore, the formed silicide thicknesses are thinner at the edge compared to the center of the line. This is because silicon is the diffuser for the titanium silicide formation reaction, as shown in Fig. 10.42. At the edges, the silicon supply is smaller when compared to the center; therefore the formed silicide is thinner at the edges. As the width gets smaller, the edge effect becomes more salient. As a result, the overall sheet resistance increases sharply as the line width shrinks. Observing this limitation, the industry started looking for alternatives to silicides with metal diffusers. With metals being the diffusers, as illustrated

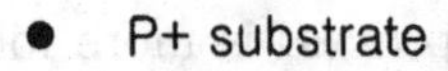

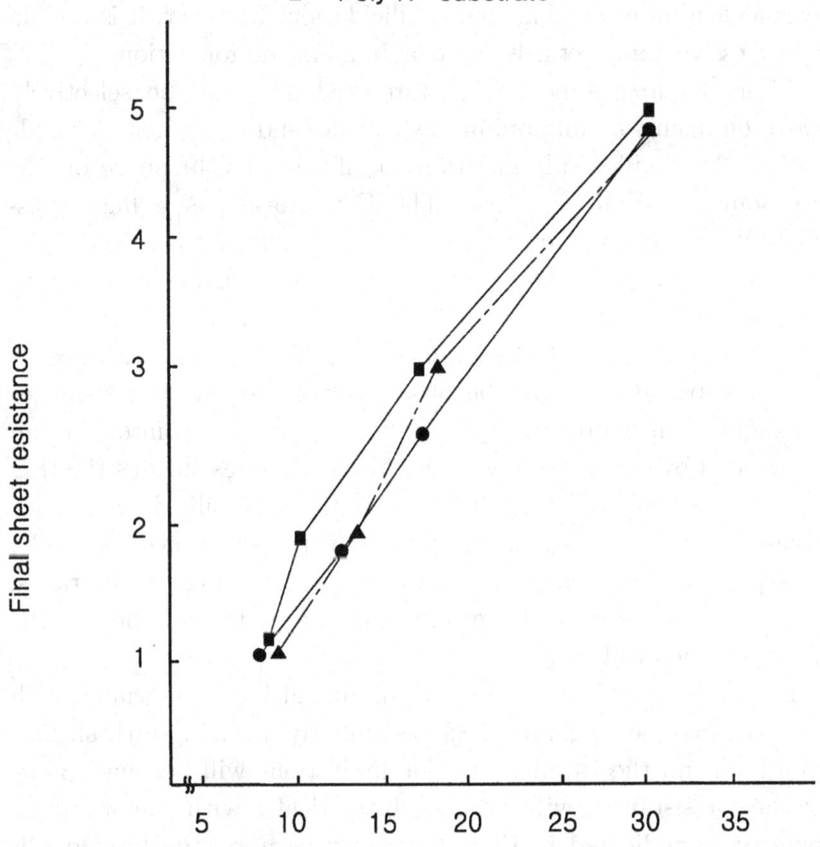

Fig. 10.39. The final sheet resistances of silicide increase with initial Ti sheet resistances (680°C/850 for 1st and 2nd anneals).

in Fig. 10.43, the silicidation reaction sees more abundant supply of metal on the edges. Therefore the edge effects disappear.

Co-salicide is the primary substitute for titanium silicide. This is mainly because Co-salicide provides no line width effects, and it requires a lower temperature for the salicidation reactions. It offers

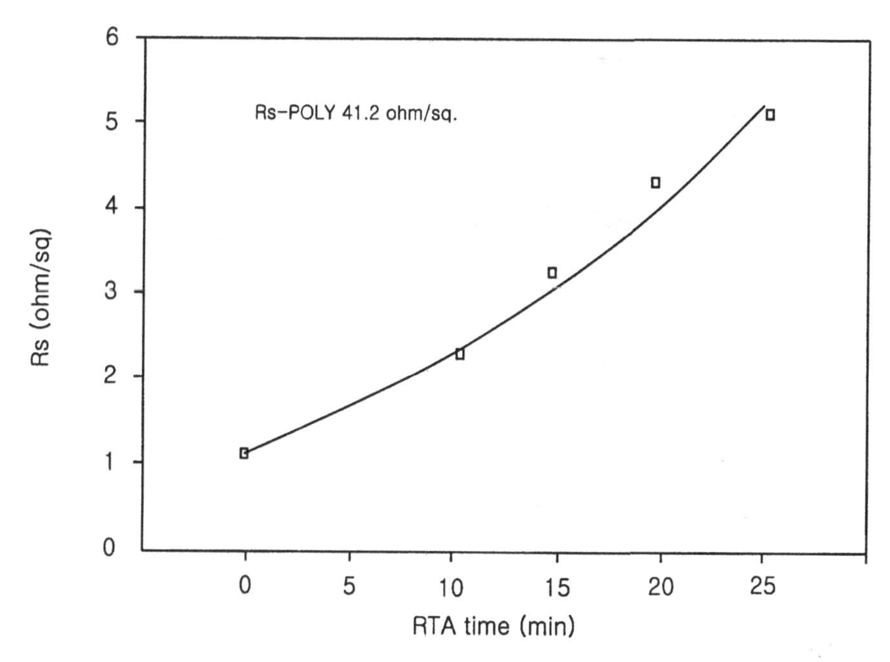

Fig. 10.40. Degradation of silicide due to excessive postsilicide thermal cycle at 850°C.

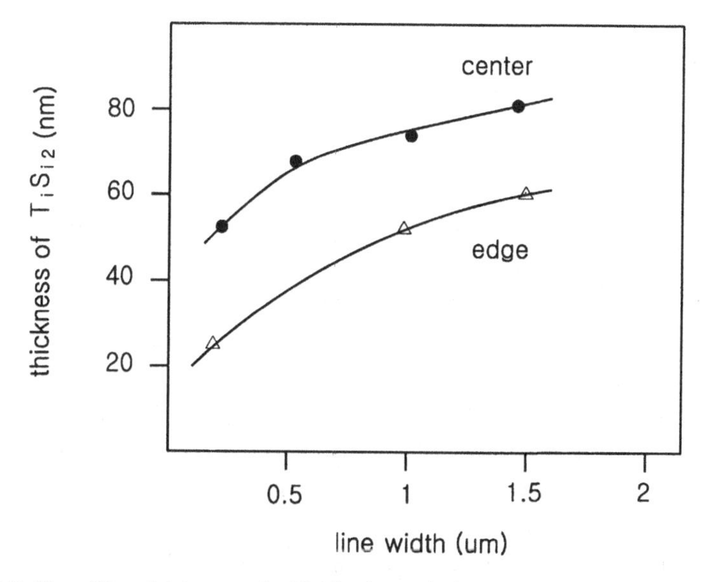

Fig. 10.41. The thickness of silicide formed decreases with decreasing line width.

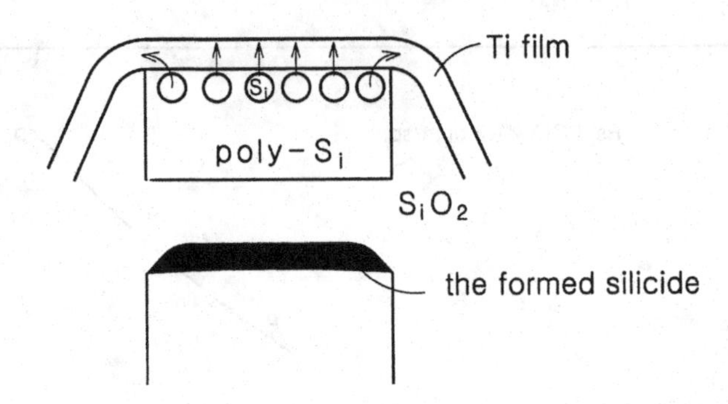

Fig. 10.42. For Ti-salicide formation, silicon is the diffuser. Silicon supply is less at the edges than at the center.

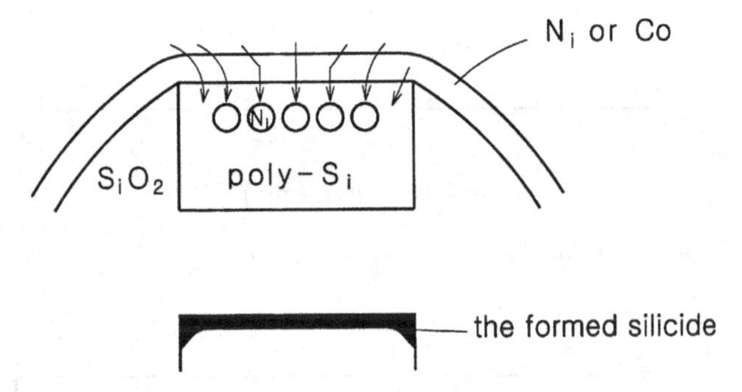

Fig. 10.43. For Ni- or Co-salicide formation, metal is the diffuser. The formed silicide is thicker at the edges.

better etching selectivity with respect to oxide etching. It can better seal off the dopant's out-diffusion since the dopant diffusivities are lower in $CoSi_x$ than in $TiSi_x$ or WSi_x. However, Co-salicide has a worse thermal stability than Ti-salicide. To form the Co-salicide, a layer of Co is sputtered on the gate and source–drain structure after spacer etching, similar to the Ti-salicide process. The first anneal is conducted at 550°C, which is a lower temperature than that for Ti-salicide. The unreacted Co is stripped off with a solution of phosphoric, acetic, and nitric acids. This is followed by a second anneal, which is performed at about 690°C.

With technology advances, the junction depths continue to be made shallower, and the use of Ti- or Co-salicide will become unacceptable since they consume significant amounts of silicon from the underlying silicon surface. Table 10.3 shows the silicon consumption ratios for various silicides. Beyond 0.1-μm technology with 150- to 100-nm junction depths, the formed silicide thickness without junction leakage for titanium and cobalt can be too thin to provide the needed sheet resistances. In comparison, Ni-salicide is a good alternative material. Furthermore, Ni-salicide can be formed at a much lower temperature (400°C) and with only a one-step anneal. This feature greatly reduces the thermal budget for the device formation. Because of this characteristic, Ni-salicide seems to have gained increased popularity in technologies beyond 0.1 μm.

Chapter 11

PLANARIZATION AND CMP TECHNOLOGY

In this chapter, we will introduce planarization technologies that essentially planarize wafer surface topography so as to facilitate the ensuing photolithography and etching processes.

Section 11.1 begins with an overall introduction to planarization, and it differentiates ILD from IMD. ILD is further discussed in Section 11.2, with a primary focus on film composition, properties, and postdeposition treatments. SOG is widely used for planarization in both ILD and IMD. It is the first material used for multimetal planarization processes. SOG material and process characterization are discussed thoroughly in Section 11.3. This is then followed by the introduction of IMD process and technology evolution in Section 11.4. CMP is considered the ultimate planarization technique to date. The CMP process, setup, and applications are discussed in Section 11.5. Finally, CMP modeling is illustrated in Section 11.6.

11.1. Introduction

In semiconductor manufacturing, planarization technology is a technique used to planarize severe topography that results from the underlying layer structures. Why are we having severe topography problems as we continue to strive for ever-shrinking feature sizes? The reason is that oftentimes, the shrinkage paths in the horizontal dimensions are more aggressive than those in the vertical shrinkage paths. This is because the horizontal shrinkage renders designers with more gross dice per wafer and therefore lower cost. There is a tendency to push as much horizontal shrinkage as possible as long as the existing processing technology allows it. On the other hand, the reason for the slower shrinkage path in the vertical dimension is twofold. First, shrinking the vertical dimensions does not yield more dice per wafer. Second, the vertical dimensions directly relate to the layer thickness, either dielectric or conducting. Reducing the thickness will result in increased resistances of conducting layers (R) and capacitance (C) of the dielectric layers; that is to say, the vertical shrinkage can degrade the circuit performance. This explains why pushing for ever-decreasing device geometries will result in ever-worsening topographies.

If severe topography is not planarized, it will cause several problems in subsequent processes. First, the conducting layers' resistance will increase. Imagine two metal lines that run between points A and B, as shown in Fig. 11.1. The one that has severe topography needs a longer metal line than the other; therefore a high metal resistance will result. Second, one is more prone to see metal stringers across topography after etching, as illustrated in Fig. 11.2. Third, the photoprinting on a highly nonplanarized surface will be out of focus. Optics theory dictates that each photolithographic process has

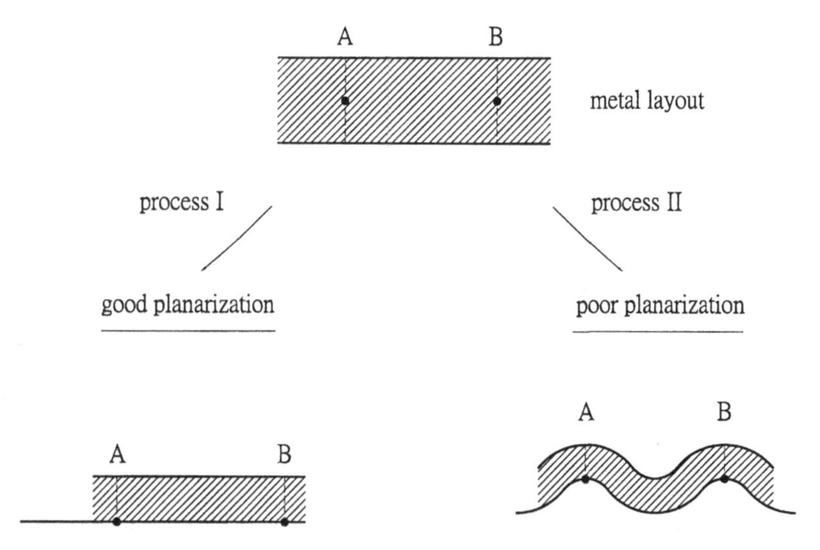

Fig. 11.1. Points A to B on a layout requires a longer metal line to connect on a poorly planarized surface.

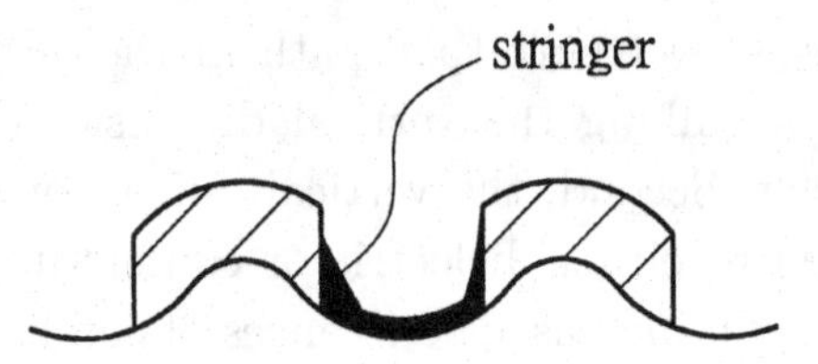

Fig. 11.2. Metal residues (stringers) are likely to form between two metal lines over severe topography.

a finite depth of focus (DOF). If the topography is so bad that the vertical step height difference between two contacts, lying at the top and bottom of the topography hills, is larger than the DOF latitude, as shown in Fig. 11.3, the contacts printed on wafers will not have the same dimensions. This will cause blind or mis-sized contacts.

Planarization technology must optimize the planarity of the topographies in terms of the extent of planarization, wafer yield, manufacturing cost, and circuit performance. Planarization is required in both ILD and IMD, as demonstrated in Fig. 11.4. ILD represents the dielectric layers between the polysilicon gate layer and the first metal layer; IMD represents the dielectric layer between two metal layers. There are two terms that are often used in the industry: local planarization and global planarization. The former refers to achieving planarity in local patterns, the latter, across patterns

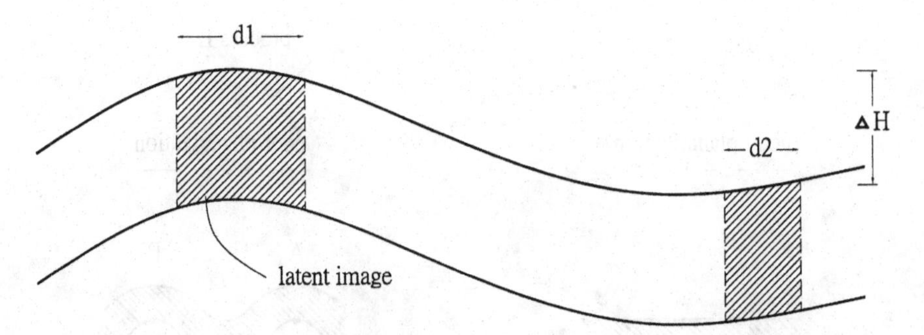

Fig. 11.3. The resist surface step height ΔH must be smaller than the depth of focus to assure good-quality printing.

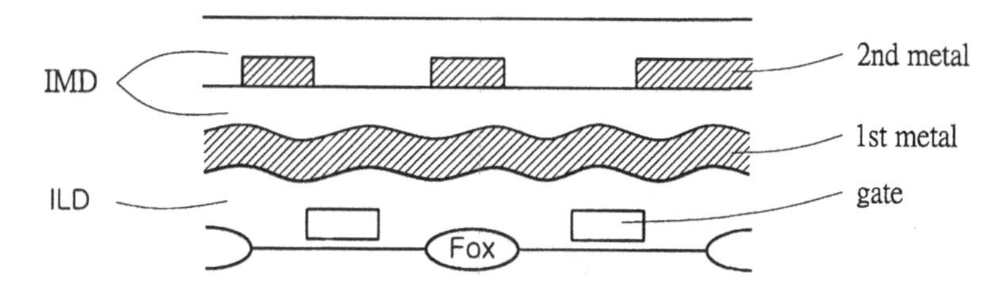

Fig. 11.4. Naming convention in semiconductor manufacturing. ILD: between gate and the first metal; IMD: between metals.

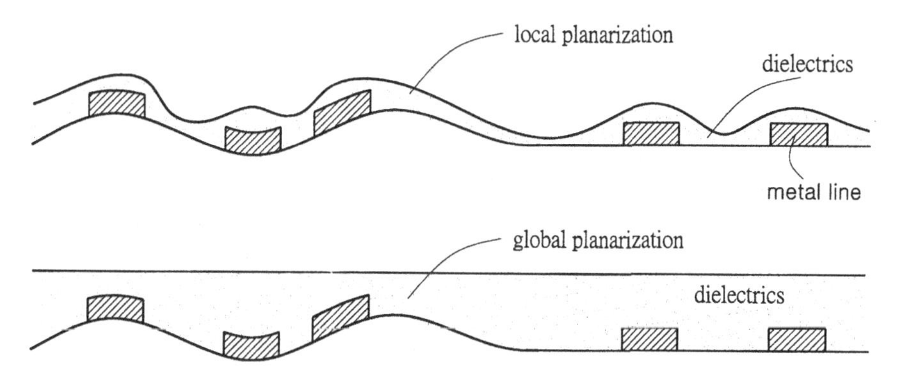

Fig. 11.5. Comparison between local and global planarization.

with different pitches over wide areas. The terms are differentiated in Fig. 11.5.

There are several approaches for measuring the planarity or the extent of planarization. Some approaches can be evaluated from measures of SEM cross section, while others are related to electrical measurements. As shown in Fig. 11.6, with SEM, the extent of planarization can be evaluated from the following equation:

Extent of planarization

$$\phi = 1 - \frac{\delta_{\text{after}}}{\delta_{\text{before}}}, \tag{11.1}$$

where δ represents the height difference. With the electrical tests, the first option is to measure the metal resistances across the

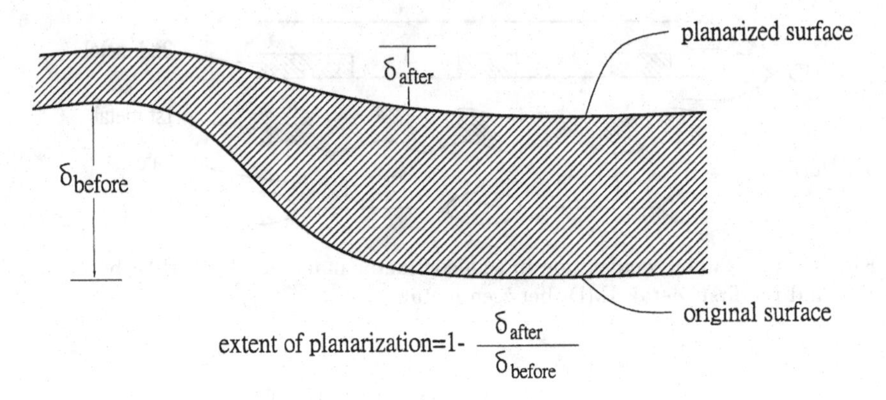

Fig. 11.6. Measuring of the extent of planarization.

topographies. The resistance of the metal line that runs between points A and B is proportional to the length, which will decrease with improved planarization, as illustrated in Fig. 11.1. The second option is to measure the resistance of the metal spacing. With a fixed etching process, the worse the topography, the higher the tendency to leave metal residues between metal lines will be. The metal resistance between two parallel metal runners will decrease with increasing metal residues.

11.2. Interlayer Dielectrics

ILD is the dielectrics stack between the polysilicon gate and the first-level metal layer. At this stage, the underlying device is formed, and therefore the thermal cycles that are required for ILD stack formation should be so low that it will not cause any further dopant diffusion, unless this was taken into account in the device engineering. The film stack should provide optimal planarization and cost-effectiveness at the designated technology node. The difficulty with planarization technology lies in the fact that few dielectric films render conformal step coverage. As a result, when filling a gap, a void tends to form as the film grows, as demonstrated in Fig. 11.7. The voids may trap moisture or chemicals, which can be squeezed out at subsequent

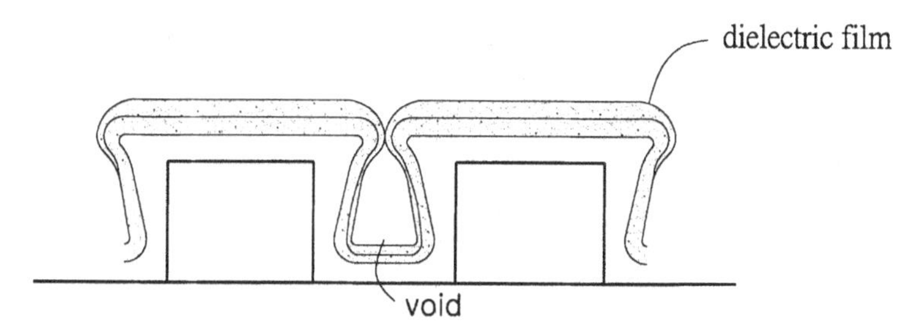

Fig. 11.7. A void formed between two conducting lines due to poor step coverage of the dielectric film deposition.

thermal cycles, such as annealing or alloy, and it may cause contamination or reliability issues. A decent planarization scheme should leave no voids.

Over the years, ILD has evolved from PSG for technology of several microns to TEOS CMP for 0.13-μm technology and beyond. PSG is a phosphorous-doped silicon dioxide film that is deposited from the following reactions at temperatures below 500°C:

$$\begin{aligned} SiH_4 + O_2 &\longrightarrow SiO_2 + 2H_2 \\ 4PH_3 + 5O_2 &\longrightarrow 2P_2O_5 + 6H_2 \,. \end{aligned} \tag{11.2}$$

Although the step coverage of thermal PSG film is not good, the as-deposited films can be reflowed. Reflow is a process step in which the as-deposited film is heated to above its glass transition temperature, causing it to flow and to planarize the underlying topography. The PSG films contain P wt% of around 6–8; the reflow temperatures can be as high as 950°C to 1100°C, depending on the P wt%, as shown in Fig. 11.8. The flow angles decrease with increasing P wt%, reflow times, and reflow temperatures. The upper P wt% is limited by contact autodoping, as shown in Fig. 11.9; basically, the phosphorous atoms in the PSG on the contact sidewall out-diffuse downward to dope the silicon substrate, and thereby they affect the device characteristics. Another negative side effect of having too much P wt% in the film is that P may react with ambient moisture to form acids, which may corrode metal lines. To protect the underlying device from being degraded, usually, an undoped silicon oxide film is deposited underneath the PSG. The film stack (USG/PSG for ILD) is often

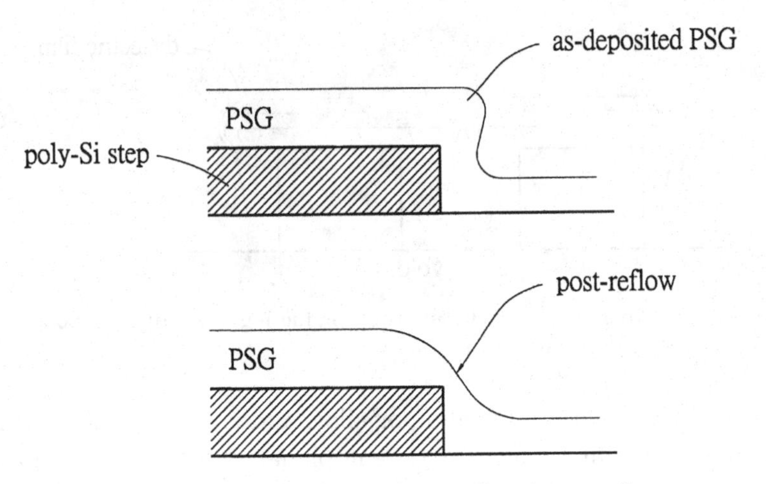

Fig. 11.8. PSG of the as-deposited and post-reflow.

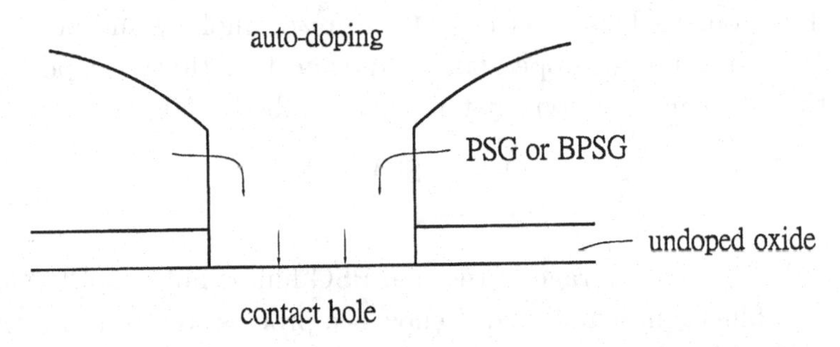

Fig. 11.9. ILD with excessively doped PSG or BPSG can induce auto-doping.

used for technologies with feature sizes of larger than 2 μm; in this case, the polysilicon gate pitch is larger than 4 μm, with a gate stack height of over 5 kÅ. This structure can be reasonably planarized by using the PSG reflow process, despite the poor step coverage of as-deposited PSG.

As the design rule moves to 1.5 μm and smaller, the required reflow temperatures of PSG become intolerably high for the underlying device. Boro phospho silicate glass (BPSG) has turned out to be a good alternative by adding B into the PSG films:

$$\begin{aligned} SiH_4 + O_2 &\longrightarrow SiO_2 + 2H_2\,, \\ 4PH_3 + 5O_2 &\longrightarrow 2P_2O_5 + 6H_2\,, \\ 2B_2H_6 + 3O_2 &\longrightarrow 2B_2O_3 + 6H_2\,. \end{aligned} \tag{11.3}$$

The reaction takes place between 400°C and 450°C. The addition of B essentially lowers the flow temperatures further, to somewhere around 850°C–900°C; this means less thermal budget compared to PSG. Typically, the B percentage by weight is around 2–4 wt%, and the P percentage by weight is around 3–6%. The B wt% is limited by the fact that high B content causes the film to be hygroscopic, which results in crystal formation on the film surface. Again, to prevent autodoping effects, BPSG films are often deposited on top of an undoped silicon dioxide film. A typical ILD thickness combination for 0.8-μm technology is about 8 kÅ of BPSG on top of 3 kÅ of undoped silicon dioxide. Too much dopant in the BPSG film can also cause the contact sidewall to bulge after contact anneal, as illustrated in Fig. 11.10. The bulging leads to metal discontinuity in Al sputtering, causing contact failure. It has been found that a delay in ambient before sending the wafers into a flow furnace or dipping the as-deposited film into the DI water can convert the surface boron oxide into boron hydroxide; this will degrade the flow ability or flow

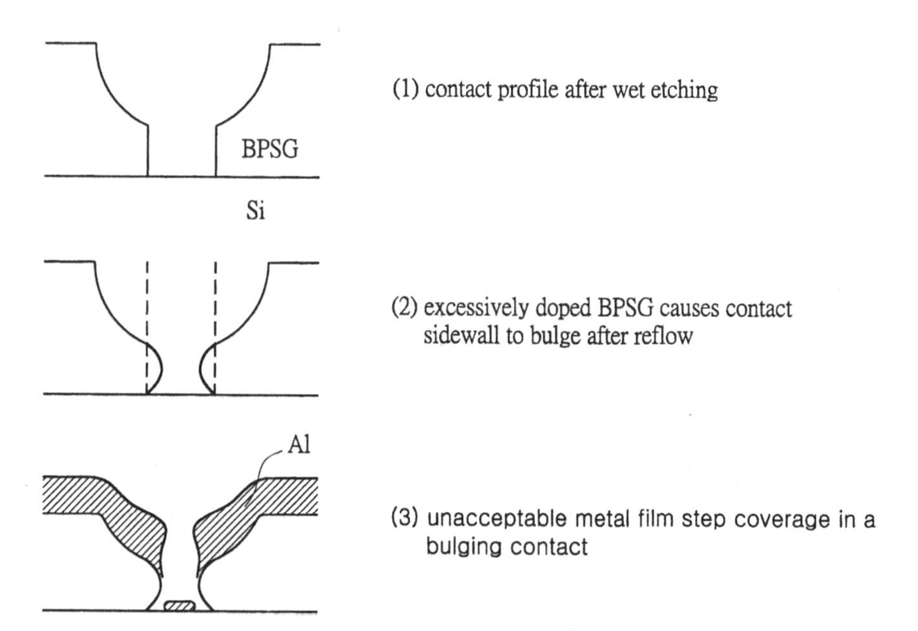

Fig. 11.10. Excessively doped BPSG causes a bulging contact sidewall and metal discontinuity in the contact.

angles. Other factors, such as flow temperatures and time, can also affect the flow angles.

As device geometries shrink to about 0.6 μm, the narrow space between the two neighboring polysilicon gates becomes very difficult to fill with BPSG since the aspect ratio (the vertical height to horizontal space ratio) is larger than 1. Furthermore, the shorter gate length allows a smaller thermal budget, and only cycles lower than 850°C are allowed for ILD flow. These two reasons drive the ILD to change from BPSG to BPTEOS, which is a TEOS oxide doped with boron and phosphorous. The reaction takes place at around 450°C. BPTEOS has a better as-deposited step coverage, and the flow temperature can be as low as 750°C–850°C. BPTEOS ILD provides excellent planarization for technologies around 0.5 μm or so. Again, the issues of autodoping, bulging, and surface hygroscopicity set the upper limits for the B and P wt% in BPTEOS.

Some advanced processes employ silicide modules to improve the circuit speed; with silicide modules, the temperature cannot exceed 750°C to retain the silicide integrity. In such circumstances, even the BPTEOS flow is considered too rigorous. A deposition-etch-deposition process can eliminate the flow thermal cycle. This process consists of PEOX or PETEOS deposition, followed by Ar sputtering, in which the sputtering rate is higher along the 45° angle. Therefore the overhang or cusp shape can be eliminated. Then a thick layer of the oxide is deposited on the top to form a very conformal final profile, as shown in Fig. 11.11. A very good planarization can be obtained by three cycles of deposition-etch-deposition to reach the final dielectric thickness. This approach can result in an even better planarization. The drawbacks of this approach are low throughput and relative high costs since several cycles of the deposition-etch-deposition and etchback processes are involved.

The extent of planarization on ILD propagates toward the back-end processes. In other words, a nonplanarized ILD can cause the first-level IMD planarization to be more difficult. This in turn will cause the second-level IMD to be even more difficult, and so on. Figure 11.12(a) illustrates an example of poor ILD planarization, while Fig. 11.12(b) shows an example of a better ILD planarization.

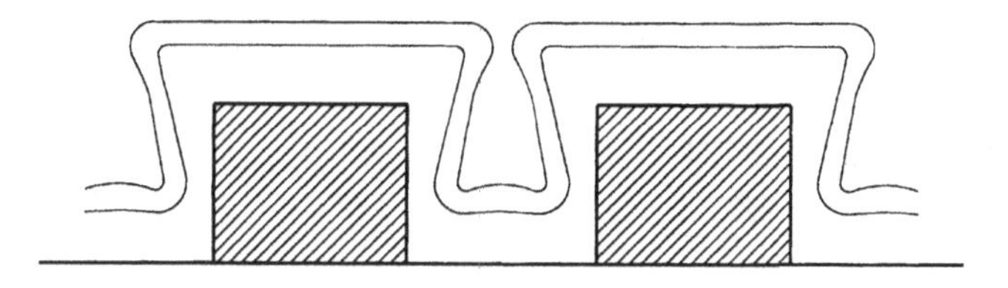

(a) continuous deposition of PEOX in a spacing leads to void

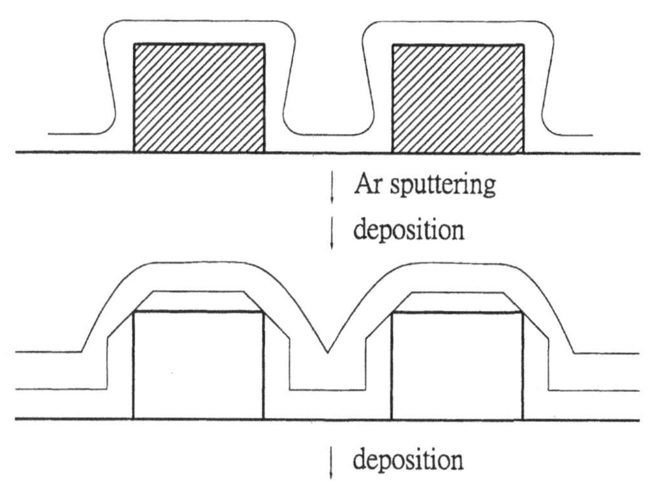

(b) the multiple dep-etch-dep gives improved planarization

Fig. 11.11. Comparison between PEOX continuous deposition and dep-etch-dep approaches.

It is obvious from Fig. 11.12(b) that it is easier to perform planarization at the second-level IMD. As device geometry migrates beyond 0.5 μm, multilevel metal structures, up to six or more metal levels, are often required. In such circumstances, a globally planarized ILD is often needed to ensure good planarization at the top IMD level. With such requirements and low thermal budget constraints, the chemical mechanical polishing is employed. This is done by depositing a thick layer of BPTEOS and a light flow at low temperatures, such as 700°C–750°C. Flow at such a low temperature regime essentially aims at making the film denser, instead of reflowing the film. It is then

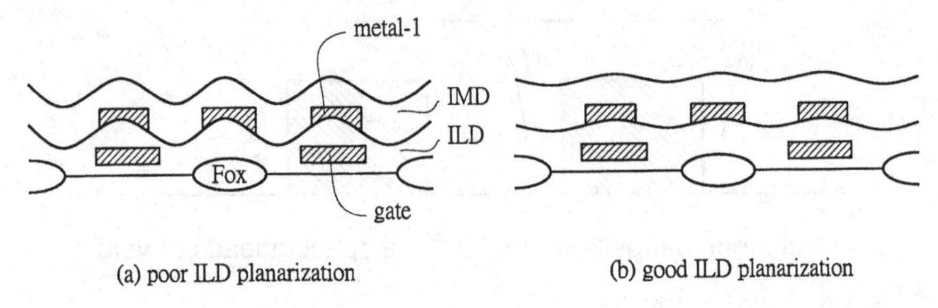

Fig. 11.12. The topograph of ILD propagates up to IMD.

followed by a chemical mechanical polishing process; the polished BPTEOS is then covered with a PETEOS layer to complete the ILD.

11.2.1. *CMP and dummy patterns*

Chemical mechanical polishing (CMP) achieves a global planarization, but it has a relatively high manufacturing cost, and its polishing rates vary with the underlying pattern density. Adding dummy patterns further improves the local and global planarization techniques. Figure 11.13(a) shows an example of a local planarization technique, such as BPTEOS flow for different pattern densities. It

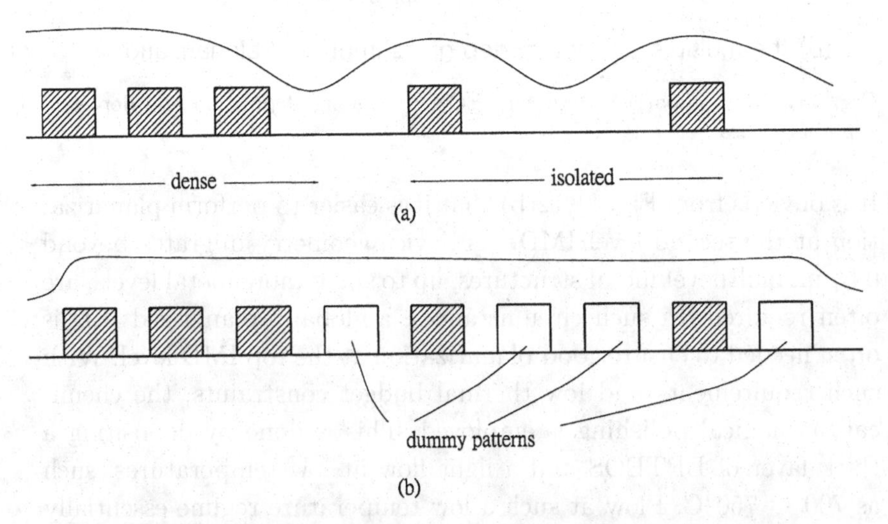

Fig. 11.13. Dummy pattern can be added to improve planarization.

can be observed that even with BPTEOS flow, the planarization is localized. Figure 11.13(b) demonstrates that by adding dummy patterns, one can actually bring the pattern densities to about the same level, and a global planarization can be achieved with CMP on flowed BPTEOS. For example, for a memory product, the polish rate on BPTEOS in the cell array (dense area) is slower than in the peripheral area (sparse area). This results in a height difference between the cell array and the peripheral areas. Therefore adding dummy patterns in the peripheral area is a great advantage in terms of achieving a more global planarization. Very likely, the CMP process, along with the addition of a dummy pattern technique, will be the ultimate global planarization technique.

11.3. Spin on Glass

Spin on glass (SOG) has been widely used for intermetal dielectrics of multilevel metal manufacturing technology, ranging from 1.2 to 0.6 μm. It is a silicon dioxide film that is formed by using the sol–gel technique. The sol–gel technique has been widely used for forming thin films and coatings, such as thin films in semiconductor manufacturing and metal or ceramic coatings for various applications. In the case of SOG, an oxide precursor in alcohol and ketones is used. The solution is sprayed over a spinning disk or substrate surface. On being spun on the substrate surface, the sol–gel is transformed into an alcogel, an oxide network that condenses in the presence of solvent. The gel is then baked at elevated temperatures, ranging from 250°C to 450°C, to drive out the solvents.

The baking leads to the formation of a hard and porous silicon oxide film. Significant volume and weight reduction are observed on baking, and yet the films remain adherent on the substrate surface. Most of the volume contraction occurs in the vertical dimension, as long as the film thickness is less than thousands of angstroms. The unique feature of the sol–gel technique is that the films can fill in narrow gaps over internal and external surfaces of complicated structures. This is why the SOG technique is widely used in semiconductor back-end planarization.

In view of the differences in precursor oxides, SOG materials can be primarily divided into two groups: the silicate and siloxane. The silicate SOG is made of $Si(OH)_4$; the siloxane SOG is made of $Si(OH)_{4-n}R_n$. R stands for light alkyl groups. The spin-coating process setup consists of an exhaust hood to carry away the outgassing solvents and a spinning chuck, on which the wafer is held, as shown in Fig. 11.14. Oftentimes, the spinning chuck rotates at a few thousand rotations per minute. A proper coating process should consist of several steps with increasing chuck rotating speeds and increasing *in situ* baking temperatures. With a fixed solid content and viscosity of SOG material, the film thickness can be altered by changing the chuck spin speed.

SOG is primarily used for IMD planarization; the curing and baking temperatures are limited up to 450°C or so, and the cured films may contain significant amounts of silanol, Si–OH. The cured SOG tends to be hygroscopic. Owing to about 8% alkyl group content in the siloxane SOG, the physical properties significantly differ from those of silicate, as illustrated in Table 11.1. After being placed in the ambient, the siloxane SOG absorbs less water than the silicate SOG. However, oxygen plasma-treated siloxane SOG tends to absorb a lot more moisture. The reason for this is that presumably,

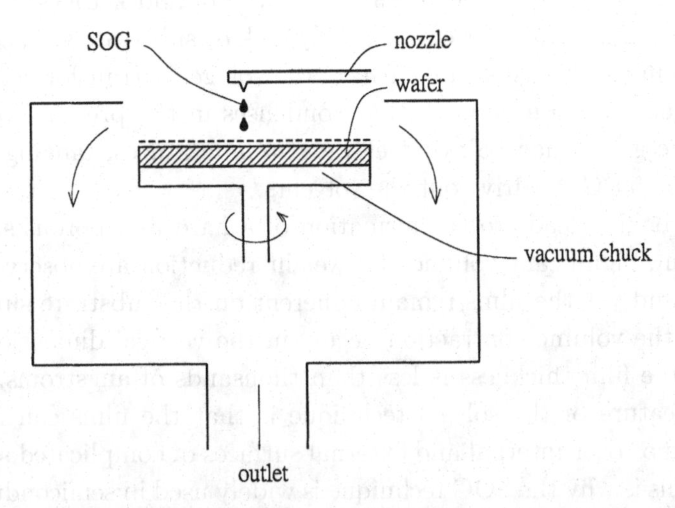

Fig. 11.14. Schematic of a SOG spin-coating system.

Table 11.1. Qualitative properties comparison of silicate and siloxane SOG.

	Silicate	Siloxane
Alkyl group	No	Yes
Stress	High	Low
Viscosity	Low	High
Thickness/coat	Thin	Thicker
Planarization	Good	Better
Prone to O_2 plasma attack	Fair	More

the oxygen plasma damages the alkyl group and possibly burns the carbons in the film structure and thereby increases its porosity. The absorbed or adsorbed moisture, if not outgassed properly, can outgas at various subsequent stages, causing fatal effects in via structures, such as poisoned vias. The poisoned via means the via resistances are abnormally high and nonuniform.

During spin coating, the surface tension of the sol–gel pushes itself into the narrow gaps; however, to achieve a good planarization, oftentimes, more than a single coating is required. It is essential to have a proper cure between each coating process to ensure that the solvent is driven out as much as possible. If this is not done, the solvent can outgas at later stages and cause disastrous metal or oxide film delamination, in addition to poisoned via problems. The coated SOG films must be stored in dry nitrogen ambient, and the queue time should be as short as possible.

Both silicate and siloxane SOG are used in a sandwich structure — undoped oxide layer/SOG/undoped oxide layer. The purpose of undoped oxide layers is twofold. First, they avoid direct contact of the SOG with metals, such as aluminum, since the residual solvent or moisture may corrode metals. Second, the undoped oxide layer functions as a stress buffer between the metal layers and the SOG material, especially during curing and baking, in which SOG undergoes significant volume shrink. It would have saved a lot of planarization efforts if the SOG material properties could be tailored such that the undoped oxide layer can be eliminated. So far, the industry has not been very successful in this aspect. For IMD applications, to maintain the underlying metal layer integrity, the postmetal process thermal cycles are often kept below 420°C for less than 30 min. For device geometries larger than 0.35 μm, the IMD technology heavily relies on SOG.

11.4. Intermetal Dielectrics

The need for IMD technology arises from the use of more than one level of metals in a device. Any IMD technology needs to fulfill two requirements: One is to provide proper electrical isolation between metal lines, and the other is to provide proper planarization for ensuing metal layer processing. In early days, with double-level metal processes with geometries of larger than 1.5 μm, one thick PECVD oxide layer served the purpose because photolithography and etching requirements could be met for such large dimensions. As technology migrated to somewhere between 1.2 μm and 0.8 μm, a planarization with SOG was often used. The silicate SOG is sandwiched between two PECVD silane-based oxides (PEOX). Oftentimes, two consecutive SOG coatings are needed to achieve good planarization. Each spin-coating process yields about 1000 Å of oxide film and is followed by a high-temperature curing process to drive out the solvents. After the double-coating SOG has been completed, a thick layer of PEOX is deposited. Via holes can then be patterned and etched on the SOG sandwich so that the nth-layer metal can be connected to the $(n + 1)$th layer.

Processes that use aluminum as the hole-filling metal require a wet–dry hole etching process to ensure good metal step coverage. Therefore the amount of wet etching must be optimized to achieve good metal step coverage and not to touch the SOG. Figure 11.15 shows that the more we wet etch on the via hole, the better cup-shaped sidewalls will be, and hence better metal step coverages will be obtained. Nonetheless, the more we wet etch, the higher the probability of SOG loss will be. This is due to the PEOX thickness and wet etch rate nonuniformity across wafers and wafer to wafer. Once the SOG is touched, the etchant will etch right through the bulk of SOG and create a hollow space; this is because the SOG wet etch

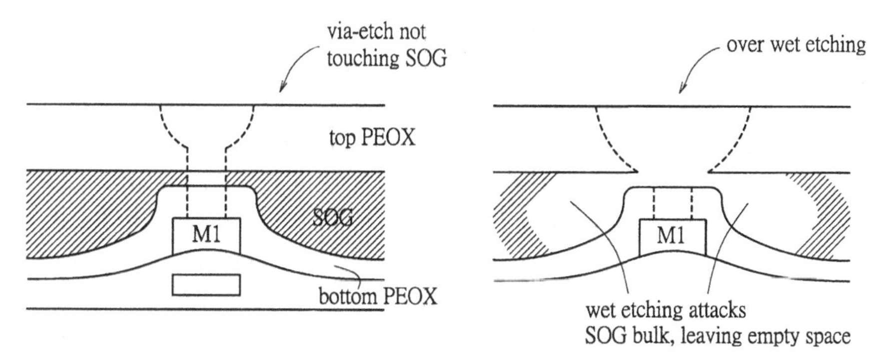

Fig. 11.15. Via wet etching. Proper amount of wet etching should stop in the upper PEOX bulk (left); excessive wet etching causes wet etchants to attack the SOG, leaving hollow space (right).

rate is much faster than that of PEOX. The hollow space can create many contamination and reliability issues. On via hole opening, it is highly recommended that the wafers not be placed in ambient for too long. The reason for this recommendation is that the SOG after curing is a dry and porous material and can easily absorb the ambient moisture. The adsorbed moisture can then outgas during via hole metal sputtering, which is conducted under high vacuum, $< 10^{-6}$ Torr. As a result, poison vias can be observed. To minimize the possibility of outgassing, a bake in the furnace is often required prior to via hole metal sputtering. Table 11.2 illustrates a typical SOG planarization process flow.

With any given SOG planarization scheme, there are ways of achieving better planarization. First, tapered metal profiles can be

Table 11.2. SOG planarization process flow.

1. Bottom PEOX deposition (5 kÅ)
2. SOG 1st coating (2 to 3 kÅ)
3. SOG baking (100°C–250°C for 2–4 min)
4. SOG 2nd coating (2–3 kÅ)
5. SOG baking
6. Top PEOX deposition (7 kÅ)
7. Via definition: photolithography and etching
8. SOG baking
9. Metal deposition

used to improve the SOG filling results since it will cause the gap openings to open up further. This requires tapered metal etching; namely, the etching chemistry erodes the resist mask as the etching proceeds. This can be achieved, but it is difficult to control in terms of metal dimensions, profiles, and stringers or residues. The second approach is to use a better step coverage bottom oxide layer, as indicated in Fig. 11.16. A better bottom oxide step coverage leads to better SOG planarization. The fourth approach is to use a thinner PEOX bottom layer, which effectively reduces the aspect ratios. Figure 11.17 illustrates that voids form in narrow metal spaces. With thicker bottoms or narrower metal spaces, it is more prone to form voids. The voids are indications of poor planarization and must be avoided to ensure good process and product reliability. The fourth approach is to form a spacer on the bottom PEOX with Ar sputtering etching or a spacer etching approach.

One can imagine that the thicker the SOG, the better the planarization results will be. However, with silicate SOG, unless triple coating is used, it is difficult to slow down the coater rotating speed to obtain a thick oxide. An alternative is to use siloxane SOG.

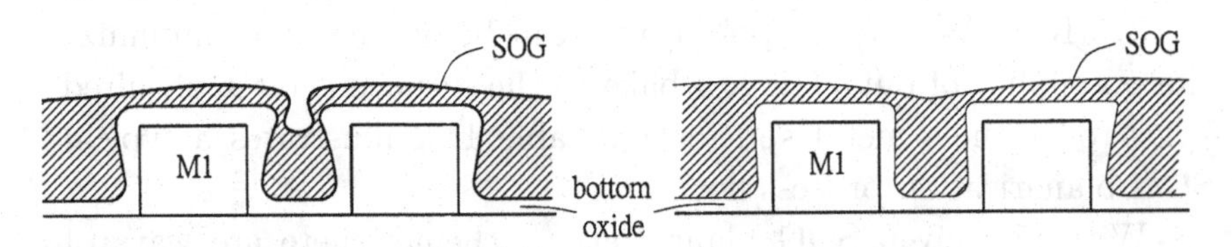

Fig. 11.16. The bottom oxide step coverage plays a critical role in SOG planarization.

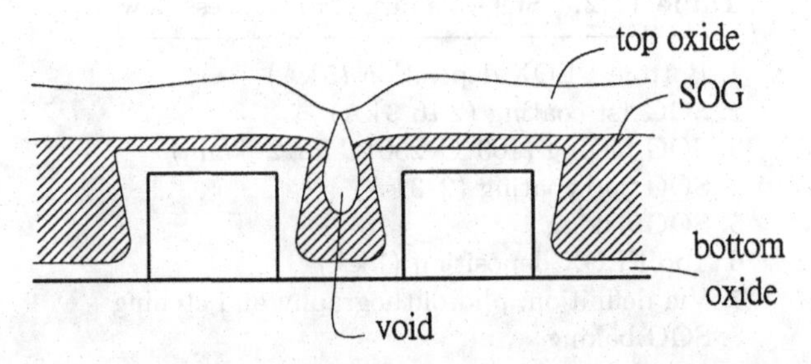

Fig. 11.17. Poor SOG gap filling in narrow gaps, leading to void formation.

With each coating, siloxane SOG provides approximately 1.6 times thicker oxides than silicate SOG does. A double-coating process can then be chosen to obtain thicker oxides. Curing after each coating is also required. A further improvement approach is to use triple coating of siloxane SOG with partial or full etchback. The partial etchback is used to partially etchback the SOG surface and not to touch the bottom PEOX, which leaves some SOG on the via sidewall. On the other hand, the full etchback stops at the bottom PEOX surface and leaves nearly no SOG on the via sidewall; therefore the poison via concerns are alleviated.

The most difficult part of the siloxane SOG process is how to ensure good quality for the deep via. This is shown in Fig. 11.18, in which the via is located between two polyrunners that form a deep valley. Deep vias tend to have more SOG on the sidewall. During via etching, if oxygen plasma is used in the course of via etching (e.g., the resist striping), the sidewall SOG will be attacked and cause the sidewall SOG to recede. The hollow space traps the residual polymers or solvents and leads to poisoned vias. One of the most important electrical measurements for SOG planarization is to measure a large number of via chain resistances and take the mean value and deviation of the resistances. Good-quality vias should have both low average via resistances and small deviations. The deviation is sometimes a good indication of via sidewall SOG outgassing. With the use of SOG, caution should be taken with the product reliability. Owing to the porous structure of the cured SOG, moisture can zip through the bulk of SOG rather quickly. The reliability tests, such as temperature humidity bias tests (THB) and pressure cook tests (PCT), have severe moisture ambient at elevated temperatures and can fail the parts if the SOG process is not optimized. Moisture can

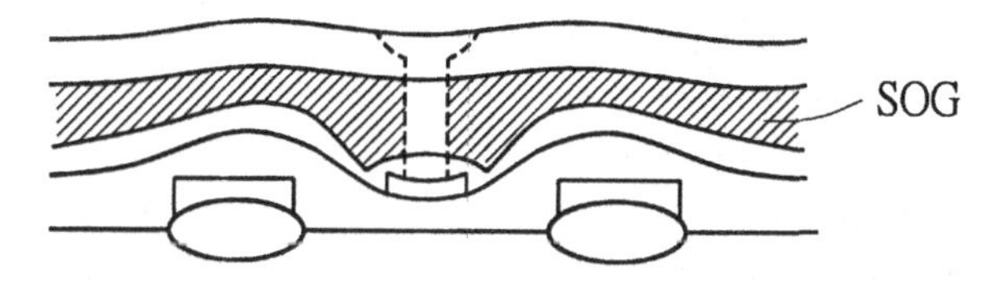

Fig. 11.18. Deep via has thick SOG on the via sidewall.

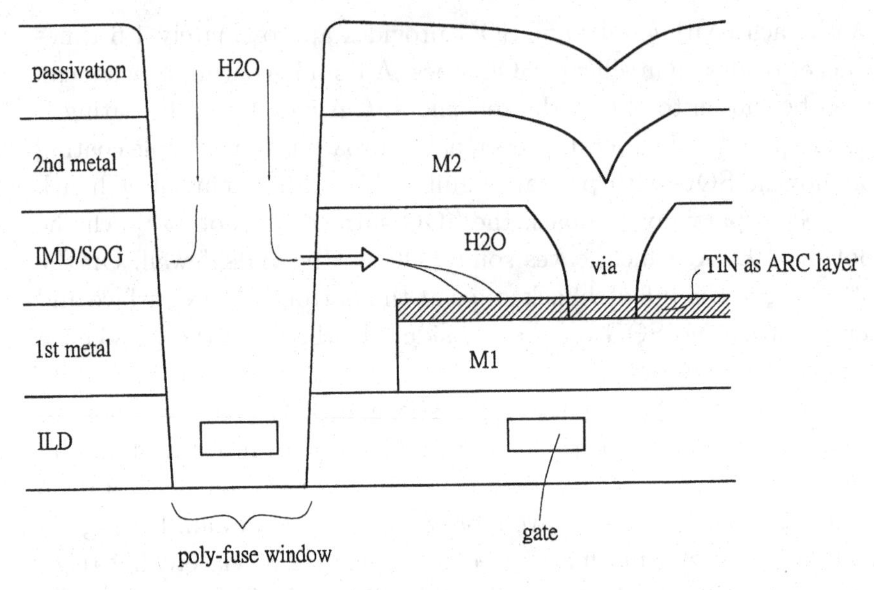

Fig. 11.19. Moisture diffuses into the circuit through the SOG sidewall to land on TiN, the top layer of the first metal, and leads to metal corrosion.

penetrate from the fuse opening into the SOG bulk and land on the TiN layer (the ARC layer of a metal line). This can cause TiN corrosion and therefore reliability failures, as shown in Fig. 11.19. In such circumstances, either the SOG full etchback process is needed to cut off the moisture diffusion channels, or the fuse window must be sealed against the moisture.

Beyond 0.25-μm technology, SOG IMD planarization becomes inappropriate for several reasons. First, W-plug is used; the IMD surface needs to be very planar for the deposited W-film to be residue-free after etchback or polishing. Second, more than four- or five-level metal schemes are required, and the poor SOG planarity can accumulate up to the top levels of the metals. There are different dielectrics that can be used to replace SOG for the IMD oxides; however, they must have one thing in common: their deposition temperatures must be below 400°C. PETEOS or PEOX oxide can be deposited with a few cycles of deposition-etch-deposition to obtain good gap filling; this is followed by a thick layer deposition. The stack is then polished back with CMP to obtain a global planarization. Via holes are then patterned and etched anisotropically to form vertical via sidewalls for W-deposition. Various oxide schemes, such as HDPCVD or SACVD oxides, can be used due to their good gap-filling capability.

11.4.1. *Planarization for passivation*

The final step in making a device is to cover the device with a protection layer (passivation layer). Passivation often consists of a double-layer structure, PEOX/PESiN, with a total thickness of over 10 KÅ. The bottom oxide is used as the stress buffer, while the nitride is the real diffusion barrier. PECVD films often do not provide good step coverage; therefore, with a fixed PEOX/PESiN thickness, as the design rules shrink, two things can happen: one possibility is that the overhangs on two neighboring metals may touch each other; the other possibility is that the corners of the nitride film may crack, as shown in Fig. 11.20. When the two overhangs touch each other, the metal space becomes an enclosed space with openings at both ends. These spaces can trap solvents or water during wet cleaning and can squeeze them out during the subsequent annealing or alloying steps, which can lead to serious contamination. The problem can be solved by either thinning down the passivation layer (especially the PEOX layer or the top metal layer's thickness) or by applying SOG between PEOX and PESiN. Of course, the manufacturing cost is increased by adding SOG.

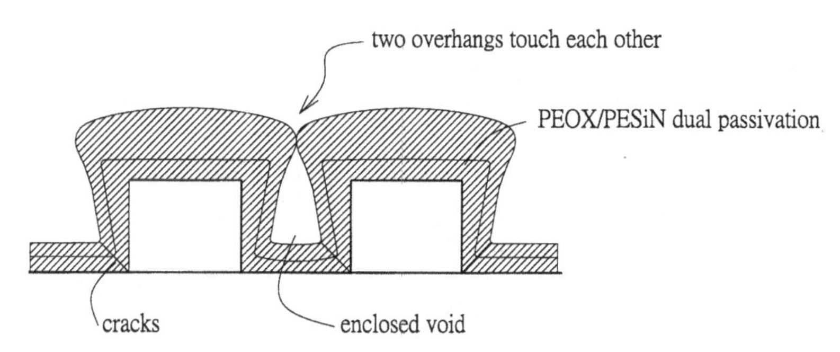

Fig. 11.20. PEOX/PESiN dual passivation can form an enclosed channel along metal lines.

11.5. Introduction to Chemical Mechanical Polishing and Its Applications

11.5.1. *Introduction to CMP*

CMP is a process that removes material from the wafer surface by means of synergistic chemical and mechanical actions. During polishing, as shown in Fig. 11.21, the wafer is held by a wafer holder and is pressed against a polishing pad on top of a platen. Pressure is exerted pneumatically to the back of the wafer. Both the wafer holder and polishing pad are set to rotate in the same direction. Slurry, which is composed of abrasive particles, deionized water, and chemicals, is continuously supplied to the center of the polishing pad. The slurry spreads and flows centrifugally as the pad rotates. The polishing pad is periodically conditioned with a diamond head to create microscratches on the pad to maintain constant and uniform polishing rates. Unlike grinding, CMP removes materials with chemical and mechanical actions taking place alternately and continuously. The wafer surface is first chemically altered (either with hydration or oxidation) by the slurry with the presence of pressure or heating from the pad mechanical forces. The altered surface layer is then removed by both chemically dissolving the slurry and mechanically dislodging it from the surface. With typical wafer CMP conditions, there is barely any material removed if only either chemical or mechanical action takes place; they work synergistically.

The polishing pad is most commonly made of organic materials such as polyurethane. It must be flexible and porous so that the pad surface can flex into the deep-lying surface and the slurry can channel

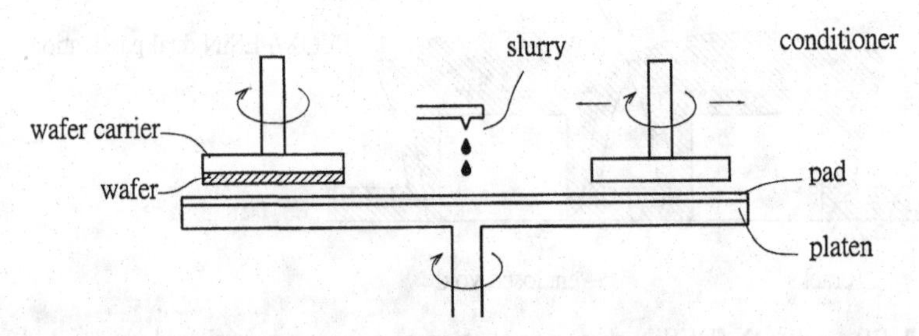

Fig. 11.21. Schematic of a chemical mechanical polishing system.

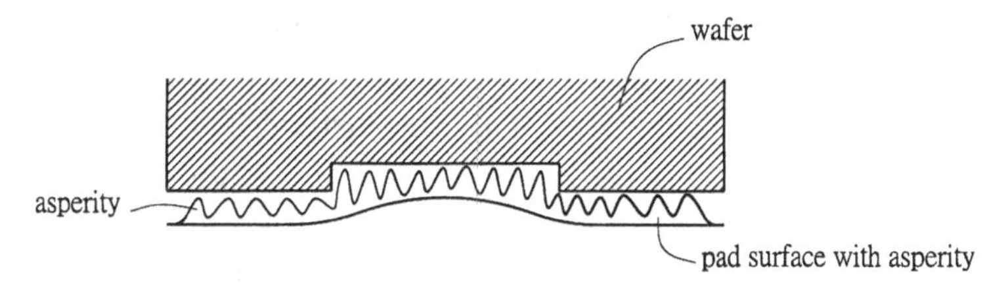

Fig. 11.22. Pad surface is flexible with asperity.

through the pad surface and reach the wafer surface, as illustrated in Fig. 11.22. In addition to the porosity and roughness, the pad surface has a few deeper trenches (tracks) that form concentric circles to facilitate the flow of the slurry in and out of the wafer surface being polished, as shown in Fig. 11.23. Its surface is rough, instead of being smooth. The roughness increases the contact surface area between the abrasive particles and the pad surface. The hardness of the pad can be altered during organic pad materials synthesis. A hard pad provides a good short-range planarity, but it is more prone to create scratches on the wafer surface. A soft pad surface allows it to easily flex in the deep-lying surface and results in a long-range planarity. The wafer is backed by a hard pad, or a carrier film, behind which the polish pressure is exerted. The wafer is fixed with a retention ring around the wafer's edge. The wafer without the carrier film tends to result in faster wafer edge polishing rates due to the bending force of the pad. By adjusting the carrier film and the polishing pad, uniform wafer polish rates can be achieved.

The system is equipped with a conditioner alongside the pad surface, which is allowed to pivot on the pad surface to create

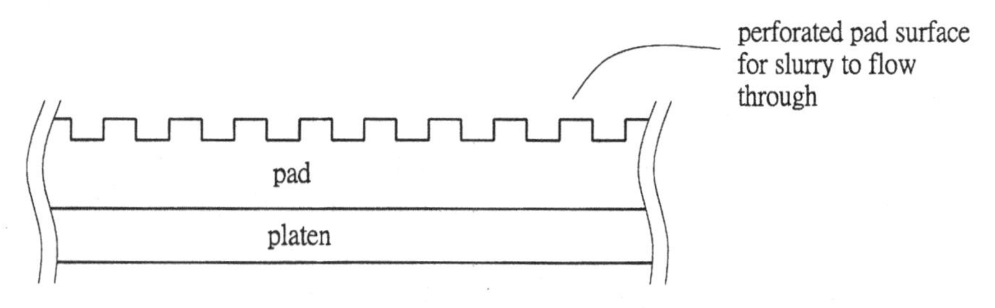

Fig. 11.23. Perforated pad surface for slurry to channel through.

microscratches on the pad surface to facilitate the delivery of slurry to the wafer surface. The head of the conditioner is cast with hard materials such as diamond or zirconium clusters. The pad surface conditions are critical to polish removal rates and uniformity. Therefore the polish pad must be continuously conditioned as polishing proceeds. Inadequate or no conditioning can lead to poor polishing rates and uniformities. This results from the fact that the pad surface pores or perforations are clogged by the slurry. As the pores on the pad surface are clogged, they lose the capability of channeling the slurry evenly to the wafer surface; as a result, the wafer polishing rates and stability are degraded. Furthermore, a used and nonconditioned pad can be heavily contaminated by cations and anions that result from polishing residues. The pad characteristics, such as porosity and hardness, are unique to a specific polishing process. They must be carefully selected and maintained to meet specific requirements.

Polishing slurries are water based, and they contain abrasive particles, oxidizing agents, and buffer agents for controlling the solution's pH values. In manufacturing, oxide and its derivatives are mostly polished in an alkaline solution with silica abrasive particles, while metal systems are polished in an acidic solution with alumina abrasive particles. The particle contents vary according to different process requirements, ranging from 5% to 20%. The abrasive particles undergo linear and rotational motions in the shear flow between the pad and wafer surfaces; therefore they assist in the dissolution of the chemically altered surface material in CMP. The polish rates increase with increasing solid contents and can be expressed as

$$\frac{R_p}{R} = 1 + \alpha \tau_w C_p \,, \tag{11.4}$$

where R_p represents the polish rate with abrasive particles, R is the polish rate without abrasive particles, α is the proportional constant, τ_w is the average shear stress on the wafer surface due to slurry, and C_p(wt%) is the abrasive particle concentrations. The constant α can be derived from the slope if the CMP operating conditions are known. Unlike grinding, which uses macroscopic particles to remove the surface material via crack propagation, CMP uses atomic-scale fine particles via bond breaking.

In CMP slurry, the particles tend to aggregate to various extents, depending on the slurry conditions such as pH values or concentrations. The aggregate sizes range from tens to hundreds of nanometers. Generally, larger particles result in higher polish rates. A uniform particle size distribution has proven to result in fewer scratches on the wafer surface during CMP. The shapes of the particles are critical parameters in CMP. The spherical particles tend to display a rolling contact mode between the pad and wafer surface. Arbitrarily shaped particles tend to display a sliding mode and thus introduce defects. Furthermore, the spherical particles result in two to three times lower polishing rates than nodular-shaped particles do.

For CMP, the chemically altered surface layer is removed via dissolution or via abrasive particles actions. To obtain a good post-CMP planarity, the dissolution must be deliberately suppressed, leaving relatively high-lying features to be removed via polishing actions. This is because the dissolution is nonselective for either high- or low-lying feature surfaces, but the polishing is selective, if the pad has a proper hardness.

11.5.2. *CMP application*

A wafer manufacturing technology migrates beyond 0.13 μm, CMP is widely used to remove or planarize back-end layers, mainly because it provides good global planarization.

11.5.2.1. *Oxide CMP*

Oxide CMP is most frequently used in advanced device manufacturing, including shallow trench isolation, and for each layer of ILD and IMD planarization. Oxide CMP is normally conducted in an alkaline solution with abrasive silica particles. Regardless of various doped oxides, the basic polishing principles are similar. Material removal in oxide CMP involves two steps. The silicon oxide surface and abrasive particles' surface are first hydrated and heated. Water molecules

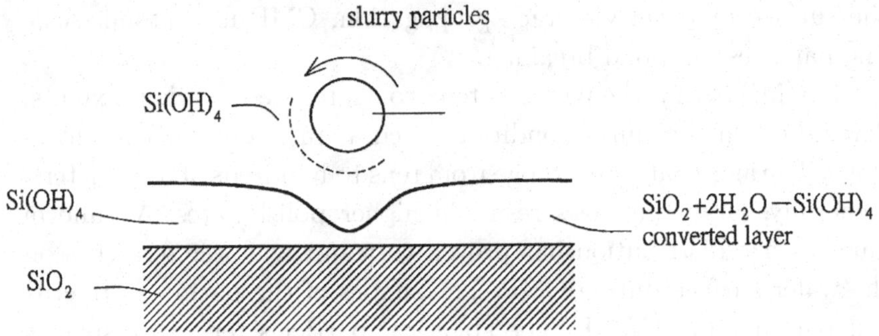

Fig. 11.24. Chemical and mechanical changes during oxide CMP.

come into contact and penetrate into the surface oxide layer, and they convert the surface SiO_2 into SiOH, as shown in Fig. 11.24:

$$SiO_2 + 2H_2O \longrightarrow Si(OH)_4 . \quad (11.5)$$

During CMP, as the pad and abrasive particles move against the wafer surface, a significant amount of heat is generated due to the friction between the abrasive particles and the wafer surface. Because the abrasive particles have a much higher overall surface area for the generated heat conduction compared to the wafer surface, the wafer surface is heated up. The heating effect further softens the hydrated surface and causes it to undergo plastic deformation. The surface SiOH bonds are then broken, and the molecules are removed by the plowing effects of the abrasive particles. It should be noted that whether the particles' plowing effect will result in surface material removal depends on whether the particles' shear force is larger than the surface bonding energy. It has been shown that the rate of oxide removal without the presence of water is significantly reduced. Furthermore, the rate of removal increases with the contents of the abrasive particles because the more abrasive particles are present, the more friction heat will be generated, and the more wafer surface heating will result. The higher the pH value of the solution (i.e., more hydroxyl ions in the solution), the higher the polish rates will be. This is partly due to hydroxylation and partly due to dissolution of silicate molecules.

The oxide polish rates depend on the film density or hardness; the thermal oxide polish rate is slower than the PECVD oxide polish rate. On the other hand, silicon nitride is harder than oxide, and the removal rate is even slower. Dopants tend to affect the polish rates as well. Similar to wet etching, the BPSG polish rates increase with increasing overall dopant concentration in BPSG. However, the polish rates on oxide with organic dopants tend to decrease with increasing organic dopant level.

Application of CMP to shallow trench isolation, STI, is shown in Fig. 11.25, in which the silicon surface is covered with a silicon oxide–nitride double layer. Pholithography and etching are carried out to pattern the trenches, followed by an oxidation to round off the sharp corners. The trenches are then filled with a CVD oxide layer, and CMP is conducted to smooth the surface. Owing to the fact that the nitride polish rate is a lot slower than the oxide layer and that the pad may flex into the trench oxide surface, a severe dishing effect may be observed. Greater dishing effect is seen in the dense pattern areas due to loading effects. Once the dishing effect is seen, it is difficult to achieve good surface planarity. A two-step approach can be employed to achieve good planarity: first, oxide CMP is used to gain a planar surface after removing a layer of oxide; then, the process is switched to a plasma isotropic etching to stop on the nitride surface.

11.5.2.2. *Tungsten CMP*

Application of tungsten is often accompanied by a tungsten etch-back or tungsten CMP process. Typically, hole patterns are defined on the oxide layer; this is followed by Ti/TiN glue layer deposition through sputtering or CVD. Then, blanket tungsten that is thicker than one half of the hole width is deposited. The blanket tungsten is then etched back to the oxide surface, leaving W-plugs in the holes. Figure 11.26 shows the process flow. Tungsten CMP is often used in lieu of the etchback process in advanced device manufacturing because the plug recess and local step height can be mitigated using the CMP process. Furthermore, the CMP process offers potentially higher yields and lower cost compared to

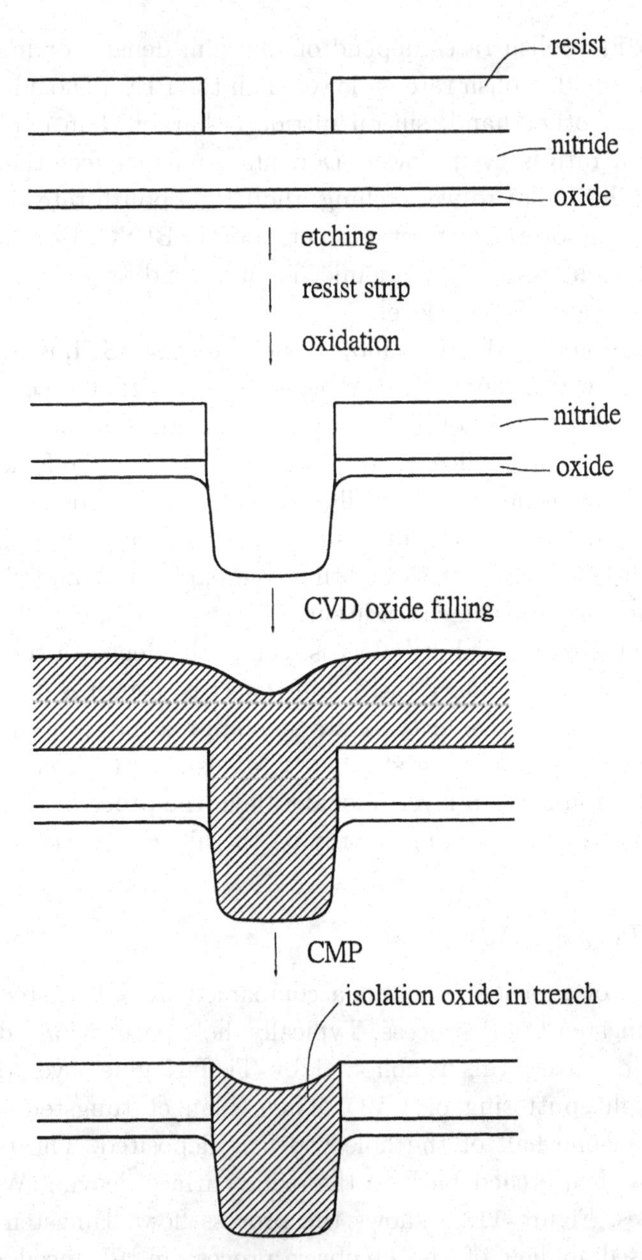

Fig. 11.25. CMP for shallow trench isolation.

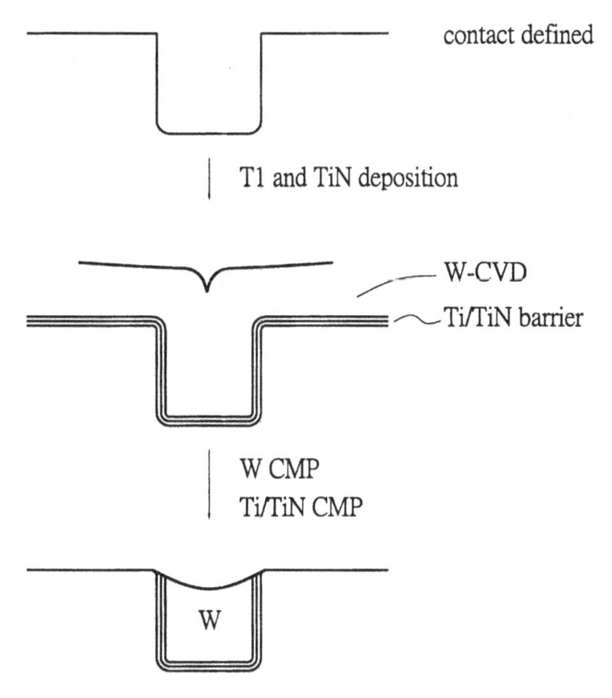

Fig. 11.26. CMP for W-plug formation.

the conventional tungsten etchback process. Tungsten CMP is often conducted in an acidic solution with alumina abrasive particles. Oxidants, such as potassium iodate, ferric nitrate, or hydrogen peroxide, can be used as well. During CMP, tungsten is oxidized with oxidants:

$$W \xrightarrow{\text{ox}} WO_x \tag{11.6}$$

$$WO_x \longrightarrow WO_n\,. \tag{11.7}$$

The initial oxide is then converted to the final oxide form, which is in turn subject to dissolution and mechanical abrasion:

$$W_{\text{bulk}} + WO_n \longrightarrow WO_{n,\text{aq}} + W \tag{11.8}$$

$$W_{\text{bulk}} + WO_n \xrightarrow{\text{abrasion}} WO_{n,\text{aq}} + W\,. \tag{11.9}$$

The first reaction represents the oxide dissolution. The second reaction represents the oxide being removed via abrasion. Both leave the tungsten surface refreshed for further oxidation. The cycle repeats continuously, until the underlying material surface emerges. For tungsten CMP, an overpolish step is often required to completely remove the glue layers. Caution should be taken as excessive overpolishing can result in plug recess. An oxide buff can be carried out to eliminate the excessive recess.

11.5.2.3. *Copper CMP*

Unlike conventional metal interconnects, copper lines are processed with CMP and damascene structures due to the fact that copper is not plasma etchable. The damascene structure starts out with an oxide layer with trenches. The barrier layer TaN/Ta is first deposited. Then, the wafer is subject to an electroplating process to coat a Cu layer on the top. Polishing of Cu can be carried out by using alumina or silica acidic slurries, with hydrogen peroxide as the oxidant. Potassium hydrogen phthalate salt can be used as the buffer if the pH is kept at 4.0. During the CMP process, Cu first undergoes mechanically enhanced oxidation with copper hydroxide, which can be further converted to copper oxide in high-peroxide slurry. Both copper hydroxide and copper oxide can dissolve into the slurry by forming a copper phthalate complex:

$$\begin{aligned}
\mathrm{Cu} &\xrightarrow{\mathrm{ox+H_2O}} \mathrm{Cu(OH)_2}\,,\\
\mathrm{Cu(OH)_2} &\xrightarrow{\mathrm{ox}} \mathrm{CuO} + \mathrm{H_2O}\,,\\
\mathrm{Cu(OH)_2} + \mathrm{RCOO^-} &\longrightarrow \mathrm{Cu^{+2}}\ \text{(phthalate complex)}\,,\\
\mathrm{CuO} + \mathrm{RCOO^-} &\longrightarrow \mathrm{Cu^{+2}}\ \text{(phthalate complex)}\,.
\end{aligned} \tag{11.10}$$

The oxides can be mechanically removed by using abrasion as well. Table 11.3 shows that the copper removal rates vary with different slurries with alumina and silica acidic solutions. The barrier layer, Ta or TaN, is harder than Cu; significant overpolishing is required to clear the barrier layer. However, as expected and illustrated in Fig. 11.27, dishing, erosion, and spacer rounding can result in dense areas due to overpolishing. The time needed to planarize a structure depends on the distribution of pattern densities on the wafer.

Table 11.3. Various types of slurries for Cu CMP. Additive is added to reduce the etch rate. Etch rate is done by immersing samples in the chemical without mechanical polishing.

Slurry	pH	Particle size (nm)	Etch rate (nm/min)	Removal rate (nm/min)
A_1	4.0	Alumina (320)	10.2	585–1109
A_2	3.6	Alumina (320)	26.6	552–948
S_1	3.5	Silica (90)	33.1	496–131
S_2	3.9	Silica (90)	2.3	156–367

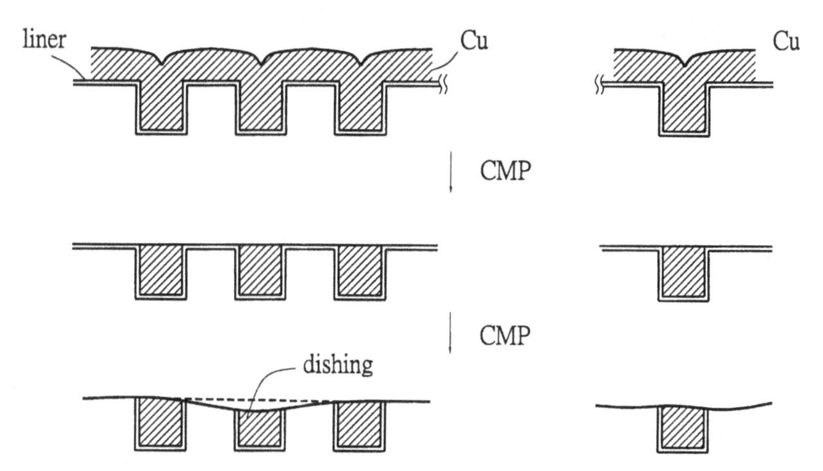

Fig. 11.27. In Cu, CMP, dishing may occur in dense areas during overpolishing.

This is because the applied pressure sensed by the high pattern density area is actually higher compared to the applied pressure in the low pattern density area; therefore the polishing rate is faster. Consequently, a certain amount of overpolishing is required to form perfect isolated Cu lines. The patterned underlying oxide erodes faster than the unpatterned underlying oxide; therefore overpolishing causes serious erosion and dishing. For Cu CMP, corrosions are found in the Ta and Cu, and they are more severe in the low pattern density areas. A passivating agent, BTA (1-H-benzotriazole), can be added to avoid corrosion. The side effect of adding BTA is that it slows down the polishing rates.

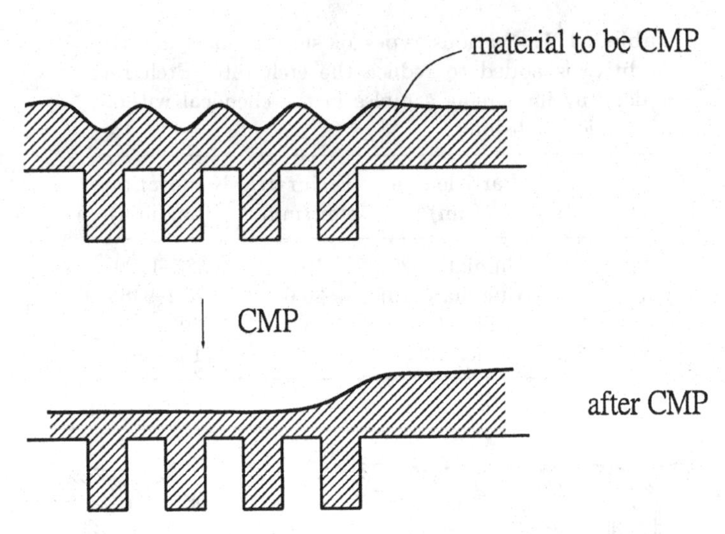

Fig. 11.28. Pattern density variation leads to step height after CMP.

The CMP local polish rates vary according to the underlying local pattern densities. The difference in polish rates is the result of effective applied pressure that is exerted on the local wafer surface. This is compounded by the fact that the pad surface is flexible and can bend along the step height and reach the low-lying surfaces. As a result, a step height still exists due to local pattern density variations, as indicated in Fig. 11.28. This step height can cause significant CD variations across the step due to the swing effect or notching (strong reflected light from the sloped surface), or it can diminish the common DOF window across the steps. Global planarization can be achieved only if caution is taken in minimizing the underlying pattern density variations. Various approaches have been taken in this regard. First, as illustrated in Fig. 11.29(a), before CMP commences, a reverse tone mask can be generated to partly etch off the overburden over the flat-lying areas to even out the overburden distribution across the wafer. Second, as also illustrated in Fig. 11.29(b), dummy patterns can be added to less dense pattern areas in the underlying patterns to achieve a pseudo-uniform pattern density distribution across the wafer.

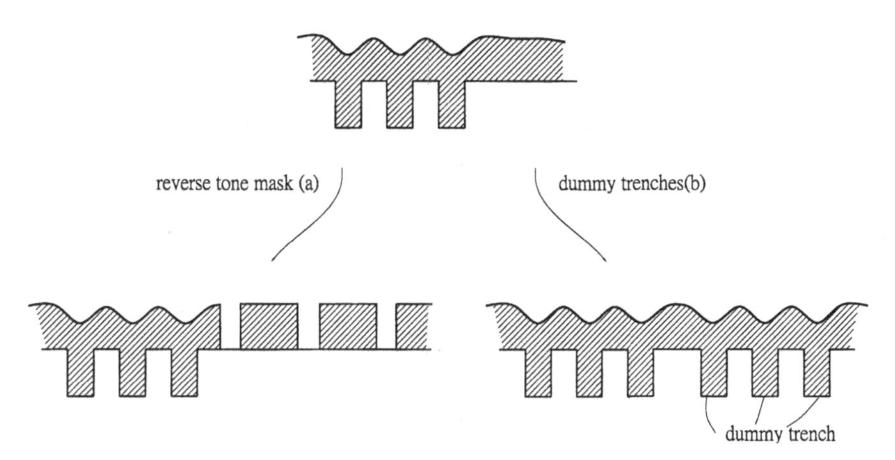

Fig. 11.29. Approaches to reduce step height due to underlying pattern density difference by (a) using a reverse tone mask and (b) adding dummy trenches.

11.6. CMP Modeling

Preston's equation is an empirical equation and has been used frequently to model the CMP process. Preston's equation correlates the applied pressure and linear velocity to the polishing material removal rates:

$$R_p = kPV\,, \tag{11.11}$$

where R_p is the material removal rate, P is the applied pressure from the rotating pad to the wafer, and V is the linear velocity of the wafer relative to the pad surface. The equation accounts for the mechanical abrasive actions through P and V and lumps all the chemical interactions into the proportional constant, k. Any operating parameter changes other than applied pressure or wafer rotating speed must be accounted for with different values of k. One of the refinements of Preston's equation is to take the shear stress of the fluid between the wafer and pad surface into account. This results in the CMP material removal rate being proportional to $(PV)^{0.5}$:

$$R_p = k'(PV)^{0.5}\,. \tag{11.12}$$

Preston's equation states no information regarding species' concentrations or polishing rate distributions across a wafer surface. To find this information, a microscopic model, including momentum, energy, and mass transfer equations, is needed. With proper assumptions on the boundary conditions, the model can be solved for the species concentration profiles across the region of interest. From the concentration profile, one can evaluate the mass flux at the surface to obtain the material removal rate.

Chapter 12

COPPER AND LOW-k

The evolution of the semiconductor industry has been striving for ever-increasing packing density, higher speed, and lower cost. Continuing with shrinking the feature sizes seem to be the main theme. This chapter explains how the back-end processes have to change to cope with the ever-shrinking technology. We will focus on metallization and the dielectric materials. These two subjects coupled with CMP form the state-of-the-art back-end processes for modern device manufacturing. Section 12.1 provides an overview of why the metal and dielectric materials need to be changed as the technology advances. Section 12.2 illustrates the Cu electroplating as well as the needed module processes. It then continues with the explanation of various low-k dielectrics and their associated processes in Section 12.3. To integrate the Cu and low-k materials is not an easy task. Section 12.4 completes this chapter by addressing several integration issues.

12.1. Back-End Processes

Let us review the signal propagation in a device. Signals are carried by the moving electrons in conducting lines. All materials have finite resistances to the moving electrons:

$$R = \frac{\rho L}{A}, \tag{12.1}$$

where R is the resistance of a conducting line with a length of L and cross-sectional area of A, as shown in Fig. 12.1. In addition, the

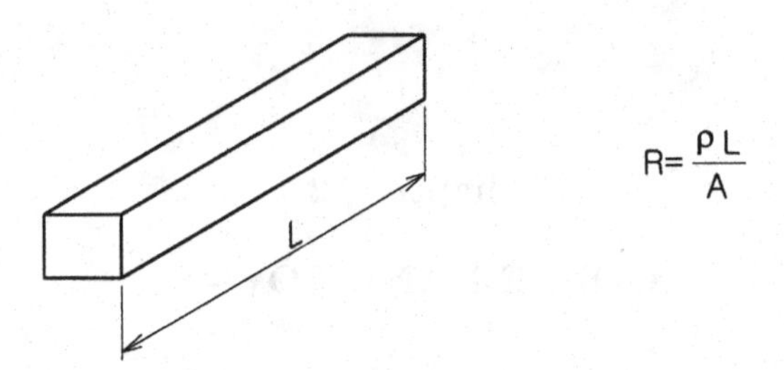

Fig. 12.1. The resistance of a conducting line is proportional to its resistivity and length and inversely proportional to its area.

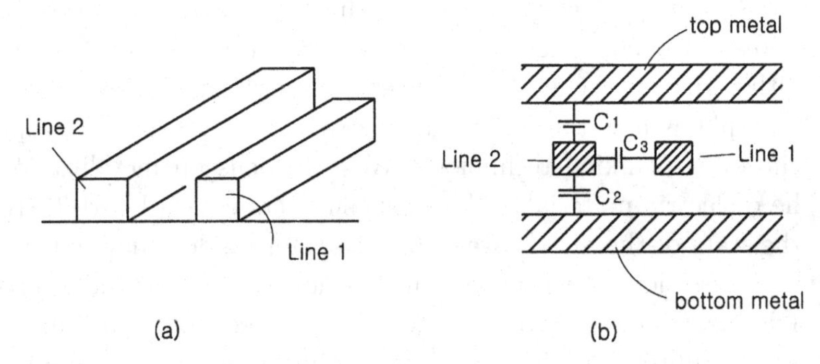

Fig. 12.2. (a) Signals propagating along a conducting line are slowed down by a neighboring line and (b) parasitic capacitors are seen among conducting line.

capacitance of a pair of capacitor plates is represented by C, and

$$C = \frac{\varepsilon A}{d}, \tag{12.2}$$

where ε is the permittivity of the medium, A is the cross-sectional area of the plates, and d represents the separation of the plates. In a device, signal propagations in a conducting line, as shown in Fig. 12.2, are slowed down by the neighboring conducting lines through RC coupling. R is the resistance of the conducting lines, and C is the capacitance of the capacitor resulting from the dielectric film between the conducting lines. The RC delay of a conduction line due to its four neighboring lines can be expressed as

$$RC = 2\rho k\varepsilon_0\left(\frac{4L^2}{p^2} + \frac{4L^2}{t^2}\right), \tag{12.3}$$

where p is the horizontal pitch, which is the sum of the line width and space, t is the vertical pitch, and k is the dielectric constant of the dielectric material that separates the conducting lines. Now, let us consider a device with an input V_{in} and an output V_{out}. The V_{out} signal will increase with a time delay after V_{in} increases. It takes some time for V_{out} to reach the maximum value, as expressed by

$$V_0 = V_{0,\max}\left(1 - \exp\left(\frac{-t}{RC}\right)\right). \tag{12.4}$$

Obviously, the larger the RC value is, the longer it will take for V_{out} to reach the maximum value. R and C are the lumped resistance and capacitance of the device, respectively. Now, as the technology shrinks, the feature sizes, p and t, are all scaled down by a factor of λ; as a result, the RC delay will increase, which will somewhat compensate the speed improvement due to transistor shrink. There are a couple of approaches to solve these issues. One approach involves using a hierarchical design that takes critical time paths and puts them on another metal layer with larger and thicker metals. This approach has been practiced in circuit design. Figure 12.3 demonstrates the trend of metal layer number with technology nodes. Increasing the metal layer number solves the speed issue but adds extra cost due to

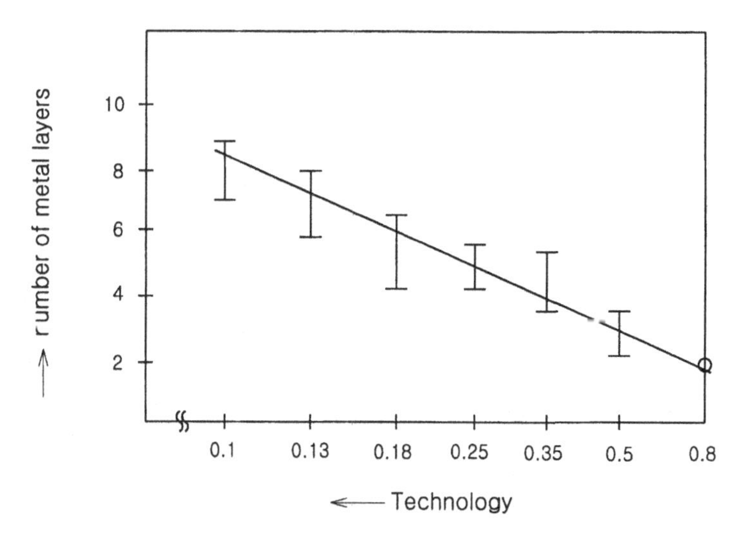

Fig. 12.3. The number of metal wires increases with advancing technology.

an increase in process steps and yield loss. Recently, a more effective approach has been introduced to volume production: to change the dielectric constants of dielectric materials and the resistivities of the conducting materials. Both κ (dielectric constant) and ρ (resistivity) are reduced by changing the materials. Table 12.1 shows the dielectric constants and resistivities for various materials. The resistivity of copper is nearly one half that of aluminum. Figure 12.4 shows the clock frequency of a microprocessor as a function of each technology feature size and interconnect technology. The frequency increases

Table 12.1. Resistivity and dielectric constant of various materials.

	Resistivity	k-values
AlCu	3.3 ($\mu\Omega$ cm)	—
Cu	1.8	—
SiO_2		4.2
F-doped		3.6
Organic low-k		2.8
C-doped		2.5–3.0
Porous silica		1.8–2.5

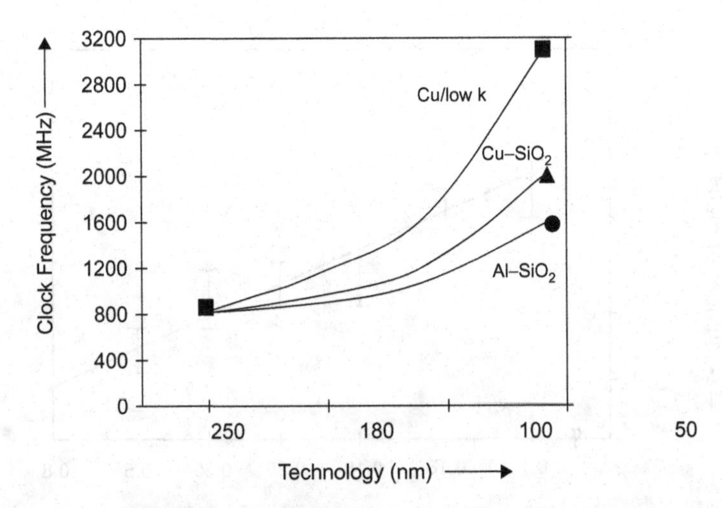

Fig. 12.4. Clock frequency achieved by a processor chip with different back-end structures.

with shrinking technology primarily due to the transistor current increases. Furthermore, the highest frequency is achieved with a combination of copper and low-k dielectrics. The lowest frequency results from using conventional oxide and aluminum structures. There are a large spectrum of dielectric materials that can be used to replace silicon oxide; however, copper seems to be the most accepted conducting material substitute. The use of copper also makes it possible for a device to have fewer conducting layers.

12.2. Cu Wiring

A low resistivity is not the only reason for replacing aluminum with copper. Cu is a superior conducting material not only in terms of lower resistivity, but also in terms of better reliability, better gap filling capability, and lower cost. Although the resistivity of Cu is slightly higher than that of Ag, it has only about 60% of Al's resistivity. The electromigration resistance of Cu is better than those of Ag and Al. Figure 12.5 shows the electromigration failures of Cu and

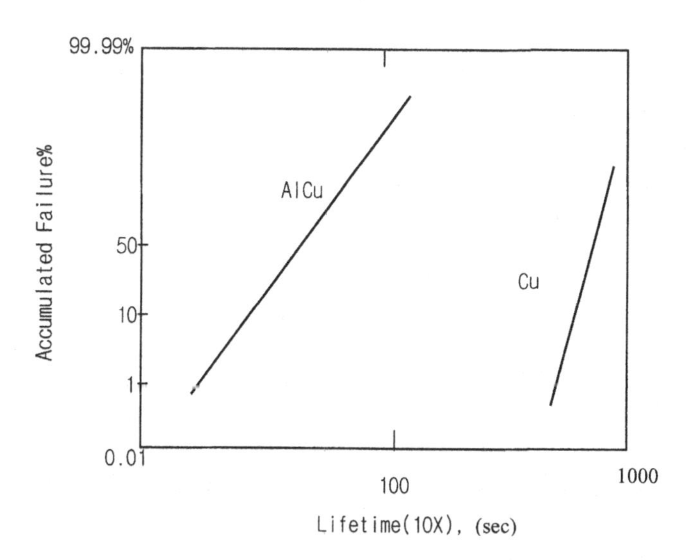

Fig. 12.5. Electromigration performance comparison between AlCu and Cu interconnects.

Al lines over time. The failure is defined as the increase in line resistance by a certain percentage, such as 30% or 50%. It can be clearly observed that the electromigration resistance of Cu is much better than that of Al. However, Cu is prone to corrosion due to the lack of self-passivated oxide. Fundamentally, the oxidant, such as moisture or oxygen, readily diffuses through the copper oxide that is formed on the surface; it reaches the fresh Cu surface and further oxidizes it. Aluminum oxide, on the other hand, performs much better in this aspect. As a result, unlike the Al process, the Cu line must be surrounded by inert materials such as silicon nitride or tantalum nitride films, the barrier layers.

Copper does not have a volatile compound at low temperature; therefore Cu is not plasma-etchable at low temperature. Cu is often employed in manufacturing in conjunction with CMP and damascene processes, in contrast with conventional etching processes for Al. Figure 12.6 illustrates the difference between the two. During conventional etching processes, Al is deposited onto an oxide surface with etched contact or via holes. Lithography is used to create patterns in the metal lines; chlorine-containing plasma is then used to etch the Al layer into the conducting lines. This is done differently for damascene processes, where the oxide is patterned and etched first, and then the Cu is deposited to fill up the trench. The excessive Cu is then polished to be removed. Another alternative is the dual damascene process, in which via and trench are etched consecutively. It is believed that the Cu dual damascene process is more cost-effective because there are less process steps.

12.2.1. *The barrier layer for copper*

Similar to the barriers for aluminum, the barriers for Cu prevent the Cu from diffusing into the silicon substrate that forms Cu_3Si and causes device failure. Several factors, such as film thickness, microstructure, density, and defects, can significantly affect the barrier properties. In general, amorphous films are better than crystalline films; the smaller the grain sizes, the better the barrier properties will be. It is also apparent that the thicker the film is,

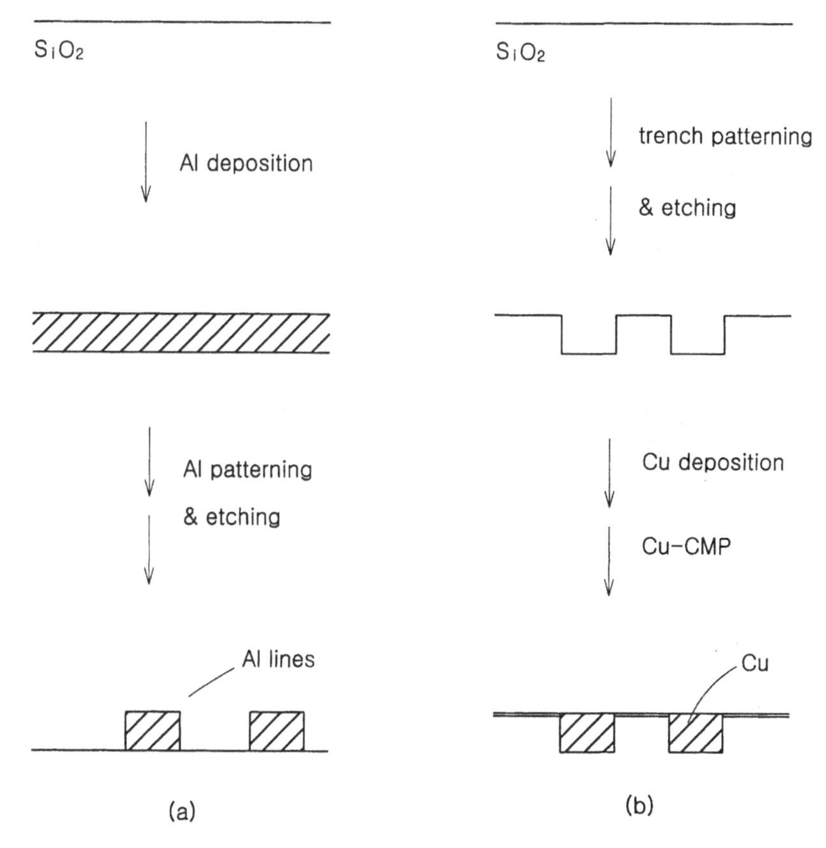

Fig. 12.6. Comparing (a) Al etching with (b) Cu damascence process.

the better the barrier properties will be. The grain boundaries of a columnar polycrystalline structure tend to render fast diffusion paths for Cu diffusion. A dense film provides better barrier properties than a porous film. Defects such as voids, pinholes, or microcracks in the barrier film degrade the barrier's performance. Table 12.2 shows various barrier layers and their resistivities. Annealing is often used to evaluate the barrier's performance. Cu barrier samples are first annealed and inspected. Transmission electron microscope (TEM) can be used to check the boundary between copper and the barrier layer. For a poor barrier, the as-deposited film structure has a well-defined boundary; however, the annealed structure tends to have Cu

Table 12.2. Resistivity of various Cu barriers.

Diffusion barrier	Resistivity ($\mu\Omega$ cm)
Ta	15 (bcc)
TaN (40% Ta, 60% N)	910 (fcc)
Ta(O) (58% Ta, 32% O)	370 (amorphous)
Ta (N, O) (80% Ta, 17% N, 3% O)	250 (amorphous)

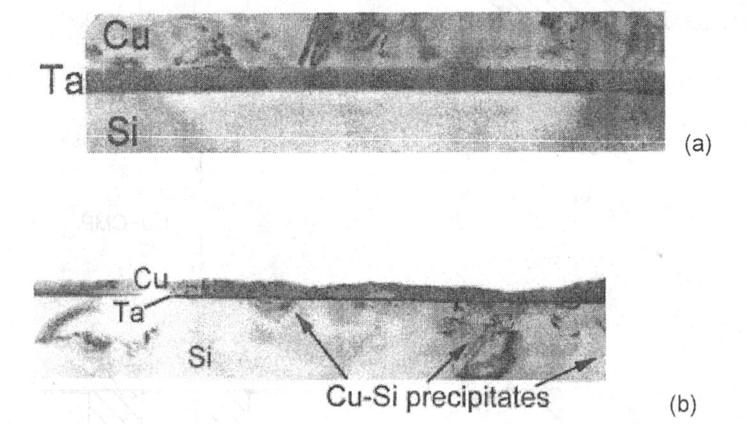

Fig. 12.7. TEM images of Cu/20 nm beta-Ta/Si: (a) as-deposited and (b) after 600°C/1 h anneal, Cu penetrates through the barrier (Stavrev *et al.*, 1999).

diffusing through the barrier and into the underlying silicon substrate. The penetrating phenomenon increases with increased anneal temperature and time. Figure 12.7 shows that for an as-deposited structure, a clearly defined boundary can be observed for Ta with respect to Cu and Si. But the annealed structure shows that the boundary is smeared, and Cu precipitates are seen in Si.

There have been several barrier candidates studied in literature. In view of various barrier requirements and commercial availability, the Ta/TaN bilayer is widely used in volume production to form a TaN/Ta/Cu structure, with Ta on the Cu side for adhesion and TaN on the dielectrics side for preventing corrosion. Both layers can be deposited with IMP PVD systems. TaN is obtained by adding nitrogen to the plasma. TaN films with various stochiometries can be obtained by varying the nitrogen flow rates to the plasma system. During the deposition, increases in the nitrogen flow rate give N-rich TaN, a higher-resistivity film with higher resistance to corrosion.

12.2.2. *The copper seed layer*

Before bulk Cu deposition, a thin Cu seed layer is required; this seed layer dictates the film structure and step coverage of the ensuing Cu bulk grown with electro chemical deposition (ECD). In addition, it functions as an electrode during ECD. Discontinuity of the seed Cu layer induces voids in Cu ECD. From an electromigration viewpoint, a $\langle 111 \rangle$ texture is preferred to achieve better performance. A strong $\langle 111 \rangle$-oriented bulk Cu can be grown on a $\langle 111 \rangle$ seed layer. Cu seed layer has a strong heteroepitaxial growth relationship with the underlying Ta. It is also critical that the barrier and seed layer depositions be carried out in a clustered tool without vacuum break. Processes with an *in situ* and vacuum break between the barrier and seed Cu film depositions show drastically different Cu textures. The difference is likely due to oxidation on the barrier surface. Figure 12.8

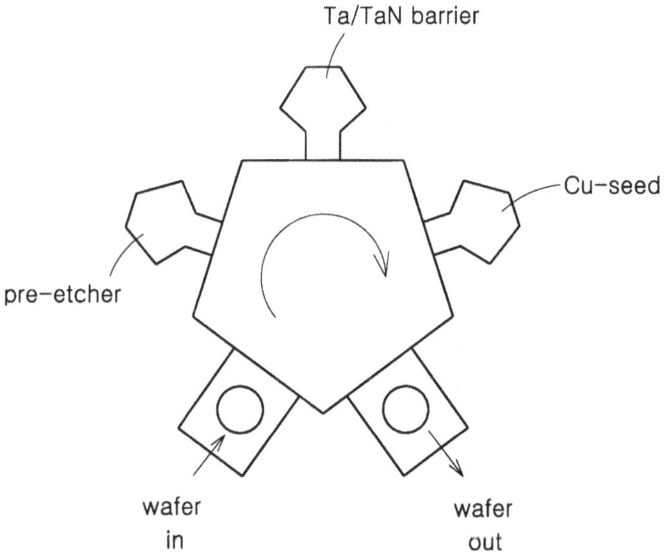

Fig. 12.8. Schematic of a clustered tool for preparing wafers for copper ECD.

shows a clustered tool that can handle degassing, precleaning, and barrier and Cu seed layer sputtering without breaking the system vacuum.

12.2.3. *The bulk copper growth*

The bulk Cu deposition can be done with MOCVD, PVD, or ECD. MOCVD has been proposed for CVD copper deposition using Cu (hfac)(tmvs) as the precursor:

$$2\text{Cu(hfac)(tmvs)} \xrightarrow{150^\circ\text{C}} \text{Cu} + \text{Cu(hfac)}_2 + 2\text{(tmvs)}\,, \qquad (12.5)$$

where Cu(hfac)(tmvs) stands for copper(I) hexafluoroacetylacetonate dihydrate, trimethylvinylsilane, as shown in Fig. 12.9. Both CVD and ECD Cu show similar ⟨111⟩ textures, but the grain size of the CVD Cu is smaller than that of ECD Cu. Therefore, as illustrated in Fig. 12.10, the CVD Cu shows inferior electromigration performance compared to ECD Cu. ECD Cu is the most accepted production approach for Cu films due to its superior step coverage capability and film quality as well as cost-effectiveness.

There are two types of ECDs: electrode-less deposition and electroplating deposition. Electrode-less deposition is a metal deposition process in which a reducing agent is added to a solution that contains the metal ions to reduce the metal ions and form metal deposits on a catalyzed or noncatalyzed surface. On the other hand, electroplating deposition is a process of passing electrical currents, as a result of

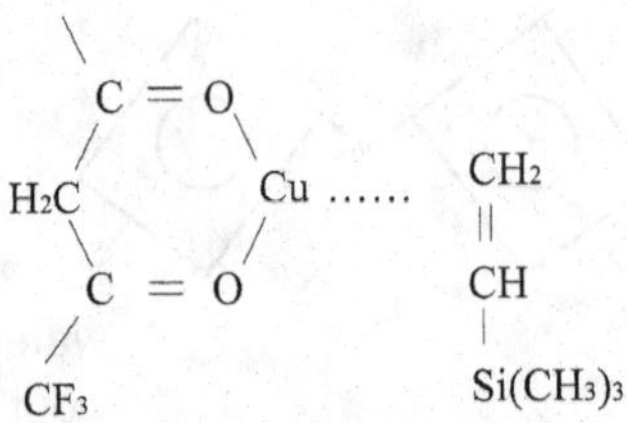

```
   CF3
     \
      C = O
     /     \
  H2C       Cu ......   CH2
     \     /             ||
      C = O              CH
     /                   |
   CF3                 Si(CH3)3
```

Fig. 12.9. The molecular structure of Cu(hfac)(tmvs).

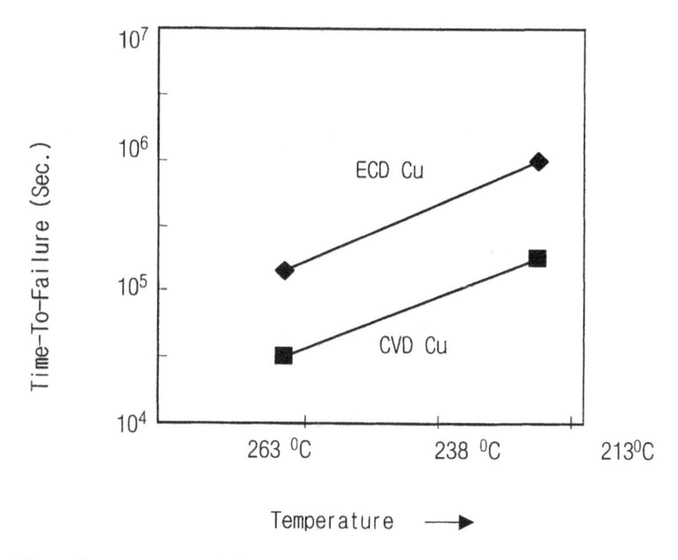

Fig. 12.10. Comparison of electromigration time to failure between CVD and ECD copper.

imposing voltage across two electrodes, through a solution that contains the metal ions. The metal ions are reduced and deposited onto the cathode surface.

Before introducing the electroplating of Cu, let us recap some fundamental concepts in electrochemistry. Consider dipping a metal rod into a solution that contains metal ions. For example, when a Cu rod has been dipped into a $CuSO_4$ solution, two reactions occur in opposite directions as soon as the rod is dipped into the solution. One reaction occurs when Cu atoms oxidize and dissolve into the solution, leaving the electrons on the metal rod surface. The metal rod is negatively charged and is surrounded by positive ions. The other reaction takes place when the copper ions deposit onto the metal rod surface. The metal rod loses two electrons to each attaching Cu^{+2} ion:

$$\begin{aligned} &\mathrm{Cu} \longrightarrow \mathrm{Cu}^{+2} + 2\mathrm{e} \\ &\mathrm{Cu}^{+2} + 2\mathrm{e} \longrightarrow \mathrm{Cu}\,. \end{aligned} \tag{12.6}$$

When the equilibrium is reached, the amount of negative charges left on the metal rod varies with different materials. For example, at equilibrium, magnesium has more negative charges on the rod than copper. The Nernst equation states that the metal ion concentration of the solution affects the metal electrode potential as

$$E = E_0 + \frac{RT}{zF} \ln a \,, \tag{12.7}$$

where R is the gas constant, F is the Faraday constant, z is the number of electrons involved in the reaction, and a is the activity of the metal ions, which can be approximated with molar concentration of the metal ions in a diluted solution. At room temperature, 25°C, the equation can be simplified to

$$E = E_0 + \frac{0.0591}{z} \ln a \,. \tag{12.8}$$

Obviously, E_0 is the standard potential of a metal electrode when its ion activity in the solution is equal to 1.

It is of little use to measure each individual metal electrode potential. They are all measured relative to a reference electrode. A standard hydrogen electrode is often used as the reference electrode, which is set to zero, as shown in Fig. 12.11. The hydrogen electrode is formed by bubbling hydrogen at 1 atm and 25°C through an inert electrode. With this setup, the voltage of different electrodes can be measured. Table 12.3 shows the standard electrode reduction potential at 25°C and 1 atm for various electrodes. The half reactions, with their corresponding standard potentials displayed from negative to positive values, indicate increasing strength for being

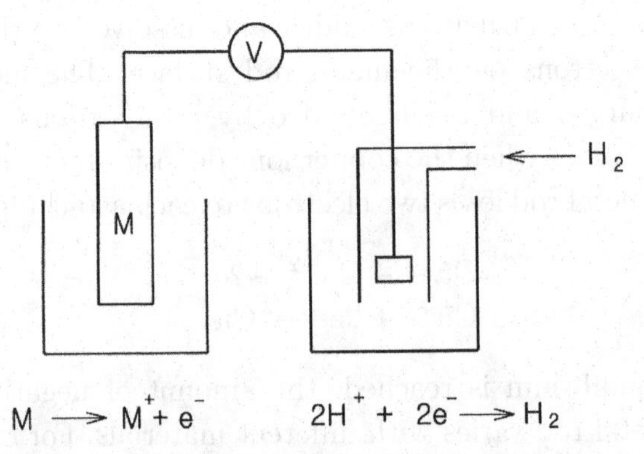

Fig. 12.11. Hydrogen electrode is used to measure the standard potential of metals or other materials.

Table 12.3. Standard electrode potential of various half reactions with solution activity of 1 at standard condition.

Half reactions	Std. potential (v)
$1/3\ Li^{+3} + e^- \rightarrow 1/3\ Li$	-3.05
$1/3\ Al^{+3} + e^- \rightarrow 1/3\ Al$	-1.71
$1/2\ Cr^{+2} + e^- \rightarrow 1/2\ Cr$	-0.71
$1/2\ Ni^{+2} + e^- \rightarrow 1/2\ Ni$	-0.23
$1/2\ Zn^{+2} + e^- \rightarrow 1/2\ Zn$	-0.76
$1/3\ Fe^{+3} + e^- \rightarrow 1/3\ Fe$	-0.05
$H^+ + e^- \rightarrow 1/2\ H_2$	0.0
$1/2\ Cu^{+2} + e^- \rightarrow 1/2\ Cu$	$+0.34$
$Hg^+ + e^- \rightarrow Hg$	$+0.8$
$1/2\ Cl_2 + e^- \rightarrow Cl^-$	$+1.37$
$1/2\ F_2 + e^- \rightarrow F^-$	$+2.8$

oxidizing agents. Now, any of the two half reactions can be combined to form an oxidation–reduction reaction. If a half reaction is reversed, the sign will also be reversed. For example,

$$2e^- + Cu^{+2} \longrightarrow Cu(s) \quad +0.34\,V \tag{12.9}$$

$$Zn(s) \longrightarrow Zn^{+2} + 2e^- \quad +0.76\,V\,. \tag{12.10}$$

The complete oxidation–reduction reaction can be shown as the following:

$$Cu^{+2} + Zn(s) \longrightarrow Zn^{+2} + Cu \quad +1.10\,V\,. \tag{12.11}$$

A positive voltage of the resulting reaction indicates that the reaction can proceed spontaneously. Of course, applying an external power can force a nonspontaneous reaction to proceed.

For copper electroplating, the cell design is set up with Cu as both the anode and cathode immersed in a copper sulfate solution. Cu ions naturally deposit on any cathodic solid surface. The external power is essentially to drive the anode Cu to dissolve into the solution, supplying Cu ions into the solution so that the plating process can be run continuously. If one immerses two Cu electrodes into a copper sulfate solution and passes a direct current through the electrodes, as one slowly increases the voltage, Cu will start to deposit onto the cathode. The potential at which the metal ions start to plate out is called deposition potential. The deposition potential depends on the material, concentration, and operating conditions. The total amount of Cu deposited onto the cathode surface can be estimated from Faraday's law:

$$W = \frac{It}{F}\frac{A}{z}, \tag{12.12}$$

where W is the accumulated weight of Cu deposits, I is the current density, t is the total elapse time, A is atomic weight of the depositing metal atom, and F is the Faraday constant (96 500 coulombs/gmole; z, the charge number of the metal ions). Assuming a uniform film thickness across the plated area, the resulting film thickness, T, can be calculated with the following equation:

$$T = \frac{W}{a\rho}, \tag{12.13}$$

where a is the plated area and ρ is the deposited film density. In an electroplating process, not all of the electrical current is utilized to deposit material on the cathode. Some of the current is wasted, for example, to heat the bath or discharge hydrogen or other impurity ions. The plating current efficiency is defined as η:

$$\eta = \frac{W}{W_c}, \tag{12.14}$$

where W is the weight of actual deposited metal and W_c is the theoretical weight of the metal.

According to Faraday's law, a plater can increase the current density and speed up the metal deposition rate. To examine if this concept can be applied to a plating bath, let us take a look at the plating mechanism. As shown in Fig. 12.12, the hydrated ions are transferred from the bulk to the cathode surface via migration, convection, and diffusion. The migration is due to ion motion under the influence of an electrical field. The convection results from fluid flow, nonuniform bath temperature, or bath agitation. Diffusion results from the ion concentration gradient. Diffusion dominates the mass transfer within the phase boundary between the electrode and electrolyte. Once the ions reach the cathode, dehydration occurs before

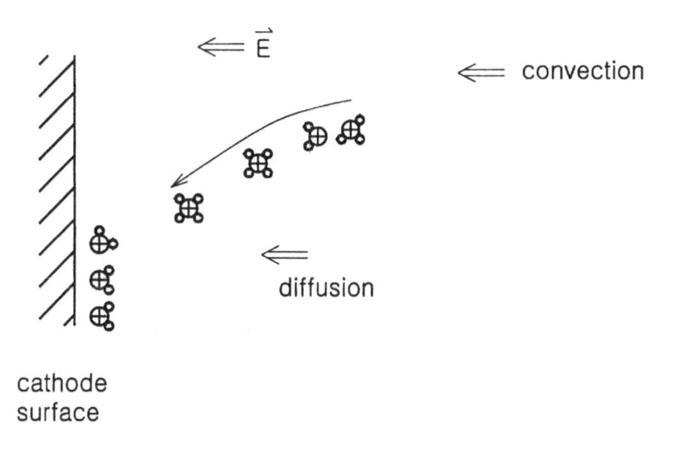

Fig. 12.12. The mass transfer of ions from bulk to the cathode surface is driven by migration, diffusion, and convection.

the ions are adsorbed on the cathode surface. The ad-ions take electrons from the cathode to form ad-atoms, which can migrate around on the surface. On reaching energetically favorable positions, they coalesce into a film. When the voltage to the plating bath is increased, the current will increase rapidly. Up to a certain point, further voltage increase does not increase the bath current. Initially, the current increase is due to a higher potential that drives a higher ion flux toward the cathode surface to be deposited (consumed). These ions must then be supplied from the bulk of the solution. As the scenario reaches the diffusion limitation, a leveling-off trend in bath current is observed. The current in this region is called the limiting current. Beyond the flat region, further increase in potential will cause an increase in the bath current again. This is due to the fact that some impurity ions start to deposit or hydrogen starts to discharge. In other words, the current efficiency will start to drop. A proper plating operation should keep the current below the limiting current to obtain high current efficiency and good film quality.

12.2.4. *The ECD system*

A schematic of Cu electroplating cell design is shown in Fig. 12.13. The electrolytes flow upward toward the wafer surface

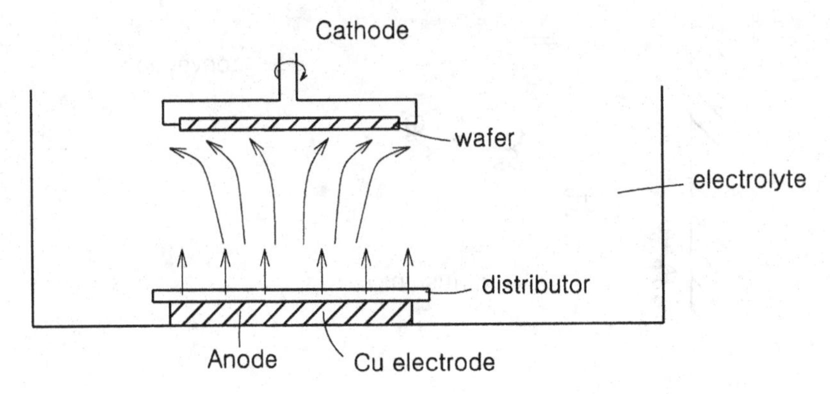

Fig. 12.13. Schematic of a copper electroplating cell.

perpendicularly. The wafers are attached to the cathode. The flow distribution must be uniform to achieve better deposited Cu film thickness uniformity. Cu is deposited on the cathode from the Cu on the anode via the copper sulfate solution. An external power is applied to complete the electrical current flow. With technologies beyond 0.13 μm, extreme requirements are imposed on the plated Cu film properties, including adhesion to underlying layers, film thickness uniformity, grain sizes, crystal orientation, chemical purity, conductivity, and stress. Several factors, such as plating current density, copper sulfate concentration, sulfuric acid concentration, additive concentration, and plating bath temperature, can influence the film deposition rates and film textures. In the absence of side reactions, the amount of deposition is proportional to the current density, as estimated from Faraday's law. However, typical plating current is kept at about 30–50% of the limiting current to avoid undesired film characteristics. An increase in the bath temperatures boosts the ion supply to the cathode, and issues related to insufficient ion supply can be mitigated. High bath temperatures also increase the nuclei growth rate and result in larger grains. However, it favors hydrogen discharge and leads to metal hydroxide precipitation on the cathode, which may be included in the film and cause poor film quality.

The plated film morphology is largely influenced by the electrode polarization. The higher the polarization, the smoother the film surface will be. The electrode polarization results when the cupric ion

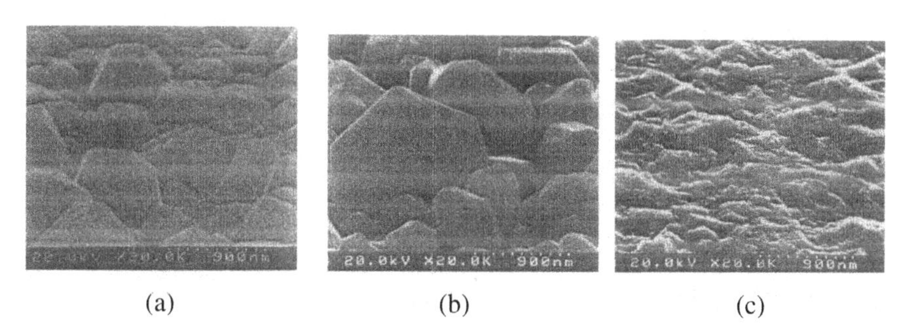

Fig. 12.14. Plated copper grain sizes decrease with increasing plating current and decreasing cupric sulfate concentration: (a) 1.6 A/dm sq., 60 g/l; (b) 2.4 A/dm sq., 150 g/l; and (c) 4 A/dm sq., 60 g/l (Gau *et al.*, 2000).

consumption due to electrode surface reactions is faster than the supply from the electrolyte. Therefore an increase in the current density by raising the applied voltage boosts the surface reaction rate and makes the cupric ions undersupplied, and thereby a polarization effect is observed. In general, a smooth Cu film, specifically a low resistive film, can be obtained by increasing the current density, as illustrated in Fig. 12.14, or by decreasing the copper sulfate concentration. Terminal effect has been found in copper plating processes. This effect is due to the fact that the plating current is fed through a contact ring around the wafer edge and is spread across the whole wafer through a thin copper seed layer. The highly resistive seed layer causes the current to be larger at the wafer's edge compared to that at the center. The terminal effect diminishes as the film thickness grows. To mitigate the terminal effect, the seed resistance can be reduced, or the electrolyte resistance can be increased.

The most challenging aspect of the Cu electroplating process is filling up the deep trenches or holes with no electrolytes trapped within. To fulfill this requirement, overhangs or conventional conformal deposition are not appropriate in the case of Cu deposition since the seam and void may trap electrolytes or impurities, as shown in Fig. 12.15. The preferred filling is shown in Fig. 12.15(c), which provides a bottom-up filling and eliminates electrolyte trapping. This filling mechanism can be achieved by including some organic or inorganic additives in the plating bath. These additives are primarily

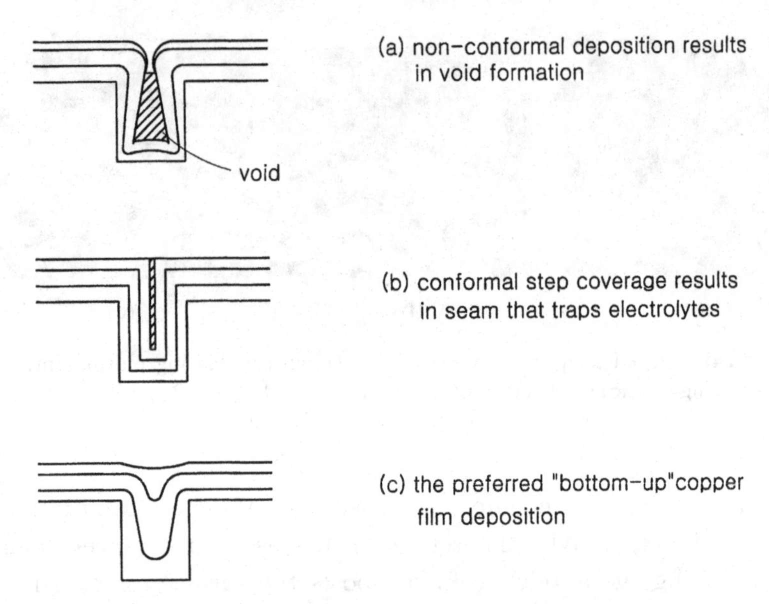

Fig. 12.15. Different film growth schemes. The preferred scheme for copper ECD is the bottom-up film growth.

organic compounds with various functional groups. Some organic additives require the existence of inorganic ions to perform properly. During the plating process, these additives are consumed by means of being incorporated into the films or breaking down and undergoing various side reactions. To achieve consistent production processes, the concentration of these additives must be kept constant. In other words, the consumed chemicals must be replenished with fresh chemicals continuously.

12.2.5. *The additives for the ECD bath*

The additives can be broadly categorized into two groups: the suppressors and the accelerators. The suppressors slow down the cupric ion adsorption rate on the cathode surface, while the accelerators enhance it. Suppressors are generally polymeric surfactants and can be further divided into carriers and levelers. Carriers adsorb on the cathode surface and form a monolayer, and thereby they offer a diffusion barrier to the cupric ions. As a result, this increases the cathodic polarization more suitably for fine grain deposition. Carriers often require the presence of chloride ions to perform properly.

Leveling is defined as the decrease in surface roughness while the plating process is taking place. Consider the diffusion of ions from the bulk of the solution to the depositing surface; they take longer to reach the recessed areas, and therefore they result in thinner deposition in those areas. Leveling facilitates an essentially flat, leveled surface after the deposition. Leveling is not a natural process, but it can be achieved with the aid of leveling agents or levelers. The leveling agent molecules, which are generally multiply charged, diffuse toward the surface and preferentially adsorb on the protruded sharp corners or edges, and thereby they inhibit the ion deposition. Therefore the growth rates on those protrusions are slower compared to those in the flat areas. As a consequence, the deposition surface can be leveled, as demonstrated in Fig. 12.16. Furthermore, the bulky leveler molecules impede their diffusion into small trenches or holes, which would otherwise increase the nonconformal deposition.

Accelerators generally are unsaturated compounds with groups that contain sulfur, oxygen, or nitrogen. One of the functions of accelerators is that they are used as brighteners, which facilitate brighter deposits. The mechanism that makes accelerators work is once again adsorption. As depositing ions arrive at the cathode, the brightener molecules quickly surround the ions, preventing them from coalescing with earlier adsorbed ions; instead, they create other nucleation sites. Consequently, fine-grained film is obtained, which looks bright. Accelerators can also decompose into intermediate hydrated molecules, which adsorb on the Cu film and provide more

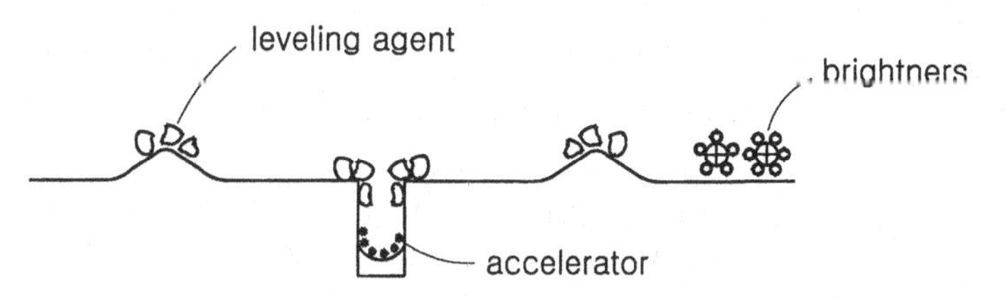

Fig. 12.16. Additives provide bottom-up film growth in trenches and uniform fine-grain copper film.

active sites for Cu adsorption, or they reduce the effect of the suppressor. In deep recessed features, as demonstrated in Fig. 12.16, the smaller accelerator's intermediate molecules adsorb on the growing Cu surface. On a concave surface, as the film grows, the surface area decreases; therefore the effective concentration (or surface coverage) of the smaller molecules increases, which causes further increase in the Cu grow rates. This renders the bottom-up growing mechanism in deep recessed areas or features. On a convex surface, the trend is reversed. Although most of the interactive reaction mechanisms are not clearly understood yet, it is known that the additives function synergistically. Engineers often need to heuristically optimize the plating operating conditions with each specific plating solution.

12.2.6. *Post-ECD anneal*

As-deposited ECD Cu film has a fine grain structure and is unstable due to excessive vacancies, dislocations, or other forms of defects. The as-deposited film recrystallizes spontaneously to eliminate such defects and releases excessive energy. The as-deposited Cu film can undergo self-annealing. The film can also be annealed by using external energy sources, such as furnace or rapid thermal annealing. With annealing, the nanocrystals grow in size, the grain boundaries and resistivity decrease, and the film stabilizes. The self-annealing proceeds naturally; the annealing rates increase with increasing film thicknesses. A thicker film takes a short time to complete the self-annealing, while a thin film can take a very long time.

For production efficiency and consistency, the annealing must proceed within a short time and in a controllable manner. Furnace annealing can be conducted with a large temperature range; the higher the temperatures, the larger the grain sizes will be. Different seed materials produce different initial grain sizes, but they end up with nearly the same annealed grain sizes. Another annealing approach is by using RTA. Certainly, the heating across the wafer must be very uniform to avoid thermal-induced stress, which may cause cracking in the wafer. Furthermore, if there is any thermal-induced film structure nonuniformity, it shows up in the CMP rate and unpredictable reliability. The annealed copper film structure can be feature-size-dependent. Copper films in shallower trenches requires more thermal energy to stabilize than those in deeper trenches. Owing to its short cycle time and consistent quality, RTA annealing is most often chosen for Cu anneal in volume production.

12.3. Low-*k* Dielectrics

As discussed earlier, *RC* delay increases with shrinking device feature size. One approach to reduce the *RC* delay is to use low-resistivity conducting materials such as copper. Another approach is to use low-*k* dielectrics, which directly reduce the capacitance. The material can be deposited with either PECVD or the spin-coating approach. Low-*k* dielectrics in general have inferior mechanical, thermal, and electrical properties when compared to thermal or CVD silicon dioxides. Ideal low-*k* dielectrics should posses the following characteristics:

(a) low dielectric constant of smaller than 4.2, which is the *k* of silicon dioxide
(b) good adhesion to metals, to avoid delamination between barrier metal layer and low-*k* materials, especially when the ambient temperature varies
(c) high thermal stability, to avoid cracks at high processing temperatures
(d) good thermal conductivity so that the heat generated as a result of power dissipation can be conducted away, to avoid device overheating
(e) low moisture adsorption: Moisture or moisture reacting with the ingredients of the low-*k* materials can form corrosive agents that attack the metal lines of Al or Cu
(f) good gap filling capability, to achieve good planarization

The dielectric constant is the medium permittivity to vacuum permittivity ratio:

$$k = \frac{\varepsilon}{\varepsilon_0}. \qquad (12.15)$$

The dielectric constant of a material indicates its capability for storing static charges between two parallel electrode plates. The space full of gases or air is very close to the vacuum. According to Debye's equation, for a given material or mixture, the dielectric constant relates to physical constants as

$$\frac{k-1}{k+2} = \frac{\rho P}{M}, \tag{12.16}$$

where k is the material's dielectric constant, ρ is the material density, P is the polarizability, and M is the molecular weight. Polarizability is the proportionality constant between the induced dipole moment and the electrical field. Obviously, to obtain a low dielectric material, one has to somehow reduce the material's density or polarizability. Starting with a host material, one can change the polarizability by doping the material with foreign atoms. On the other hand, the material density change can be done by introducing some bulky organic functional groups into silicon oxide. It can enlarge the free volume of the oxide; this is because the molecular packing density decreases. Alternatively, it can be accomplished by purposely creating pores in the material bulk.

FSG (SiO_xF_y), fluorine-doped silicate glass, is probably the first low-k dielectric material that was used in production, starting from the 0.13-μm technology node. Either HDP or PECVD methods can be used for deposition. For HDP CVD, SiF_4, silane, or TEOS can be used for reaction with CF_4. For PECVD, TEOS or silane can be used for reaction with C_2F_6, NF_3, or SiF_4. Reactions take place between 200°C and 400°C. Fluorine replaces some of the oxygen atoms and results in forming Si–F bonds, which have lower electrical polarizability, thereby giving rise to lower dielectric constants. FSG typically has a dielectric constant of about 3.6, compared to 4.2 for silicon dioxide. The dielectric constant for fluorine-doped oxide decreases with increasing fluorine concentrations. However, fluorine concentration should be kept below 5% to maintain film stability. The major drawback of FSG is that the F in FSG can react with moisture and lead to HF acid. HF can corrode Ta or other metal layers in the Cu metallization scheme.

Carbon-doped silica (CDS) glass is formed by introducing the alkyl group (CH_3-) into silicon oxide. Because the alkyl group is bulky and creates free volume, it thereby reduces the material density. Low material densities imply low k values. Therefore, as more and more methyl groups are incorporated into the film, the film density decreases, and so does the film dielectric constant. CDS can be obtained by using different techniques such as spin on dielectrics (SOD), CVD, or PECVD. One example of the SOD techniques is the methyl silses quioxanes (MSQ) formation via polymerization of methyltrialkylsilane, $CH_3Si(OH)_3$. MSQ has a k value of 2.7–3.1, and it is stable up to 475°C. Other carbon-containing chemicals, such as methylsilane, trimethylsilane (3MS), or tetramethylsilane (4MS), can be oxidized by N_2O or hydrogen peroxide to form CDS.

Some inorganic and organic polymers have been extensively studied and commercialized as low-k dielectric materials. Inorganic materials have O–Si–O backbones. One of the commercial examples is hydrogen silses quioxanes (HSQ), a siloxane-based polymer in which silicon is directly attached to hydrogen and oxygen, as shown in Fig. 12.17. The silicon at either end contains an hydroxyl group. On annealing, two Si–OH groups undergo condensation reactions to form Si–O–Si. After 400°C curing, the film ends up with a dielectric constant of 3.0. One good commercial example of an organic polymer is a cross-linked polyaromatic polymer; after heat treatment, its k value can be as low as 2.65, and it is stable up to 400°C.

Materials with k values of smaller than 2.5 are called ultra-low k materials. The best approach to achieve ultra-low k is to generate pores in the materials. Porous low-k materials are generally

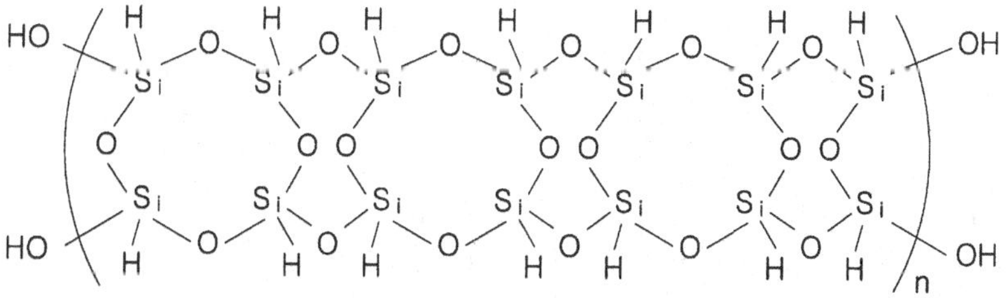

Fig. 12.17. The ladder structure HSQ.

obtained with a spin-coating technique, followed by a baking step around 350°C–400°C. How can the porosity of a material change its dielectric constant? Let us look at a material with a starting k value of k_{solid}. As porosity p is generated in the material, the Matthiesen's model provides an approximation of the effective modulus as

$$E_{\text{eff}} = (1 - p)E_{\text{solid}} + pE_{\text{air}}\,, \tag{12.17}$$

where E_{eff} is the predicted modulus of the porous material and E_{solid} is the modulus of the starting material. For predicting the effective dielectric constant of a porous material, the model becomes

$$k_{\text{eff}} = (1 - p)k_{\text{solid}} + pk_{\text{air}}\,. \tag{12.18}$$

It can be observed that the effective dielectric constant of a porous material depends on the k value of the starting material, k_{solid}, and porosity. Since 1 is the lowest possible k value for all media, the higher the porosity, the lower the dielectric constant will be. However, the porous material's thermal conductivity, hardness, and elasticity also decrease with increasing porosity. In view of these characteristics, to achieve a certain low k, better material properties can be obtained when starting with a low-k material compared to a high-k material. For example, one needs to create 60% porosity in a solid for a k of equal to 4.0 to obtain an effective k of 2.2. With such a high porosity, it is difficult to maintain good material properties. However, 25% porosity is needed when starting with a material with a k of equal to 2.6; namely, starting with a low-k material often results in an ultra-low-k material with more acceptable properties. One more novel example of creating porosity is TEOS reacting with water:

$$(C_2H_5O)_4Si + 2H_2O \longrightarrow SiO_2 + 4C_2H_5OH\,. \tag{12.19}$$

The product is silica with ethanol embedded in it. If the product is left dried naturally, it forms a high-density aerogel, which is called xerogel, and it contains 60–90% air. If the product is dried at high pressures and temperatures (supercritically), it forms aerogel, which contains 90–99% air in the silica. Proper control over starting material and porosity allows one to obtain various porous silicas with effective dielectric constants ranging from 3.0 to nearly 1.0. Low-k materials are generally dry and porous. Caution should be taken to prevent them from absorbing moisture since moisture leads to an increase in dielectric constant and device reliability concerns.

12.4. Integration of Copper and Low-k Materials

Integrating copper and low-k dielectric materials poses far more challenges than the conventional aluminum and silicon oxide processes. Copper is a fast diffuser in oxide. Once it gets into silicon, it forms deep trapping centers and fails device functions. Therefore copper interconnects or plugs must be completely sealed from the rest of the structures or layers. As discussed earlier, a conventional etching approach cannot be employed to form copper interconnects as copper compounds are relatively nonvolatile at low temperatures. Copper interconnects are formed with damascene and CMP processes and they are often integrated with low-k materials. Low-k materials are more vulnerable to plasma and wet etching chemical attacks. Furthermore, it may crack during the ensuing CMP or annealing steps. Caution must be taken to successfully integrate the copper and low-k materials.

In manufacturing, the damascene process is often carried out in the form of dual damascene, that is, to form the trench and via simultaneously. With this approach, there are two options available: the via-first approach and the trench-first approach. Silicon nitride film is extensively used in the low-k damascene process since it provides protection for the low-k materials against harsh process conditions such as plasma etching or ashing. It can also act as an etching stopper with respect to low-k dielectrics etching. Figure 12.18 shows the process flow for the trench-first approach. The trench is first patterned and etched, which stops at the middle nitride layer. Then, the second

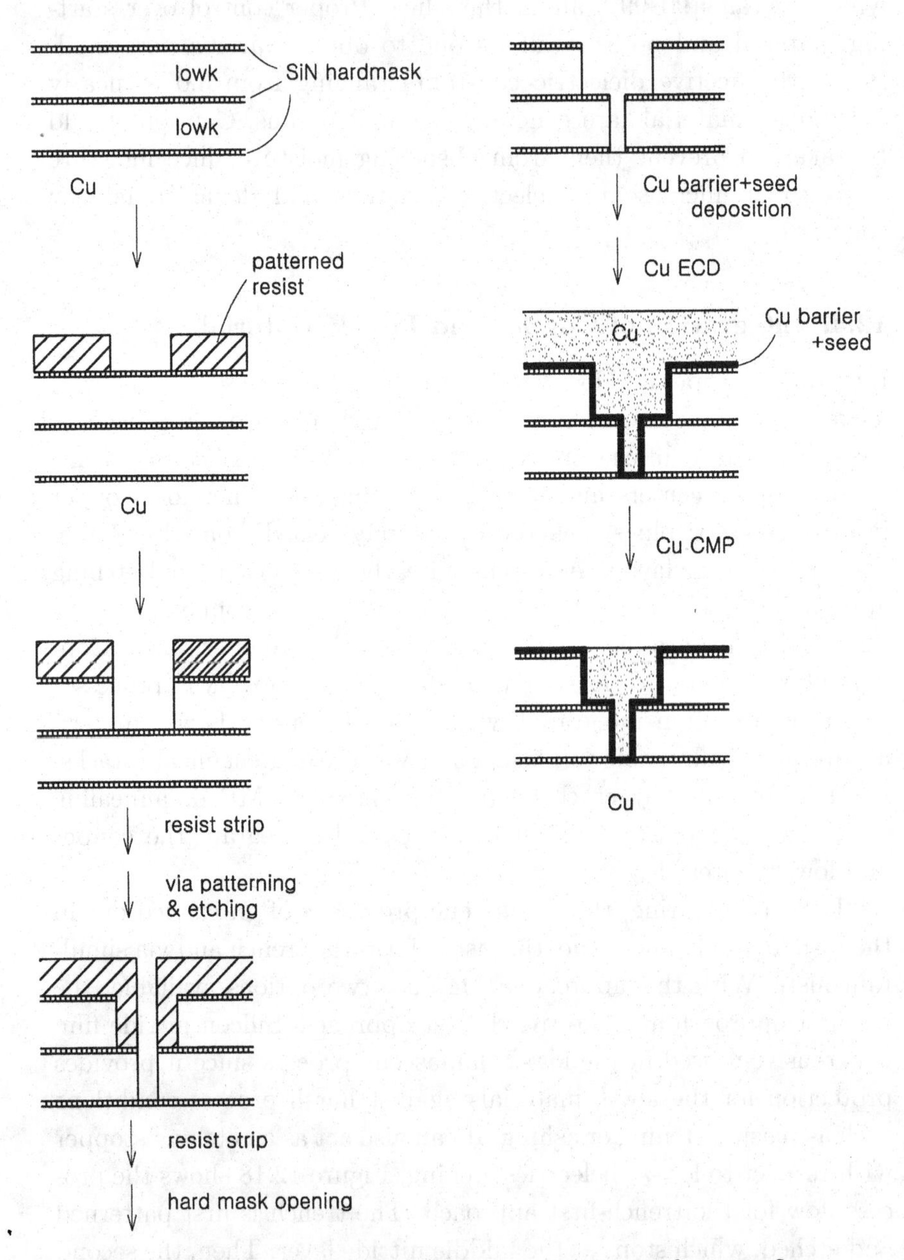

Fig. 12.18. Trench-first dual damascene structure and process flow.

patterning is accomplished by defining the via within the trench. Via etching is carried out, and it is stopped right at the top of the bottom nitride hard mask layer. Several issues can be noted in this situation. First, the resist is quite thick when the via in the trench is being defined. This makes lithography difficult since a thick resist leads to poor resolution, especially when defining small via sizes. Second, the via etching should stop at the bottom nitride surface and switch to a milder recipe so as not to sputter out the underlying copper atoms, which may cause massive contamination. Third, during the resist strip, either with ashing or wet stripping, the recipes or chemicals must be properly selected so that they will not attack the fragile low-k material on the sidewall. The copper barrier (typically Ta/TaN) and seed layer are sputtered into the trench and via using the IMP approach. It is important to have a uniform thickness across the bottom and the sidewall; otherwise, copper can diffuse readily in the dielectric bulk and cause contamination. On the other hand, if the barrier does not completely seal copper, the copper can diffuse along the copper lines and squeeze out at the barrier opening area, and thereby cause failures. ECD copper is deposited onto the structure with significant amounts of overgrowth to ensure that holes are properly filled. Finally, the copper CMP is carried out. At this point, special attention should be paid to the shear force that is exerted from the pad on the wafer since the low-k material is fragile. Excessive pressure or shear force can lead to cracks.

The other damascene approach is the via-first approach, as indicated in Fig. 12.19. Via is defined and etched through the bulk of the low-k dielectric, and it stops at the bottom nitride surface. Again, to prevent copper contamination, the bottom nitride remains. A second photolithography is employed to open up the trench area and stop at the middle nitride surface. It should be noted that some resist is left behind to cover the via hole, which is beneficial. It protects the sidewall low-k material and the bottom nitride. After the damascene structure is formed, the bottom nitride is opened up, and the copper deposition process is carried out and followed by a copper CMP. With the issues involving the trench-first approach,

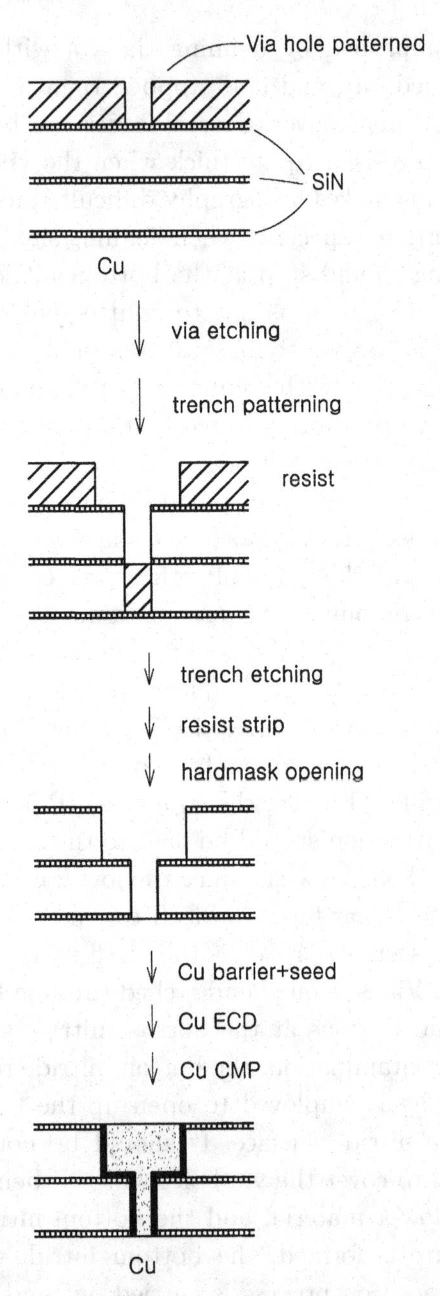

Fig. 12.19. The via-first dual damascene structure and process flow.

as discussed earlier, the via-first approach seems to gain more popularity in production. Furthermore, as the k values of the dielectrics continue to be driven lower, the nitride cap tends to be replaced with SiC, SiCN, or SiCOH films for low total effective dielectric constants.

Glossary

acceptor a dopant that has three valence electrons; it creates a hole when covalent bonding with neighboring silicons.

ambipolar diffusion model an assumption stating that in a plasma system, the electrons and ions move collectively with the same diffusivity.

anisotropic etching an etching process that is characterized by having a preferential direction in which the etching proceeds.

anti-punch-through (APT) implant a counter doping implant placed beneath the transistor channel region to improve the device window.

antireflective coating (ARC) a coating on top of a shining substrate, such as Al, to reduce its reflectivity.

atmospheric pressure CVD (APCVD) a CVD process that operates near atmospheric pressure.

barrier metal a thin layer of metal, such as Ti, TiW, or TiN, used underneath the Al to ensure the contact integrity.

bird's beak length (BBL) the length of the oxide underneath the nitride edge after the local oxidation.

binary mask a mask pattern consisting of opaque and transparent areas, as defined using Cr absorber.

borophosphorous doped TEOS (BPTEOS) TEOS oxide film doped with B and P for lowering its flow temperature.

borophosphosilicate glass (BPSG) a silicon dioxide doped with B and P.

capacitively coupled plasma plasma that is generated with a parallel plate electrode configuration.

carbon doped silicate glass (CDS) C-doped silicon oxide, having a dielectric constant raging from 2.7 to 3.1.

charge to breakdown (QBD) an oxide quality test methodology that basically tests the total charges that an oxide can stand before breaking down.

chemical amplified resist (CAR) a resist that involves acid catalytic reactions in its deprotection step.

chemical mechanical polishing (CMP) a material removing process using synergistic chemical and mechanical actions.

chemical vapor deposition (CVD) a gas mixture passes over a heated substrate, and chemical reactions are initiated in the vicinity of the substrate surface to form a nonvolatile solid film.

cold-wall CVD reactor a CVD reactor in which the reactor wall temperature is lower than wafer temperature.

collimated sputtering a sputtering process with a collimator that filters the randomly traveling atoms, leaving only the vertical flux onto wafer surface.

complementary MOS (CMOS) circuits a circuit that consists of both NMOS and PMOS transistors.

critical dimension (CD) the dimensions of critical portions of a circuit.

Cu(hfac)(tmvs) a chemical source for Cu CVD.

Czechralski crystal growth a widely used process for growing silicon ingots, which are then made into wafers.

damascene process a trench is formed first and filled with metals, such as Cu, which are then polished with CMP to form metal interconnects.

deionized water (DI water) water that contains no ions.

depth of focus (DOF) the vertical distance within which the imaging quality is acceptable.

dimethyl aluminum hydride (DMAH) a precursor for CVD aluminum.

diazonaphthoquinone (DAQ) a type of photoactive compound.

die-to-database inspection a mask inspection that compares the transmitted image of a die to its own processed design image.

die-to-die inspection a mask inspection that compares the transmitted images of two neighboring dies.

diffusion a process that introduces the precursor dopants to the substrate surface; this is followed by a drive-in diffusion at high temperature to allow the dopants to spread isotropically into the substrate bulk.

dissociation a molecule being impacted by an energetic electron and dissociated into radicals.

donor a dopant that has five valence electrons; it provides an extra electron when covalent-bonding with neighboring silicons.

doping a process of adding various atoms into a silicon substrate to alter its resistivity and type.

double-level metal (DLM) a device with two levels of metals to increase its integration level.

drive-in diffusion the second step of a diffusion process in which the predeposited dopants are further driven into the substrate.

dry oxidation silicon is oxidized with oxygen or nitrous oxide.

dual passivation a passivation layer composed of PECVD oxide–PECVD nitride layer structure.

dummy patterns patterns that are not electrically active in the circuit.

dynamic random access memory (DRAM) a type of memory chip; its stored information disappears if the power is turned off.

e-beam the electron beam that is used to expose resist-coated blanks to make masks.

electrical field to breakdown (EBD) an oxide quality test approach that basically tests what electrical field the oxide can stand without experiencing breakdown.

electrically programmable read-only memory (EPROM) a type of memory chip; its stored information does not disappear when the power is turned off, and memory can be programmed electrically; the stored information can be erased with UV exposure.

electrochemical deposition (ECD) a deposition process occurring in an electrolyte involving electrical and chemical actions.

electrodeless deposition a deposition process occurring in a solution using reducing agents in the absence of electricity.

electromigration material movement in a conducting line caused by the electrical current.

electron impact cross section a quantity that describes the probability of an electron impact that results in a certain event occurring; the event could be ionization, dissociation, excitation, and so on.

etchants the etching chemicals.

etching selectivity the etching rate ratio of two materials.

fan filtering unit (FFU) an apparatus used in semiconductor clean room for air circulation.

Faraday's law an equation stating that the total metal deposits in ECD are proportional to the total electrical charges passing through the solution times the atomic weight divided by the charges of the ions.

Fick's law the diffusion law stating that the species diffusion flux is proportional to the species concentration gradient.

field oxide a thick oxide island grown on silicon surface to isolate a device from its neighboring devices.

fluorine-doped silicate glass (FSG) F-doped silicon oxide having a dielectric constant around 3.6.

focus ion beam (FIB) repair a mask defect repair that uses a focused ion beam.

free radicals atoms with unpaired electrons that are chemically active and can initiate high-temperature reactions at low ambient temperatures.

front-end device manufacturing the device manufacturing flow from silicon oxidation to the step before contact formation.

gas-assisted etching (GAE) an etching reaction used for removing defects on masks.

global planarization a planarization technique that provides a long-range planarization over patterns of all pitches.

glue layer a thin metal layer (TiN or TiW) used between tungsten and the underlying oxide to improve tungsten's adhesion on oxide surface.

HALO implant a counterdoping approach that places the dopants of opposite type around the source and drain junctions with ion implants.

hard mask a film, rather than a resist, that is used for an etching process in which the etchant does not attack the hard mask.

hexamethyldisililazane (HMDS) a chemical that promotes the resist adhesion on the silicon surface.

high-density plasma (HDP) CVD a type of PECVD with magnetic field enhancement such that the ion flux toward the substrate surface is higher than the radical flux.

hot-wall CVD reactor a CVD reactor in which the reactor wall temperature is close to the wafer temperature; hot-wall reactors are often used for diffusion and low-temperature CVD.

inductively coupled plasma (ICP) a plasma that is generated inductively.

integrated circuit (IC) a circuit that is composed of various components, such as resistors, capacitors, junctions, transistors, and interconnects, to fulfill designated functions.

integrated device manufacturer (IDM) a semiconductor company that has its own product line, including product definition, design, wafer manufacturing, and final chip, device, or system delivery.

inter layer dielectrics (ILD) dielectric materials used in the front end (before contact formation) of device manufacturing.

inter metal dielectrics (IMD) the dielectric films used in a device between metal layers.

intrinsic semiconductor a semiconductor material without any impurities.

ion implant a doping process in which the dopants are ionized and then accelerated to impinge onto the substrate surface and penetrate into desired depths.

ionized metal plasma (IMP) sputtering a sputtering with sputtered atoms being ionized to enhance the vertical flux.

isopropyl alcohol (IPA) a chemical widely used in semiconductor factories for cleaning.

isotropic etching an etching process that is characterized by having equal etching rates in all directions.

large-angle tilted implant drain (LATID) an implant with a large angle, at around 30° to normal direction, for optimizing a transistor.

Larmor frequency the frequency at which a charged particle circles around the influencing magnetic field lines.

laser repair mask defect repair that uses a laser beam.

lightly doped drain (LDD) a device source–drain structure in which the junction concentration profile is graded, instead of being sharp.

local oxidation (LOCOS) a local oxidation approach to form field oxide.

local planarization a planarization technique that provides a short-range planarization over certain pitches.

low-pressure CVD (LPCVD) a CVD process that operates under low pressures to increase the species' diffusivities.

low-K material material with a dielectric constant smaller than silicon oxide.

majority carrier the carrier in a doped semiconductor that is larger in population.

mask critical dimension uniformity (CDU) the uniformity of the critical dimensions on a mask layer, often expressed in terms of range (max–min) or 3σ.

mask error enhancement factor (MEEF) the CD difference on the mask divided by the CD difference on the wafer divided by the reduction ratio of photolithography.

mask (photomask) a quartz blank with patterned chrome that can be printed onto a wafer surface; a complete IC design consists of n layers of masks.

mean free path the average distance that a particle travels without colliding with others.

metal oxide semiconductor (MOS) a device made on silicon surface that is composed of a source, drain, and a gate oxide/gate structure above the channel area.

methyl silses quioxanes (MSQ) carbon-doped silicon oxide.

Miller indices an index used for identifying crystal orientation.

minority carrier the carrier in a doped semiconductor that is smaller in population.

molecular beam epitaxy (MBE) a non-CVD epitaxy growth using an evaporation approach.

negative resist a type of resist for which the exposed area of the resist turns insoluble in developing solution.

next generation lithography (NGL) photolithography beyond 157 nm wavelength.

NMOS an MOS transistor with its source and drain doped with n-type dopants.

n-type dopant dopants with five valence electrons.

off-axis illumination (OAI) photolithography using nonnormal incident light for exposure.

optical proximity correction (OPC) a technique of altering the mask patterns to obtain good pattern transfer fidelity between the designed patterns and the printed patterns on wafers.

optical proximity effect (OPE) a phenomenon in which the printing CD of a feature depends on its proximity.

optical transfer function (OTF) an approach that relates the input of an optical system to its output in the frequency domain.

overlay the alignment between two related layers.

particle inspection a mask inspection that detects the particles on the mask by using the information of transmitted and reflected light.

passivation the final dielectric layer or protection layer on a device.

pellicle an organic film used to seal the Cr pattern such that particles will not fall on the mask surface, causing defects.

phase shifting mask a mask using a phase shifting material, such as $MoSiO_x$, to form patterns; the shifter enhances the resolution by phase shifting.

phosphorous-doped silicate glass (PSG) silicon oxide doped with phosphorous.

photo acid generator (PAG) a component of a chemical amplified resist that releases acid on the occurrence of photochemical reactions.

photo active compound (PAC) a component of a resist that undergoes chemical reactions on receiving photons.

photolithography a process step that prints the photomask pattern onto the wafer surface.

physical vapor deposition (PVD) a process in which the target material is vaporized or sputtered and then condensed onto a substrate surface.

planarization a technique to smooth out the topography.

plasma a partially ionized gas system that consists of neutral particles, electrons, and ions.

plasma-enhanced chemical vapor deposition (PECVD) a CVD process in which the reactions are initiated by energetic electrons in the plasma.

plasma-enhanced TEOS (PETEOS) a TEOS oxide film grown with the PECVD approach.

plasma etching an etching process initiated by energetic electrons in plasma.

plasma sheath in a plasma system, a region depleted of electrons that forms next to the electrode or an object surface due to an electron flux toward the surface; there is a voltage drop across the sheath.

plating current efficiency the ratio of actual to theoretical weight of the deposits in an ECD process.

polymethylmethacrylate (PMMA) an old-generation resist for e-beam lithography.

PMOS an MOS transistor with its source and drain doped with *p*-type dopants.

poisoned via resistances of vias that are nonuniform and abnormally high.

polybutene-1-sulfone (PBS) a positive tone e-beam resist with a high throughput but lacking plasma etching resistance.

polymethyl-α-chloroacrylate-co-α-methylstyrene (ZEP) a positive tone resist for e-beam lithography suitable for plasma etching applications.

polybuffered LOCOS (PBLOCOS) a variation of LOCOS with a polysilicon buffer layer between the oxide and nitride layers.

positive resist a resist that is characterized by having its exposed area turning soluble in developing solution.

post coating delay (PCD) the delay time between the resist coating and the exposure.

post exposure bake (PEB) a baking step after exposure to remove the standing wave effect.

post exposure delay (PED) the delay time between the exposure and postexposure bake (PEB).

post-PEB delay (PPD) the delay time between the PEB and developing.

pressure cooker test (PCT) a reliability test item with high temperature and humidity ambient.

priming a pretreating step for the silicon surface that improves the resist adhesion.

***p*-type dopant** dopants that have three valence electrons.

pyrolysis thermal decomposition in the absence of oxygen.

quick down rinse (QDR) a wafer rinse process in which the water level rises above the wafers and drains very quickly to rinse away particles and contaminations.

Raman backscattering spectroscopy (RBS) a type of material composition analysis instrument: The underlying principle is to measure the energy of the ion that impinges into a sample and the energy of the ion that is scattered so as to identify various elements in the sample.

rapid thermal processor (RTP) a heating apparatus that provides a very fast temperature ramp up and ramp down in seconds.

reactive ion etching (RIE) an etching process that is enhanced with ion bombardment.

reflow a high-temperature annealing step that enables an as-deposited oxide film to flow.

remote plasma CVD a type of PECVD in which free radicals are extracted out of plasma bulk for deposition; such a process can be free of ion bombardment.

resolution the finest pattern that can be printed or imaged with high fidelity on an exposure system.

resolution-enhanced technique (RET) a technique used to enhance the photolithography resolution.

SC-1 clean RCA standard clean-1 consists of ammonium hydroxide and hydrogen peroxide and removes organic residues and metallic particles.

self-aligned silicide (salicide) a silicide formed and self-aligned to source, drain, and gate.

semiconductor a material that has a resistivity ranging between those of conductors and insulators.

silicate SOG inorganic spin on glass (oxide).

silicide a compound composed of refractory metal or noble metal silicides that provides about 10 times lower resistance than polysilicon or silicon.

silicon the second richest element on earth; it is made into silicon wafer to be used as a raw material for semiconductor manufacturing.

siloxane SOG SOG with organic groups.

silses quioxanes (HSQ) a siloxane-based polymer in which silicon is directly attached to hydrogen and oxygen.

soft baking a baking step after resist coating to drive solvents in the resist.

spacer an oxide sidewall that is needed for forming the LDD.

spin on glass (SOG) a solgel that can be spin-coated onto the wafer surface to form silicon oxide film.

sputtering a process that uses energetic heavy ions to bombard target material to sputter out target atoms, which then fall on substrate surface to form a thin film.

standing wave effect the interference effect between the incident and reflected lights that causes the resist edge profile to wiggle.

static random access memory (SRAM) a type of memory chip; its cell is composed of six or four transistors, and it has a much faster access time but larger cell size compared to DRAM.

step coverage a measure of the deposited film thickness difference across a step or a hole structure; it equals the ratio of the film thickness at the bottom corner to that on the field region.

shallow trench isolation (STI) a process of forming isolation by etching trenches on silicon surface and filling it up with oxide.

stoichiometry the relative atomic ratio of elements in a compound.

stress migration material migration caused by stress.

sub-resolution assisting feature (SRAF) nonprintable assisting feature (much smaller than the main features) that is added on a mask to improve the photolithography process window.

substitutional diffusion diffusing atoms moving from one vacancy to the other.

substractive etching the conventional etching process; patterns are defined with photolithography and thon otchod.

temperature humidity bias test (THB test) a reliability test item with high temperature, humidity, and electrical biases for chips that involve high temperature and high humidity.

tetraethyloxylsilane (TEOS) a chemical, $Si(OC_2H_5)_4$, that can be thermally decomposed to form oxide film.

tetrakisdiethyl amido titanium (TDEAT) a precursor for CVD TiN.

tetrakisdimethyl amido titanium (TDMAT) a precursor for CVD TiN.

tetramethylsilane (4MS) a precursor containing four methyl groups for making CDS.

thermal budget the maximum tolerable value of the summation of diffusivity times the duration time of high-temperature steps in a process flow.

thermal oxidation a process that employs oxidants to oxidize a bare silicon surface to silicon dioxide at elevated temperatures.

time-dependent dielectric breakdown (TDDB) a method of testing the oxide quality by stressing the oxide at a constant voltage and measuring the time when the oxide breaks down.

trichloroethane (TCA) a chlorine-containing species that is often used as chlorine source in silicon oxidation.

trimethylsilane (3MS) a precursor containing three methyl groups for making CDS.

tungsten polycide a combined word for "tungsten silicide on polysilicon."

ultra large scale integration (ULSI) the integration level of a chip that has over 10 million transistors.

wet oxidation silicon oxidation with water molecules.

white ribbon effect, or the Kooi effect a ribbon around the bird's beak edge after the local oxidation process; the ribbon is composed of a mixture of silicon oxide and silicon nitride.

References

Chapter 1

Boylestad, R. and L. Nashelsky, *Electronic Devices and Circuit Theory*, 4th ed. (Prentice Hall, 1987).

Hoerni, J. A., UA Patent No. 3,025,589, 20 March 1962.

Larrabee, G. B., A challenge to chemical engineers — Microelectronics, *Chem. Eng. Prog.* **6**, 51 (1985).

Morgan, D. V., K. Board, and R. H. Cockrum, *An Introduction to Microelectronics Technology* (John Wiley, 1985).

Ohmi, T., *Ultra-Clean Technology Handbook* (Marcel Dekker, 1993).

Reid, T. R., *The Chip* (Simon and Schuster, 1984).

Sasaki, H., Multimedia: Future and impact for semiconductor technology, *IEDM*, 111 (1997).

Sienko, M. J. and R. A. Plane, *Chemistry Principles and Properties* (McGraw-Hill, 1974).

Tolliver, D. L., Plasma processing in microelectronics — Past, present and future, *Solid State Technol.*, 99 (1980).

Van Zant, P., *Microchip Fabrication*, 4th ed. (McGraw-Hill, 2000).

Weste, N. and K. Eshraghian, *Principles of CMOS VLSI Design, A System Perspective* (Addison-Wesley, 1985).

Chapter 2

Allen, P. E. and D. R. Holberg, *CMOS Analog Circuit Design* (Saunders College, 1987).

Dorf, R. C. (ed.), *The Electronic Engineering Handbook* (CRC Press, 1993).

Grove, A., *Physics and Technology of Semiconductor Devices* (John Wiley, 1967).

Muller, R. S. and T. I. Kamins, *Device Electronics for Integrated Circuits* (John Wiley, 1977).

Ong, D. G., *Modern MOS Technology — Processes, Devices and Design* (McGraw-Hill, 1984).

Prince, B., *Semiconductor Memories — A Handbook of Design, Manufacturing, and Applications*, 2nd ed. (John Wiley, 1991).

Wolf, S., *Silicon Processing for the VLSI Era*, Vol. 3 (Lattice Press, 1995).

Chapter 3

Bhat, M., H. H. Jia, and D. L. Kwong, Growth kinetics of oxides during furnace oxidation of Si in N_2O ambient, *J. Appl. Phys.* **78**(4), 15 (1995).

Buchanan, D. A., Scaling the gate dielectric: Materials, integration and reliability, *IBM J. Res. Develop.* **43**(3), 245 (1999).

Chang, C. P., C. S. Pai, F. H. Baumann, C. T. Liu, C. S. Rafferty, M. R. Pinto, E. J. Lloyd, M. Bude, F. P. Miner, K. P. Cheung, J. I. Colonell, W. Y. C. Lai, H. Vaidya, S. J. Hillenius, R. C. Liu, and J. T. Clemens, A highly manufacturable corner rounding solution for 0.18 μm shallow trench isolation, *IEDM*, 27.2.1 (1997).

Chen, J. Y., *CMOS Devices and Technology for VLSI* (Prentice Hall, 1989).

Gusev, E. P., H. C. Lu, E. L. Garfunkel, T. Gustafsson, and M. L. Green, Growth and characterization of ultrathin nitrided silicon oxide films, *IBM J. Res. Develop.* **43**(3), 265 (1999).

Hook, T. B., J. S. Burnham, and R. J. Bolam, Nitrided gate oxides for 3.3 V logic applications: Reliability and device design considerations, *IBM J. Res. Develop.* **43**(3), 393 (1999).

Kern, W., *Handbook of Semiconductor Wafer Cleaning Technology* (Noyes, 1993).

Kooi, E., *The Invention of LOCOS* (IEEE, 1991).

Lin, T. H., N. S. Tsai, and C. S. Yoo, Twin-white-ribbon effect and pits formation mechanism in PBLOCOS, *J. Electrochem. Soc.* **138**(7), 2145 (1991).

Liu, H. I., D. K. Biegelsen, N. M. Johnson, F. A. Ponce, and R. F. W. Pease, Self-limiting oxidation of Si nanowires, *J. Vac. Sci. Technol. B* **11**(6), 2532 (1993).

Luo, T. Y., Effect of hydrogen content on reliability of ultra thin *in situ* steam generation (ISSG) SiO_2, *IEEE Electron Dev. Lett.* **21**, 430 (2000).

Ngau, J. L., P. B. Griffin, and J. D. Plummer, Silicon orientation effects in the initial regime of wet oxidation, *J. Electrochem. Soc.* **149**(8), F98 (2002).

Okorn-Schmidt, H. F., Characterization of silicon surface preparation processes for advanced gate dielectrics, *IBM J. Res. Develop.* **43**(3), 351 (1999).

Osburn, C. M. and D. W. Ormond, Processing for advanced devices with hot-wall furnace RTP, *J. Electrochem. Soc.* **119**(5), 591 (1972).

Roenigk, K. F. and K. F. Jensen, Low pressure CVD of silicon nitride, *J. Electrochem. Soc.* **134**(7), 1777 (1987).

Shum, D. P., J. M. Higman, M. G. Khazhinsky, K. Y. Wu, S. Kao, J. D. Burnett, and C. T. Swift, Corner field effect on the CMP oxide recess in shallow trench isolation technology for high density flash memories, *IEDM*, 27.3.1 (1997).

Thakur, R. P. S., P. J. Timans, and S. P. Tay, RTP technology for tomorrow, *Solid State Technol.*, 171 (June 1998).

Chapter 4

Carter, G. and G. S. Callingan, *Ion Bombardment of Solids* (Elsevier, 1969).

Cote, D. R., S. V. Nguyen, A. K. Stamper, D. S. Armbrust, D. Tobben, R. A. Conti, and G. Y. Lee, Plasma assisted chemical vapor deposition of dielectric films for ULSI semiconductor circuits, *IBM J. Res. Develop.* **43**(1/2), 5 (1999).

Frost, L. S. and A. V. Phelps, Rotational excitation and momentum transfer cross-section for electrons in hydrogen and nitrogen from transport coefficients, *Phys. Rev.* **127**, 1621 (1962).

Haller, I., Importance of chain reactions in the plasma deposition of hydrogenated amorphous silicon, *J. Vac. Sci. Technol. A* **1**, 1376 (1983).

Holland, J. R. and A. T. Bell, *Technique and Applications of Plasma Chemistry* (John Wiley, 1974).

Lieberman, M. A. and A. J. Licgtenberg, *Principles of Plasma Discharge and Materials Processing* (John Wiley, 1994).

Mamikonyan, E. R., L. S. Polak, and D. I. Slovetskii, Dissociation of nitrogen molecules from electrically excited state (electron impact dissociation), *Khim. Vys. Energ.* **6**, 483 (1972).

Perrin, J., J. P. M. Schmidt, G. deRosney, B. Oevillan, and A. Lioret, Dissociation cross section of silane by electron impact, *Chem. Phys.* **73**, 383 (1982).

Rapp, D. and P. Englander-Golden, Total cross section for ionization and attachment in gases by electron impact, *J. Chem. Phys. I. Positive Ionization* **43**, 1464 (1965).

Schuegraf, K. K., *Handbook of Thin-Film Deposition Processes and Techniques* (Noyes, 1988).

Slovetskii, D. I. and A. D. Urbas, Cross section of the dissociation of ammonia by electron impact, *Khim. Vys. Energ.* **13**, 475 (1979).

Turban, G., Basic phenomena in reactive low pressure plasmas used for deposition and etching, *Pure Appl. Chem.* **56**(2), 214 (1984).

Winter, H. F., *Topic in Current Chemistry: Plasma Chemistry III*, eds. S. Verpek and M. Vanugopulan (Springer, 1980).

Yoo, C. S., Modelling and experimentally study of plasma enhanced chemical vapor deposition of silicon nitride, PhD thesis, Chem. Eng. Dept., Worcester Polytechnic Institute, Mass (1988).

Chapter 5

Bloem, J., Y. S. Oei, H. H. C. de Moor, J. H. L. Hanssen, and L. J. Giling, Epitaxial growth of silicon by CVD in a hot-wall furnace, *J. Electrochem. Soc.* **132**(8), 1973 (1985).

Creighton, J. R., A mechanism for selectivity loss during tungsten CVD, *J. Electrochem. Soc.* **136**(1), 271 (1989).

Farrow, R. F. C., The kinetics of silicon deposition on silicon by pyrolysis of silane, *J. Electrochem. Soc.* **121**(7), 899 (1974).

Kern, W. and V. S. Ban, Chemical vapor deposition of inorganic thin films, *Thin Film Processes* (Academic, 1978).

Kern, W. and V. S. Ban, *Chemical Vapor Deposition of Inorganic Thin Films* (Academic, 1978).

Lanford, W. A. and M. J. Rand, The hydrogen content of plasma-deposited silicon nitride, *J. Appl. Phys.* **49**, 2473 (1978).

McConica, C. M. and K. Krishnamani, The kinetics of LPCVD tungsten deposition in a single wafer reactor, *J. Electrochem. Soc.* **133**, 2542 (1986).

Nguyen, V. S., W. A. Landford, and A. L. Rieger, Variation of hydrogen bonding, depth profile, and spin density in plasma-deposited silicon nitride and oxynitride film with deposition mechanism, *J. Elelctrochem. Soc.* **133**(5), 970 (1986).

Roenigk, K. F. and K. F. Jensen, Low pressure CVD of silicon nitride, *J. Electrochem. Soc.* **134**(7), 1777 (1987).

Schmitz, J. E. J., *CVD of Tungsten and Tungsten Silicide* (Noyes, 1992).

Shareef, I. A., G. W. Rubloff, M. Anderle, W. N. Gill, J. Cotte, and D. H. Kim, Subatmospheric chemical vapor deposition ozone/TEOS process for SiO_2 trench filling, *J. Vac. Sci. Technol. B* **13**, 1888 (1995).

Sinha, A. K. and E. Lugujjo, Lorentz–Lorenz correlation for reactively plasma deposition SiN films, *Appl. Phys. Lett.* **245**, 15 (1978).

Takahashi, R., Y. Koga, and K. Sugawara, Gas flow pattern and mass transfer analysis in a horizontal flow reactor for chemical vapor deposition, *J. Electrochem. Soc.* **119**(10), 1406 (1972).

Tsai, Y. H., C. S. Yoo, Y. C. Chao, and S. L. Hsu, BPSG film post-growth treatments, *Thin Solid Films*, 371 (1989).

Wilke, T. E., K. A. Turner, and C. G. Takoudis, Chemical vapor deposition of silicon under reduced pressure in hot-wall reactors, *Chem. Eng. Sci.* **41**, 643 (1986).

Yoo, C. S., T. H. Lin, and N. S. Tsai, Si/W ratio changes and film peeling during polycide annealing, *Jpn. J. Appl. Phys.* **29**, 2535 (1990).

Yoo, C. S., Y. S. Tsai, and Y. C. Chao, Silicon diffusion mechanism in polycide oxidation, *J. Electrochem. Soc.* **138**(1), 321 (1991).

Yoo, C.-S. and A. G. Dixon, Modeling and simulation of a novel multi-radial-flow CVD reactor, *J. Cryst. Growth* **97**, 337 (1989).

Chapter 6

Armacost, M., P. D. Hoh, R. Wise, W. Yan, J. J. Brown, J. H. Keller, G. A. Kaplita, S. D. Halle, K. P. Muller, M. D. Naeem, S. Srinivasan, H. Y. Ng, M. Gutsche, A. Gutmann, and B. Spuler, Plasma-etching processes for ULSI semiconductor circuits, *IBM J. Res. Develop.* **43**(1/2), 39 (1999).

Awadelkarim, O. O., T. Gu, R. A. Ditizio, P. I. Mikulan, S. J. Fonash, J. F. Rembetski, and Y. D. Chan, Creation of deep gap states in Si during Cl_2 or HBr plasma etch exposure, *J. Vac. Sci. Technol. A* **11**(4), 1332 (1993).

Beale, D., S. Siu, and R. Patrick, Trends in aluminum etch rate uniformity in a commercial inductively coupled plasma etch system, *J. Vac. Sci. Technol. B* **16**(3), 1059 (1998).

Chang, K. M., T. H. Yeh, I. C. Deng, and H. C. Lin, Highly selective etching for polysilicon and etch-induced damage to gate oxide with halogen-bearing ECR plasma, *J. Appl. Phys.* **80**(5), 3048 (1996).

Chin, B. L. and E. P. van de Ven, Plasma TEOS for interlayer dielectric applications, *Solid State Technol.*, 119 (1988).

Coburn, J. W. and H. Winters, Plasma etch — A discussion of mechanisms, *J. Vac. Sci. Technol. A* **16**(2), 391 (1979).

Egitto, F. D., F. Emmi, R. S. Horwath, and V. Vukanovic, Plasma etching of organic materials I. Polyimide in O_2/CF_4, *J. Vac. Sci. Technol. B* **3**, 893 (1985).

Kang, S. Y., K. H. Kwon, S. I. Kim, S. K. Lee, M. Y. Jung, Y. R. Cho, Y. H. Song, J. H. Lee, and K. I. Cho, Etch characteristics of Cr by using Cl_2/O_2 gas mixture with ECR plasma, *J. Electrochem. Soc.* **148**(5), G237 (2001).

Lee, S. and Y. Kuo, Chlorine plasma/copper reaction in a new copper dry etching process, *J. Electrochem. Soc.* **148**(9), G524 (2001).

Matsuda, T., M. Shaprio, and S. Nguyen, Dual frequency plasma CVD fluorosilicate glass deposition for 0.25 μm interlevel dielectrics, *Proc. 1st Int. DUMIC*, 22–28 (1995).

Park, S. K. and D. J. Economou, A mathematical model for a plasma-assisted downstream etching reactor, *J. Appl. Phys.* **66**(7), 3256 (1989).

Rana, V. V. S., A. Gupta, S. Hong, D. Cheung, and P. Lee, Low dielectric constant fluorine-doped TEOS films, *MRS Spring Meeting*, paper no. N5.1 (1997).

Samukawa, S., M. Sasaki, and Y. Suzuki, Perfect selective and highly anisotropic ECR plasma etching for WSi_x/poly-Si at electron cyclotron resonance position, *J. Vac. Sci. Technol. B* **8**(5), 1062 (1990).

Schuegraf, K. K., *Handbook of Thin Film Deposition Processes and Techniques* (Noyes, 1988).

Shanon, V. L. and M. Z. Karim, Study of material properties and suitability of plasma-deposited fluorine-doped silicon dioxide for low dielectic constant interlevel dielectrics, *Thin Solid Films* **270**, 489 (1995).

Stamper, A. K., V. McGahay, M. Shapiro, L. A. Miller, X. Tian, A. Byrant, and L. A. Serianni, Optimization of AlCu wiring delay in advanced CMOS technology, *Proc. MRS Symposium on Low-Dielectric Constant Materials II*, Boston, 183–188 (2–3 Dec. 1996).

Suzuki, T., H. Kitagawa, K. Yamada, and M. Nagoshi, Residue formation and elimination in chlorine-based plasma etching of Al–Si–Cu interconnects, *J. Vac. Sci. Technol. B* **10**(2), 596 (1992).

Wang, C. K., T. L. Ying, C. S. Wei, L. M. Liu, H. C. Cheng, and M. S. Lin, Investigation of a high quality and UV transparent PECVD silicon nitride film for non-volatile memory applications, *Jpn. J. Appl. Phys.* **34**, 4736 (1995).

Westerheim, A. C., R. D. Jones, P. J. Mager, J. H. Dubash, T. J. Dalton, M. G. Goss, S. K. Baum, and S. K. Dass, High-density, inductively coupled plasma etch of sub-half-micron critical layers: Transistor polysilicon gate definition and contact formation, *J. Vac. Sci. Technol. B* **16**(5), 2699 (1998).

Yoo, C. S. and A. G. Dixon, Plasma deposition of silicon nitride films in a radial flow reactor, *AIChE J.* **35**(6), 995 (1989).

Yoo, C. S., Modeling and experimental study of plasma enhanced chemical vapor deposition of silicon nitride, PhD thesis, Chem. Eng. Dept., Worcester Polytechnic Institute, Mass (1988).

Zhang, D., S. Rauf, T. G. Sparks, and P. L. G. Ventzek, Integrated equipment-feature modeling investigation of fluorocarbon plasma etching of SiO_2 and photoresist, *J. Vac. Sci. Technol. B* **21**(2), 828 (2003).

Chapter 7

Born, M. and E. Wolf, *Principles of Optics*, 7th ed. (Cambridge University Press, 1999).

Dammel, R., *Diazonaphthoquinone-Based Resists* (SPIE Optical Engineering Press, 1993).

Glendinning, W. B. and J. N. Helbert, *Handbook of VLSI Microlithography: Principles, Technology and Applications* (Noyes, 1991).

Guenther, B. D., *Modern Optics* (John Wiley, 1992).

Holmes, S. J., P. H. Mitchell, and M. C. Hakey, Manufacturing with DUV lithography, *IBM J. Res. Develop.* **41**(1/2), 7 (1997).

Ito, H., Deep UV resist: Evolution and status, *Solid State Technol.*, 164 (1996).

Ito, H., Dissolution behavior of chemical amplified resist polymers for 248-, 193-, and 157-nm lithography, *IBM J. Res. Develop.* **45**(5), 683 (2001).

Kirchauer, H., Photolithography simulation, PhD thesis, Institute for Microelectronics, Technical University of Vienna (1998).

Williams, T. L., *The Optical Transfer Function of Imaging Systems* (IOP, 1999).

Wilson, C. G., Organic resist materials, in *Introduction to Microlithography*, 2nd ed., eds. L. F. Thompson, C. G. Wilson, and M. J. Bowden (ACS, 1994).

Wolf, S., *Silicon Processing for the VLSI Era* (Lattice Press, 1995).

Young, I. T., R. Zagers, L. J. Van Vliet, J. Mullikin, F. Boddeke, and H. Netten, Depth-of-focus in microscopy, SCIA '93, *Proc. 8th Scandinavian Conference on Image Analysis*, Tromso, Norway, 493–498 (1993).

Chapter 8

Brunner, T. A., Rim phase shifter mask combined with off-axis illumination, *Proc. SPIE*, 1927, Optical/Laser Microlithography VI, p. 54 (1993).

Chen, J. F., T. Laiding, and R. Coldwell, Practical method for full chip optical proximity correction, *Proc. SPIE* **3051**, 790 (1997).

Inoue, T., M. Nakada, T. Uozumi, and E. Sugata, Growth and surface properties of Lanthanum hexaboride crystal, *J. Vac. Sci. Technol. A* **21**(4), 952 (1982).

Ito, H., Chemical amplified resists: Past, present and future, *SPIE* **3678**, 2 (1999).

Liebmann, L., S. Mansfield, J. Bruce, M. Cross, I. Graur, A. McGuire, J. Krueger, and D. Sunderling, Optimizing style options for sub-resolution assist features, *Proc. SPIE*, **4346**, 141 (2001).

Lin, B. J., Off-axis illumination — Working principles and comparison with alternating PSM, *Proc. SPIE*, 1927, Optical/Laser Microlithography VI, p. 890 (1993).

Mack, C. A., Optimization of the spatial properties of illumination for improved lithography response, *Proc. SPIE*, 1927, Optical/Laser Microlithography VI, p. 125 (1993).

Mauer, J. L., H. C. Pfeiffer, and W. Stickel, Electron optics of an electron beam lithography system, *IBM J. Res. Develop.*, 514 (1997).

Mederios, D. R., A. Aviram, C. R. Guarnieri, W. S. Huang, R. Kwong, C. K. Magg, A. P. Mahorowala, W. M. Moreau, K. E. Petrillo, and M. Angelopoulos, Recent progress in electron beam resists for advanced mask-making, *IBM J. Res. Develop.* **45**(5), 639 (2001).

Melngailis, J., Focus ion beam technology and applications, *J. Vac. Sci. Technol. B* **5**(2), 469 (1987).

Nakasuji, M. and H. Wada, Single crystal LaB_6 electron gun for variably shaped electron beam optics, *J. Vac. Sci. Technol. A* **17**(6), 1367 (1980).

Pfeiffer, H. C., Variable spot shaping for electron beam lithography, *J. Vac. Sci. Technol. A* **15**, 887 (1978).

Prewett, P. D. and P. J. Heard, Repair of opaque defects in photomasks using focused ion beam, *J. Phys. D: Appl. Phys.* **20**, 1207 (1987).

Rai-Choudhury, P. (ed.), *Handbook of Microlithography, Micromachining and Microfabrication* (SPIE Optical Engineering Press, 1997).

Ronse, K., R. Jonckheere, C. Juffermas, and L. Van Denhov, Comparison of various phase shift strategies and applications to 0.35 μm, *Proc. SPIE*, 1927, Optical/Laser Microlithography VI, p. 2 (1993).

Saitou, N., S. Ozasa, and T. Komoda, Variable shaped electron beam lithography system, EB55: II electron optics, *J. Vac. Sci. Technol.* **19**(4), 1087 (1981).

Siegel, J., K. Ettrich, E. Welsch, and E. Mattias, UV-laser ablation of ductile and brittle metal films, *Appl. Phys.* **213**, 64 (1997).

Sukgiura, E., H. Watanabe, T. Saito, T. Imoriya, Y. Todokoro, and M. Inoue, The application of phase-shifting mask to DRAM cell capacitor fabrication, *Proc. SPIE*, 1927, Optical/Laser Microlithography VI, p. 79 (1993).

Tison, J. K. and M. G. Cohen, Lasers in mask repair, *Solid State Technol.*, 164 (February 1996).

Uhl, A., J. Bending, L. Leistner, U. Jagdhold, and J. Bauer, E-beam and deep UV exposure of PMMA based resists — Identical or different?, *SPIE* **3333**, 1452 (1986).

Volk, W. W., J. N. Wiley, C. C. Hung, S. Khanna, C. S. Yoo, S. Biellak, C. H. Lin, and D. Wang, Reticle blank inspection and its role in zero defect manufacturing, *Proc. SPIE*, **4409**, 520 (2001).

Wong, K. K., Resolution enhancement techniques in optical lithography, SPIE, International Society for Optical Engineering (2001).

Chapter 9

2001 ITRS Roadmap (www.public.itrs.net).

Griffin, P. B., M. Cao, P. Vande Voorde, Y. L. Chang, and W. M. Greene, Indium transient enhanced diffusion, *Appl. Phys. Lett.* **73**, 2986 (1998).

Mahan, J. E. and A. Vantomme, A simplified collisional model of sputtering in the linear cascade regime, *J. Vac. Sci. Technol. A* **15**(4), 1976 (1997).

Narajan, J. and O. W. Holland, Formation of metastable supersaturated solid solutions in ion implanted silicon during solid phase crystallization, *Appl. Phys. Lett.* **41**(3), 239 (1982).

Rücker, H., B. Heinemann, R. Barth, D. Bolze, V. Melnik, D. Krüger, and R. Kurps, Formation of shallow source/drain extensions for metal-oxide-semiconductor field effect transistors by antimony implantation, *Appl. Phys. Lett.* **82**, 826 (2003).

Shibahara, K., M. Mifuji, K. Kawabata, T. Kugimiya, H. Furumoto, M. Tsuno, S. Yokohama, M. Nagata, S. Miyazaki, and M. Hirose, Low resistive ultra shallow junction for sub $0.1\,\mu$m MOSFETs formed by Sb implantation, *IEDM*, 21.7.1 (1996).

Solmi, S. and D. Nobili, High concentration diffusivity and clustering of arsenic and phosphorous in silicon, *J. Appl. Phys.* **83**, 2484 (1998).

Tuck, B., *Introduction to Diffusion in Semiconductors* (Peter Peregrinus, 1974).

Wolf, S., *Silicon Processing for the VLSI Era* (Lattice Press, 1995).

Ziegler, J. F. (ed.), *Handbook of Ion Implantation Technology* (Elsevier, 1992).

Chapter 10

Barnes, M. S., J. C. Forster, and J. H. Keller, Apparatus for depositing material into high aspect ratio holes, US Patent 5,178,739, 12 January 1993.

Cheng, H., M. Chiang, and M. Hon, Growth characteristics and properties of TiN coating by CVD, *J. Electrochem. Soc.* **142**(5), 1573 (1995).

Chiang, C., M. Lee, D. Fraser, L. C. Yip, S. Mittal, and K. Wu, Interaction of metal with underlying dielectric films in multilevel interconnect systems, *VMIC Conf.*, p. 470 (1989).

Joshi, R. V. and S. Brodsky, Collimated sputtering of TiN/Ti liners into sub-half-micrometer high aspect ratio contacts lines, *Appl. Phys. Lett.* **61**(21), 2613 (1992).

Kittl, J. A., W. T. Shiau, D. Miles, K. E. Violette, J. C. Hu, and Q. Z. Hong, Salicide and alternative technologies for future ICs: Part 1, *Solid State Technol.*, 81 (1999).

Mahan, J. E. and A. Vantomme, A simplified collisional model of sputtering in the linear cascade regime, *J. Vac. Sci. Technol. A* **15**, 1976 (1997).

Miller, N. E. and I. Beinglass, CVD tungsten interconnect and contact barrier technology for VLSI, *Solid State Technol.*, 85 (1982).

Nicolet, M. A. and M. Bartur, Diffusion barriers in layered contact structures, *J. Vac. Sci. Technol. A* **19**(3), 786 (1981).

Ohguro, T., S. Nakamura, M. Kioke, T. Morimoto, A. Nishiyama, Y. Ushiku, T. Yoshitomi, M. Ono, M. Saito, and H. Iwai, Analysis of resistance behavior in T- and Ni-salicided polysilicon films, *IEEE Trans. Electron. Develop.* **41**(12), 2305 (1994).

Polignano, M. L. and N. Circelli, The impact of the metallization technology on junction behavior, *J. Appl. Phys.* **68**(4), 1869 (1990).

Roberts, B., A. Harrus, and R. L. Jackson, Interconnect metallization for future device generations, *Solid State Technol.*, 69 (1995).

Rossnagel, S. M. and J. Hopwood, Magnetron sputter deposition with levels of metal ionization, *Appl. Phys. Lett.* **63**, 3285 (1993).

Rossnagel, S. M., Sputter deposition for semiconductor manufacturing, *IBM J. Res. Develop.* **43**(1/2) (1999).

Sherman, A., Growth and properties of LPCVD TiN as diffusion barrier for silicon device technology, *J. Electrochem. Soc.* **137**(1), 1892 (1990).

Steinbruchel, C., On the sputtering yield of molecular ions, *J. Vac. Sci. Technol. A* **3**, 1913 (1985).

Thornton, J., Magnetron sputtering: Basic physics and application to cylindrical magnetrons, *J. Vac. Sci. Technol. A* **15**(2), 171 (1978).

Woratschek, B., P. Carey, M. Stolz, and F. Bachmann, Improved excimer laser planarization of AlSi with addition of Ti and Cu, *VMIC Conf.*, p. 309 (1989).

Yamashita, M., Fundamental characteristics of a built-in high frequency coil-type sputtering apparatus, *J. Vac. Sci. Technol. A* **7**, 151 (1989).

Yoo, C. S. and J. C. Jans, Thickness measurement of combined a-Si and Ti on C-Si using a monochromatic ellipsometer, *SPIE Conf.*, p. 1231 (1991).

Yoo, C. S., T. H. Lin, N. S. Tsai, and L. J. Ijzendoorn, Si/W ratio changes and film peeling during polycide annealing, *Jpn. J. Appl. Phys.* **29**, 2535 (1990).

Yoo, W. S., A. J. Atanos, and J. Daviet, Cobalt silicide processing in a susceptor-based LP-RTP system, *Solid State Technol.*, 125 (1999).

Yun, J. Y., M. Y. Park, and S. W. Rhee, Comparison of TDMAT and TDMET as precursors for metallorganic CVD of titanium nitride, *J. Electrochem. Soc.* **146**(5), 1804 (1999).

Chapter 11

Case, C. B., C. J. Case, G. P. Schwartz, and N. M. Rutherfod, Gap-fill for sub-0.5 μm technologies: Process and cost considerations for spin-on-glass, *VMIC Conf.*, p. 114 (1994).

Cote, D. R., S. V. Guyen, W. J. Cote, S. L. Pennington, A. K. Stamper, and D. V. Podlesnik, Low-temperature chemical vapor deposition processes and dielectrics for microelectronic circuit manufacturing at IBM, *IBM J. Res. Develop.* **39**(4), 437 (1995).

Forester, L., T. Collins, G. van den Bosch, H. Meynen, B. Coenegrachts, and L. Van Den Hove, Interlevel dielectric engineering for improved device performance in half-micron CMOS, *VMIC Conf.*, p. 111 (1994).

Homma, T., M. Suzuki, and Y. Murao, A fully planarized multilevel interconnection technology using semi-selective TEOS-ozone CVD at atmospheric pressure, *J. Electrochem. Soc.* **140**(12), 3591 (1993).

Lee, S., K. Lee, H. Oh, C. Oh, Y. W. Kim, D. Kim, and B. Kim, Multilevel metallization for ASIC technology, *VMIC Conf.*, p. 59 (1994).

Machida, K. and H. Oikawa, SiO_2 planarization technology with biasing and ECR plasma deposition for submicron interconnections, *J. Vac. Sci. Technol. B* **4**(4), 818 (1986).

Park, T. S., Y. G. Shin, H. S. Lee, M. H. Park, S. D. Kwon, H. K. Kang, Y. B. Koh, and M. Y. Lee, Correlation between gate oxide reliability and the profile of the trench top corner in STI, *IEDM*, 96 (1996).

Saito, M., M. Hirasawa, K. Kato, N. Kobayashi, and A. Takamatsu, Improved SOG gap-filling capability through the control of organic components, *VMIC Conf.*, p. 620 (1996).

Stamper, A. K., V. McGahay, T. J. Hartswick, D. R. Cote, L. A. Miller, and E. G. Walton, Advanced interconnect dielectric gapfill processes for high aspect ratio CMOS technologies, *VMIC Conf.*, p. 617 (1996).

Sun, Y., N. Rayji, and S. Cagnina, 0.8 μm double level metal technology with SOG filled tungsten plug, *VMIC Conf.*, p. 51 (1991).

Sze, S. M., *VLSI Technology* (McGraw-Hill, 1983).

Takata, Y., A. Ishii, M. Matsuura, A. Ohsaki, M. Iwasaki, J. Miyazaki, N. Fujiwara, J. Komori, T. Katayama, S. Nakao, and H. Kotabni, A highly reliable multilevel interconnection process for 0.6 μm CMOD device, *VMIC Conf.*, p. 13 (1991).

Yu, C., S. Poon, Y. Limb, T. K. Yu, and J. Klein, Improved multilevel metallization technology using chemical mechanical polishing of W plugs and interconnects, *VMIC Conf.*, p. 144 (1994).

Chapter 12

Baker, R., Topography control using sacrificial capping layers, *Solid State Technol.*, 33 (2004).

Bessor, P., A. Marathe, L. Zhao, M. Herrick, C. Capasso, and H. Kawasaki, Optimizing the electromigration resistance of copper interconnects, *IEDM*, 6.1.1 (2000).

Bohr, M. T., Interconnect scaling — The real limiter to high performance LSI, *IEDM*, 241 (1995).

Cote, D. R., S. V. Nguyen, A. K. Stamper, D. S. Armbrust, D. Tobben, R. A. Conti, and G. Y. Lee, Plasma-assisted chemical vapor deposition of dielectric thin films for ULSI semiconductor circuits, *IBM J. Res. Develop.* **43**(1/2), 5 (1999).

Gau, W. C., T. C. Chang, Y. S. Lin, J. C. Hu, L. J. Chen, C. Y. Chang, and C. L. Cheng, Copper electroplating for future ultralarge scale integration interconnects, *J. Vac. Sci. Technol. A* **18**(2), 656 (2000).

Golden, J. H., C. J. Hawker, and P. S. Ho, Designing porous low-k dielectrics, *Semiconductor Int.* **5**, 1 (2001).

Graham, L., T. Ritzdorf, D. Clarke, and R. Thakur, Thermally driven recrystallization of electroplated copper, in *Semiconductor Fabtech.* 11th ed. (2000).

Gray, W. D. and M. J. Loboda, New barrier layers can help Cu/low-k integration, *Solid State Technol.*, 37 (2002).

Haveman, R. and J. A. Hutchby, High performance interconnects: An integration overview, *Proc. IEEE* **89**(5) (2001).

Jordan, K. G. and C. C. W. Tobias, The effect of inhibitor transport on leveling in electrodeposition, *J. Electrochem. Soc.* **138**(5), 1251 (1991).

Lanzi, O. and U. Landau, Terminal effect at a resistive electrode under Tafel kinetics, *J. Electrochem. Soc.* **137**(4), 1139 (1990).

Maenhoudt, M., I. Pollentier, V. Wiaux, D. Vangoidsenhoven, and K. Ronse, 248 nm and 193 nm lithography for damascene patterning, *Solid State Technol.*, s15 (2001).

Roha, D. and U. Landau, Mass transport of leveling agent in plating: Steady-state model for blocking additives, *J. Electrochem. Soc.* **137**(3), 824 (1990).

Seah, H., S. Mridha, and L. H. Chan, Annealing of copper electrodeposits, *J. Vac. Sci. Technol. A* **17**(4), 1963 (1999).

Stavrev, M. and D. Fischer, Behavior of thin Ta-based films in the Cu/barrier/Si system, *J. Vac. Sci. Technol. A* **17**, 993 (1999).

Thomas, M. E. and N. Iwamoto, Transport phenomena in porous low-k dielectrics, *Semiconductor Int.* **6**, 1 (2002).

Index